Bioremediation in Latin America

Analía Alvarez • Marta Alejandra Polti
Editors

Bioremediation in Latin America

Current Research and Perspectives

Springer

Editors
Analía Alvarez
Pilot Plant of Industrial and Microbiology
National Council of Scientific
and Techniques
Natural Sciences College and Miguel
Lillo Institute
Tucumán, Argentina

Marta Alejandra Polti
Pilot Plant of Industrial and Microbiology
National Council of Scientific
and Techniques
Natural Sciences College and Miguel
Lillo Institute
Tucumán, Argentina

ISBN 978-3-319-05737-8 ISBN 978-3-319-05738-5 (eBook)
DOI 10.1007/978-3-319-05738-5
Springer Cham Heidelberg New York Dordrecht London

Library of Congress Control Number: 2014948191

Printed on acid-free paper

Springer is part of Springer Science+Business Media (www.springer.com)

Contents

Contributors

A. Alvarez Pilot Plant of Industrial and Microbiology, National Council of Scientific and Techniques, Natural Sciences College and Miguel Lillo Institute, Tucumán, Argentina

H.M. Alvarez Centro Regional de Investigación y Desarrollo Científico Tecnológico, Universidad Nacional de Patagonia San Juan Bosco, Comodoro Rivadavia, Chubut, Argentina

G.B. Álvarez Universidad de Buenos Aires, Buenos Aires, Argentina

M.J. Amoroso Planta Piloto de Procesos Industriales Microbiológicos (PROIMI), CONICET, Av. Belgrano y Pasaje Caseros, Tucumán, Argentina

Facultad de Bioquímica, Química y Farmacia, Universidad Nacional de Tucumán, Tucumán, Argentina

Universidad del Norte Santo Tomás de Aquino (UNSTA), Tucumán, Argentina

J.D. Aparicio Planta Piloto de Procesos Industriales y Microbiológicos, Consejo Nacional de Investigaciones Científicas y Técnicas, Tucumán, Argentina

Facultad de Bioquímica, Química y Farmacia, Universidad Nacional de Tucumán, Tucumán, Argentina

M.C. Atjián Planta Piloto de Procesos Industriales Microbiológicos (PROIMI), CONICET, Av. Belgrano y Pasaje Caseros, Tucumán, Argentina

C.S. Benimeli Planta Piloto de Procesos Industriales Microbiológicos (PROIMI), CONICET, Av. Belgrano y Pasaje Caseros, Tucumán, Argentina

Universidad del Norte Santo Tomás de Aquino (UNSTA), Tucumán, Argentina

L. Benítez Centro de Investigación y Desarrollo en Fermentaciones Industriales, Centro Científico Tecnológico La Plata, Consejo Nacional de Investigaciones Científicas y Técnicas, Universidad Nacional de La Plata, La Plata, Buenos Aires, Argentina

G. Briceño Departamento de Ingeniería Química, Núcleo Científico y Tecnológico en Biorecursos, Universidad de La Frontera, Temuco, Chile

Departamento de Ciencias Químicas y Recursos Naturales, Núcleo Científico y Tecnológico en Biorecursos, Universidad de La Frontera, Temuco, Chile

L.I.C de Figueroa Planta Piloto de Procesos Industriales y Microbiológicos, Consejo Nacional de Investigaciones Científicas y Técnicas, Universidad Nacional de Tucumán, Tucumán, Argentina

M.L. Castrejón-Godínez Programa de Gestión Ambiental Universitario, Universidad Autónoma del Estado de Morelos, Cuernavaca, México

J.P. Cazón Centro de Investigación y Desarrollo en Fermentaciones Industriales, Centro Científico Tecnológico La Plata, Consejo Nacional de Investigaciones Científicas y Técnicas, Universidad Nacional de La Plata, La Plata, Buenos Aires, Argentina

V.L. Colin Planta Piloto de Procesos Industriales y Microbiológicos (PROIMI), CONICET, Av. Belgrano y Pasaje Caseros, Tucumán, Argentina

Universidad de San Pablo-Tucumán, Tucumán, Argentina

A.J. Cortes Planta Piloto de Procesos Industriales y Microbiológicos (PROIMI), CONICET, Av. Belgrano y Pje. Caseros, Tucumán, Argentina

E.L. Cruz Planta Piloto de Procesos Industriales y Microbiológicos, Consejo Nacional de Investigaciones Científicas y Técnicas, Tucumán, Argentina

S.A. Cuozzo Planta Piloto de Procesos Industriales y Microbiológicos, Consejo Nacional de Investigaciones Científicas y Técnicas, Universidad Nacional de Tucumán, Tucumán, Argentina

M.C. Diez Departamento de Ingeniería Química, Núcleo Científico y Tecnológico en Biorecursos, Universidad de La Frontera, Temuco, Chile

E. Donati Centro de Investigación y Desarrollo en Fermentaciones Industriales, Centro Científico Tecnológico La Plata, Consejo Nacional de Investigaciones Científicas y Técnicas, Universidad Nacional de La Plata, La Plata, Buenos Aires, Argentina

N. Durán Institute of Chemistry, Biological Chemistry Laboratory – NanoBioss, State University of Campinas, Campinas, Brazil

Center of Natural and Human Sciences, Federal University of ABC, Sao Paulo, SP, Brazil

P.M. Fernandez Planta Piloto de Procesos Industriales Microbiológicos PROIMI-CONICET, Av. Belgrano y Caseros, Tucumán, Argentina

M.S. Fuentes Planta Piloto de Procesos Industriales y Microbiológicos, Consejo Nacional de Investigaciones Científicas y Técnicas, Tucumán, Argentina

M.E.R. Girault Universidad de Buenos Aires, Buenos Aires, Argentina

M. Hilal Universidad Nacional de Tucumán, Tucumán, Argentina

V. Irazusta Planta Piloto de Procesos Industriales y Microbiológicos, Consejo Nacional de Investigaciones Científicas y Técnicas, Tucumán, Argentina

M.P. Lanfranconi Centro Regional de Investigación y Desarrollo Científico Tecnológico, Universidad Nacional de Patagonia San Juan Bosco, Comodoro Rivadavia, Chubut, Argentina

D.S.T. Martinez Brazilian National Nanotechnology Laboratory, National Center for Research on Energy and Materials, Campus CNPEM, Campinas, Sao Paulo, Brazil

M.A. Martinez Planta Piloto de Procesos Industriales y Microbiológicos (PROIMI), CONICET, Av. Belgrano y Pasaje Caseros, Tucumán, Argentina

Facultad de Ciencias Exactas y Tecnología, Universidad Nacional de Tucumán, Tucumán, Argentina

M.M. Martorell PROIMI-CONICET (Planta Piloto de Procesos Industriales Microbiológicos), Av. Belgrano y Caseros, Tucumán, Argentina

Química Biológica, Calidad de productos IV, Gestión de Empresas Agroalimentarias, Universidad de San Pablo, Tucumán, Argentina

N.B. Moraga Instituto de Investigaciones para la Industria Química (INIQUI), Consejo Nacional de Investigaciones Científicas y Técnicas (CONICET); Universidad Nacional de Salta (UNSa), Salta, Argentina

M.L. Ortiz-Hernández Centro de Investigación en Biotecnología, Programa de Gestión Ambiental Universitario, Universidad Autónoma del Estado de Morelos, Cuernavaca, México

E. Pagano Universidad de Buenos Aires, Buenos Aires, Argentina

H.F. Pajot PROIMI-CONICET (Planta Piloto de Procesos Industriales Microbiológicos), Av. Belgrano y Caseros, Tucumán, Argentina

Química Biológica, Calidad de productos IV, Gestión de Empresas Agroalimentarias, Universidad de San Pablo, Tucumán, Argentina

G. Palma Departamento de Ciencias Químicas y Recursos Naturales, Núcleo Científico y Tecnológico en Biorecursos, Universidad de La Frontera, Temuco, Chile

A.J. Paula Department of Physics, Federal University of Ceara, Fortaleza, CE, Brazil

M.A. Polti Pilot Plant of Industrial and Microbiology, National Council of Scientific and Techniques, Natural Sciences College and Miguel Lillo Institute, Tucumán, Argentina

G.W. Poratti Centro de Investigación y Desarrollo en Fermentaciones Industriales, Centro Científico Tecnológico La Plata, Consejo Nacional de Investigaciones Científicas y Técnicas, Universidad Nacional de La Plata, La Plata, Buenos Aires, Argentina

C. Prado Universidad Nacional de Tucumán, Tucumán, Argentina

F. Prado Universidad Nacional de Tucumán, Tucumán, Argentina

V.B. Rajal Instituto de Investigaciones para la Industria Química (INIQUI), Consejo Nacional de Investigaciones Científicas y Técnicas (CONICET); Universidad Nacional de Salta (UNSa), Salta, Argentina

A.S. Ridolfi Universidad de Buenos Aires, Buenos Aires, Argentina

A. Rodriguez Planta Piloto de Procesos Industriales y Microbiológicos, Consejo Nacional de Investigaciones Científicas y Técnicas, Tucumán, Argentina

A. Rodriguez Centro de Investigación en Biotecnología, Universidad Autónoma del Estado de Morelos, Cuernavaca, México

N. Romano-Armada Universidad Nacional de Salta, Salta, Argentina

M. Rosa Universidad Nacional de Tucumán, Tucumán, Argentina

O. Rubilar Departamento de Ingeniería Química, Núcleo Científico y Tecnológico en Biorecursos, Universidad de La Frontera, Temuco, Chile

J.M. Saez Planta Piloto de Procesos Industriales y Microbiológicos, Consejo Nacional de Investigaciones Científicas y Técnicas, Tucumán, Argentina

E. Sánchez-Salinas Centro de Investigación en Biotecnología, Universidad Autónoma del Estado de Morelos, Cuernavaca, México

A.B. Seabra Exact and Earth Sciences Department, Federal University of Sao Paulo, Diadema, SP, Brazil

P. Sineli Planta Piloto de Procesos Industriales y Microbiológicos, Consejo Nacional de Investigaciones Científicas y Técnicas, Universidad Nacional de Tucumán, Tucumán, Argentina

María Z. Simón Solá Planta Piloto de Procesos Industriales y Microbiológicos, Consejo Nacional de Investigaciones Científicas y Técnicas, Universidad Nacional de Tucumán, Tucumán, Argentina

G. Tortella Departamento de Ingeniería Química, Núcleo Científico y Tecnológico en Biorecursos, Universidad de La Frontera, Temuco, Chile

M. Viera Centro de Investigación y Desarrollo en Fermentaciones Industriales, Centro Científico Tecnológico La Plata, Consejo Nacional de Investigaciones Científicas y Técnicas, Universidad Nacional de La Plata, La Plata, Buenos Aires, Argentina

L.B. Villegas Consejo Nacional de Investigaciones Científicas y Técnicas, San Luis, Argentina

N. Yagnentovsky Centro de Investigación y Desarrollo en Fermentaciones Industriales, Centro Científico Tecnológico La Plata, Consejo Nacional de Investigaciones Científicas y Técnicas, Universidad Nacional de La Plata, La Plata, Buenos Aires, Argentina

Chapter 1
Impacts of Agriculture in Latin America: Problems and Solutions

Neli Romano-Armada, María Julia Amoroso, and Verónica B. Rajal

Abstract According to the Merriam-Webster dictionary definition, agriculture is "the science, art, or practice of cultivating the soil, producing crops, and raising livestock, and in varying degrees the preparation and marketing of the resulting products." As it is, even though this definition is very comprehensive, there is more than meets the eye about the universe of activities involved in agronomic practices. Agriculture is witness to many changes since its dawning, evolving from animal and seed domestication to genetic modification of organisms through molecular biology techniques to better suit worldwide demand. These changes influenced land ownership as well as technology development. Different climatic and anthropological realities generate diverse productive systems that adapt to every situation. Developing countries with less appropriate technologies for industrialization have large rural populations where peasant economies are sometimes critical for subsistence. Despite all the benefits of agriculture, it sometimes faces drawbacks for proper production. Other times agriculture brings along sociopolitical and environmental problems. Cultural practices have detrimental impacts on soil quality and water availability. Moreover the environment is the ultimate recipient of pesticides, herbicides, fertilizers, and other by-products. Nonetheless bioremediation techniques involving plants and microorganisms are in constant development to try to decrease the negative effects of farming, bringing these emerging processes new options for a better management in agriculture.

N. Romano-Armada • V.B. Rajal (✉)
Instituto de Investigaciones para la Industria Química (INIQUI),
Consejo Nacional de Investigaciones Científicas y Técnicas (CONICET);
Universidad Nacional de Salta (UNSa), Av. Bolivia 5150, 4400, Salta, Argentina

Universidad Nacional de Salta (UNSa),
Av. Bolivia 5150, 4400, Salta, Argentina
e-mail: vbrajal@gmail.com

M.J. Amoroso
Planta Piloto de Procesos Industriales Microbiológicos (PROIMI),
CONICET, Av. Belgrano y Pasaje Caseros, 4000 Tucumán, Argentina

Facultad de Bioquímica, Química y Farmacia,
Universidad Nacional de Tucumán, 4400 Tucumán, Argentina

Universidad del Norte Santo Tomás de Aquino (UNSTA),
4000 Tucumán, 4400 Tucumán, Argentina

A. Alvarez, M.A. Polti (eds.), *Bioremediation in Latin America: Current Research and Perspectives*, DOI 10.1007/978-3-319-05738-5_1

1.1 Introduction

Since the beginning of times, humankind has sought ways to satisfy a perpetual need for food and improve quality of life. The hunter-gatherer lifestyle demanded a large amount of time and energy to meet these needs. Domestication of seeds and livestock meant the possibility of settling for nomadic populations by providing a means of a reliable food supply. It is an undeniable fact that since people were able to produce surplus food that could be stored or traded, more time was available for activities other than farming. This development of agriculture was a substantial trigger for the evolution of today's civilization.

Several theories attempt to explain the origin and evolution of agriculture. Regarding geographical location, the concept of "Centers of Origin of Cultivated Plants" first posed in the mid 1920s by Nikolai Vavilov (1992) had an important effect on the subsequent generations who attempted to demonstrate where the sites of origin of agriculture were. Vavilov equated crop diversity centers with countries of origin of agriculture, inducing, perhaps, a conceptual confusion between both phenomena for most authors who based their theories around his original work (Harris 1990).

There are also several theories regarding how anthropological behavior drove agriculture evolution, but there is little consensus as to what propels the evolution of culture. Rosenberg (1990) harshly criticized the theories of human domestication of species, stating that agriculture was not only dependent on the evolution of domesticates, but also on the evolution of human behavior. Whether domestication was consciously or unconsciously sought, those species growing under the protected environments of humans were more successful, while those without protection were not suitable for agriculture, wane, and rapidly lost (Thrall et al. 2010).

The cause and effect of what led people to start farming in different parts of the world is still unclear. Did climate conditions at the end of the last Ice Age cause favorable environs that favor wild cereals and other annual plants in the Near East? Perhaps a demographic rise in East Asia required increased food resources to sustain the population, forcing people to find homegrown solutions.

Whether it was livestock on the Fertile Crescent or cereals on the Near East, it was in the Neolithic period that domestication dawned. The dates for the domestication of animals such as cattle, sheep, pigs, and goat range from 13,000 to 11,000 years ago, about the same period as squash cultivation in Mexico. Around 9,000 years ago, the wild progenitors of wheat, barley, and peas were grown in Syria, simultaneously; the first mutations of teosinte were starting to see light as maize in Mexico (Beadle 1980), while rice and millet were farmed in China. The presence of grinding stones for grain in early Neolithic villages' houses marks the shifting from nomadic to a settled way of life. Appearing 5,000 years ago, the youngest domesticates of this agronomic evolutionary race were the Andean potato and the North American sunflower (Harlan 1971).

Regardless the reasons for its independent origins, as agriculture developed, the environmental conditions between domesticated and wild species became markedly different. Over the years, domestication, selection, hybridization, and induced mutation have led to changes in the plants and animals regarding their appearance, nutritional value, yield, and even less obvious traits, such as patterns of social behavior. We could infer that domestication promotes a human-driven process of selection towards those desirable traits on the species to breed.

Among agronomic practices such as husbandry, or tilling and crop rotation, introduced for fertility reasons, developments in the understanding of evolution and genetics clearly reflect its influence on animal and plant improvement. The advances in molecular biology have speeded up this process by applying genetic modification technologies, enabling modification of specific genes or gene pathways to express desired characteristics such as disease control or resistance to particular chemical compounds or water stress, among others (Thrall et al. 2010).

Thus, agriculture is the cause and effect of a wide scope of selective forces. Modern agriculture is the consequence of continuous change. Today's agriculture will benefit in the near future not only from the application of more efficient conventional breeding strategies, but also from the advances in genetic manipulation of species. For the arising of new plants and animals, ecological and evolutionary principles must be considered. New organisms will need to adapt to the current and future scenario, where climate change and sustainability will drive management and development of agronomic practices.

1.2 Agriculture's World Reaches

According to FAOSTAT [Statistical online Food and Agriculture Organization from the United Nations (FAO) database], the current world population is approximately 7.1 billion people, and Latin America accounts for 9 % of that total.

In the early 1960s almost 67 % of the world's population was, according to FAO categorization, rural population. This proportion decreased over the years, to a present 48 %, setting the trend to an estimated projection of 31 % for the 2050s. Nevertheless not all the rural population is involved in farming activities. In the 1980s, while 60 % of the world's population classified as rural population, only 50 % was considered to be agricultural population. The latter has been decreasing over the years with a tendency to keep doing so over the future at an alarming rate (FAOSTAT 2013).

Despite the decline of rural residents, the demographic projection is for the world's population to continually expand in the next decades; hence there will be a growing need to convert additional lands to farming to keep up with that demand. In the last five decades the agricultural area has grown 10 % in Latin America, going from 0.56 to 0.74 billion hectares. According to the current trend in agricultural lands, one could imagine expansion to its full potential over the years (FAO 2012a; FAOSTAT 2013).

It is a global concern how the land conversion to agriculture does not compensate the demographic climbing. In Latin America alone the ratio of arable land has descended from 0.39 to 0.28 ha per person in the last five decades. This unbalanced situation ultimately causes the population to overtake the resources where agricultural lands are limited. Under this intense pressure, the productive capacity of the land is surpassed.

At the moment, in the world there are more than 1.5 billion hectares used for agricultural purposes (i.e., arable land and those destined to permanent crops). Mainly 90 % of the potentially accessible agricultural land is located in Latin America and sub-Saharan Africa, and half of that is shared by only seven countries, among which are Brazil, Argentina, Colombia, and the Plurinational State of Bolivia. This uneven distribution of land potentially available for agriculture limits the agronomic border expansion around the world, leaving in some places virtually no spare land. According to FAO (Lindquist et al. 2012), despite the presence of considerable amounts of land potentially suitable for agriculture, much of it is covered by forests, protected for environmental reasons, or employed for urban settlements. Nevertheless, in developing countries, generally in order to satisfy the growing population demand, agricultural border expansion occurs at the expense of lands with environmental covers such as forest or grasslands.

1.3 Drawbacks of Agricultural Development

To accomplish benefits such as food security and undernourishment abatement, developing countries face important problems that hinder the rise of productivity to its potential level in agriculture. The root of these problems is usually related to environmental or sociopolitical issues.

A worldwide consideration is the intertwined dependence of agriculture and nature. Climate change and agriculture are interrelated processes, both of which take place on a global scale. Global warming projects have significant impacts on conditions affecting agriculture, including temperature, precipitation, and glacial runoff. These conditions determine the carrying capacity of the biosphere to produce enough food for the human population and domesticated animals. Rising carbon dioxide levels will also have effects, both detrimental and beneficial, on crop yields (McGrath and Lobell 2013).

The overall effect of climate change on agriculture will depend on the balance of these and other effects. Assessment of the latter might help to properly anticipate and adapt farming to maximize agricultural production. Also disturbed soil conditions such as acidification, salinization, sedimentation, nutrient depletion, among others destroy the productive capital of the valuable soil resource (Roy et al. 2006).

Feeble governance implies improper management strategies and a lack of resources for investment in transportation infrastructure. This lack of government support along with the low primary production price can be felt as a hurdles race to many farmers. To abolish famine in a sustainable manner, it is necessary to plan and appropriate significant increases on agronomic investments. In spite of their limitations, farmers in developing countries invest every year, in productive assets, four times their governments' investment on agriculture. This overwhelming presence of investments from particular farmers outshines the public investment in agriculture, noting the critical role farmers should have on any strategy that aims to improve the amount and efficacy of agronomic investments and production (FAO 2012b).

1.4 Agronomic Systems Characteristics

The many variants of agricultural systems that farmers apply to the land can be classified according to a diverse spectrum of criteria. The three major classification criteria used for this are water source, target market and use of technology, and magnitude of land covered and profitability.

Many other criteria might be used to classify productive systems, such as net profitability, soil sustainability, use of chemical compounds, among others.

Rain-fed and irrigated agriculture. Two types of systems can be distinguished when considering water availability and basically what makes the difference is who provides the water.

In the rain-fed system rain is the source of water. In this way the species to be cultivated are limited to those for which the optimal growth conditions are those ruling the wet season. In the world, 80 % of the arable lands are rain fed and 90 % of them are located in Latin America. Although this is the major production system in the world, it is highly influenced by the climatic variability, questioning the potential production levels of the land (FAO 2012a).

On the other hand, in the irrigated system, the source of water is always man-made; the farmers water the crops by artificial means. These include groundwater extraction, surface water uptake and channeling, and dam or reservoir constructions to store water and ensure the continuous supply for irrigation. As such, the species grown become independent of the season, and are not defined by rain but by other climatic factors such as temperature and photoperiod.

Subsistence and industrial agriculture. Target market and technology applied divide as well two major systems, subsistence and industrial agriculture. The first involves production of sufficient amount of food and produce to meet the needs of the farmer's family group, and warehouse a small amount for the family, with little surplus to market. The technical level is primitive, often employing hand tools because of topography demands, or simply for lacking of mechanization. The application of chemicals is also made with little or no regard to human health and environment protection. Because of more permissive laws and bad governance toxic products banned in developed countries are of frequent use.

Opposed to this situation is the industrial agriculture system. Based on satisfying consumer demand with no intention of self-consumption, the aim is to profit from the agronomic production. The generation of large quantities of products is possible as a consequence of mechanization of agriculture which saves time and money, because of the higher yields agronomic specialization allows, and the quick transportation which favors commercialization, profitability increase. This is a common system in industrialized and in some developing countries; it is also frequently seen in poor countries with the introduction and utilization of foreign capital investments.

Intensive and extensive agriculture. Intensive and extensive agronomic systems refer to the amounts of labor and capital relative to land area. The intensive system produces greater yields per unit of land than extensive agriculture but it also takes higher financial inputs. In other terms, a farm using extensive agriculture will require more land than an intensive agricultural one to produce a similar profit. The latter requires intensive labor for the application of fertilizer and pesticides, harvesting, and other manual tasks.

Intensive animal farming involves large numbers of livestock raised, confined, or on limited land, requiring important quantities of food, water, and medical input, whereas extensive farming commonly refers to sheep and cattle farming in vast areas with low agricultural productivity. Nomadic herding is a common practice among Latin American rural populations that benefit from extensive farming. The herders move the flock to feed from winter to summer pastures, and vice versa.

The rising of landgrabbing in Latin America during the past decade has a negative impact on local population. Landgrabbing is a polemic activity of intensive industrialized agriculture in developing countries. The food crisis of 2007–2008, caused by the competition between foods versus fuel, led developed countries to seek control over large tracts of lands overseas. These contentious large-scale land acquisitions pose a threat to the land's fare distribution and ownership by reconcentration of land and capital (Borras et al. 2012).

1.5 Environmental Impacts of Agriculture

The land enabled every year is not enough to compensate the demographic growth; hence there is a continuous search for yield enhancement. For this purpose several practices are applied to agriculture, ranging from animal and plant improvement by genetic engineering to the spreading of agrochemical. Thus, not only benefits come from higher yield, but also collateral effects.

The impacts of agriculture vary according to the different cultural practices, type of crop, and prevailing environmental factors.

1.5.1 Deforestation

The purposes of clearing forest lands can be diverse. The reasons range from conversion of forest land to farmlands and ranches, as well as urban areas, and hydrocarbon's exploitation, among others. At the end, the spirit of deforestation is the removal of a forest or stand of trees. Contrary to an intuitive rationalization, wood extraction for logging and fuel production is responsible for only 19 % of the world deforestation, while industrial agriculture and subsistence farming account for 32 and 49 %, respectively (UNFCCC 2007).

When forests are replaced by agricultural land, either for crops or cattle, there is a decrease on atmospheric carbon sequestration. Even if the former forests are completely covered by crops, its carbon immobilization capacity is much lower than that of trees provoking the shrinkage of long-term carbon sinks around the world. There is certain controversy about how developed and highly industrialized countries benefit economically from their long ago deforested lands, and environmentally from carbon sinks of forests in developing countries. Needless to say, the environmental service forests provide (like biodiversity and soil conservation, underground water recharge) is difficult to compare to the monetary profit obtained from agriculture because the value system for each is different. Environmental services are more of long-term value, while market goods are immediate but for a short term.

1.5.2 Climate Change

The earth's surface absorbs the sun radiation and this energy is then redistributed by the atmospheric and oceanic circulation, being then irradiated back into space in longer wavelengths. Therefore, factors disrupting the entrance of radiation from the sun, the energy irradiated to space by the earth, or that alters the energy redistribution in the atmosphere or between it and the lands or oceans have the potential to change the global weather (Serio 2013).

The presence of humans and their continuous consuming and transformational activities have an impact on the weather. Climate change and agriculture have reciprocal effects upon each other. Agronomic activities in general do its bit to the phenomenon of global warming by emission of greenhouse gases to the atmosphere and changes on the land cover which modify the surface's albedo and alter the radiation balance.

Greenhouse gas emission. Greenhouse gases on the atmosphere absorb part of the infrared energy irradiated by the earth, creating optimal conditions for life on the planet. When greenhouse gases accumulate in the atmosphere, they absorb more terrestrial radiation and a fraction of it is emitted back into the planet's surface creating a radiative forcing. Thus the earth's efficiency to irradiate energy to space decreases.

In 1997, the Kyoto Protocol was adopted by Parties to the United Nations Framework Convention on Climate Change (UNFCCC) and entered into force in 2005. It sat binding obligations on industrialized countries to reduce emissions of greenhouse gases; however, developing countries are parties without binding targets.

Agriculture is responsible for the emission of 9.4 carbon dioxide (CO_2) equivalent petagrams ($Pg = 10^{15}$ g) in the world (FAOSTAT 2013). Agricultural activities such as managing livestock, creating and maintaining rice paddies, sugar cane burning, and soil bacterial nitrification-denitrification produce methane, CO_2, and nitrous oxide gases.

Cattle enteric fermentation and rice cultivation are the major sources of atmospheric methane in the world, adding up to 2.5 CO_2 equivalent Pg. Latin America's contribution alone is 26 and 3 % of the world total for cattle and rice, respectively (FAOSTAT 2013). The waterlogged soil of paddy fields provides optimal condition for methanogenesis, releasing methane to the atmosphere. Recent research shows that the latter emissions could be reduced by drainage; this aerates the soil interrupting methane production, and increases the rice yield (Ma et al. 2013; Yang et al. 2013).

Autotrophic and heterotrophic respiration emit carbon to the atmosphere, but biomass burning along with fossil fuel used in agriculture activities adds up to 4.7 Pg of CO_2 emitted worldwide with 18 % attributable to countries in Latin America (FAOSTAT 2013). In some countries, pasture burning is a practice to promote bud sprouting at the end of the dry season, releasing significant amounts of carbon to the atmosphere. "Slash and burn" deforestation also results in a multiple contribution to the overall total carbon emissions. It not only eliminates the carbon reuptake source but also, when the cut vegetation burns, it releases back into the atmosphere the carbon that was immobilized.

Nitrous oxide (N_2O) is a greenhouse gas that when compared with CO_2 has 310 times the ability per molecule of gas to trap heat in the atmosphere. Research studies suggest that it also causes ozone depletion (Ravishankara et al. 2009). Because of natural soil microbial processes from nitrification and denitrification, N_2O naturally forms beneath the ground surface. Agricultural soils provide a large source of N_2O into the atmosphere, especially lands where intensive farming drives high applications of many greenhouse gases causing products, among them manure and nitrogen fertilizers. During the year 2010, agriculture released a total of 2.0 CO_2 equivalent Pg of N_2O into the atmosphere (FAOSTAT 2013).

Land covers transformation. Changes on the world population demand translate into changes in land use. Some changes are programed and others unwanted, but most of them bring about negative impacts like biodiversity losses, climatic change, and soil erosion among others (Marlenko 2003). These changes on land use lead to land cover transformation.

The earth's surface reflects, absorbs, and transmits its own energy and that of the sun, giving to the many land covers in the world a characteristic spectral reflectance patterns or spectral signature (i.e., a specific combination of emitted, reflected, or absorbed electromagnetic radiation at varying wavelengths which can uniquely identify an object). Bare soil and vegetation spectral signatures are very different. The reflectance of vegetation is higher on the visible and close infrared regions of the electromagnetic spectrum, whereas, for soil, it is higher on the intermediate infrared region. Implementation of different agronomic practices can change the energy emitted by the soils for better or worse. Higher contents of water in the soil reduce the reflectance while saline conditions increase it (Navone and Maggi 2013).

1.5.3 Soil Degradation

The quality of a soil is a combination of the physical, chemical, and biological properties which determine its capacity to increase crop and livestock productivity. It is dynamic and can change on the short run according to specific soil characteristics, environmental conditions, land use, and agricultural practices applied (Albanesi et al. 2013).

Many are the faces of soil degradation. Desertization is an extreme outcome of a combination of forms where fertile land turns into barren land or desert. Processes like overgrazing, cover elimination, topsoil loss, salinization, and nutrient depletion prevent plant life at desert boundaries.

Nutrient depletion. In natural ecosystems the nutrients recycle, having little external input from rain or atmospheric dust, whereas agro ecosystems are open systems that go through the continuous extraction of nutrients from the land. If these nutrients are not replenished, the soil impoverishment endangers its sustainability (Álvarez and Rubio 2009; Lavado 2012).

A decline in organic matter shows that the constant agronomic use of the land decreases the carbon content in the soil (Álvarez and Steinbach 2012). Some species can mine the soil's nutrients, needing high subsidies from fertilizers. Monoculture and excessive watering have a high nutrient removal capacity. A negative balance where nutrients have been exhausted can only be tolerated for short periods without leading to soil deterioration. Since improving depleted soils is expensive, farmers may shift to other fields or deforest new lands for farming (Roy et al. 2006).

Soil acidification. Acid soils exist by natural and anthropogenic causes. The latter can be a product of acid rain deposition, negligible in soils with high buffer capacity, or by farming. Two are the main agronomic factors which alter the acidity of the soils, aggravating, if applicable, their natural acidic conditions. High nutrient extraction leads to a decline of bases with a consequent pH reduction. Also, the application of nitrogen fertilizers that produce ammonia such as urea, ammonium nitrate, ammonium sulfate, or anhydrous ammonia leads to acidification by the proton liberation that occurs when microbial oxidation turns ammonia to nitrate (Roy et al. 2006; Álvarez and Rubio 2009). Thus, with the constant application and use of high doses of these fertilizers, the possibility exists of negative impacts on the soil in the long run.

Acidic conditions allow aluminum to dissolve, making it bioavailable for plants. This element, along with manganese, inhibits plant growth. Furthermore, the hydrogen ion has negative effects on plants in two ways: first by competing with other ions it interferes with membrane functioning and second, the high energy cost the plant must expend to maintain the pH balance in the root cells. All this leads to the need of amendment application to correct the pH (Álvarez and Rubio 2009).

Soil erosion. Erosion is the outcome of topsoil removal. Topsoil is at risk whenever plant cover is removed from the surface. Overgrazed and nutrient-exhausted lands that prevent the proper growth of vegetation leave the surface exposed to erosion (Roy et al. 2006). Depending on the region, the transport agent can be wind and/or water; either one can be highly damaging. Runoff sedimentation can aggradate downstream water bodies and even lower lands by upstream removed material deposits.

1.5.4 Pollution

Agricultural practices have by-products of a biotic or abiotic nature. Certain practices diminish environmental quality and can harm humans' health or economic interests via contamination. Contamination may come from point to non-point sources, and its impact can go from local to global. Good management practices and techniques are necessary to control the amount and fate of pollutants.

Fertilizers. Nitrogen (N) and phosphorus (P) are essential plant nutrients. They can be supplied to agricultural land by synthetic and organic fertilizers. To estimate the dose of fertilizer a specific crop requires it is necessary to consider the nutrient removal from the cultivated species to yield a given production, the nutrient availability on the soil, and the proportion of nutrient the plant absorbs per kilo of fertilizer applied to the soil. In other words, the balance is achieved with a combination of the nutritional demand and utilization efficiency of the crop, and the nutrient supply from the soil (Benintende and Benintende 2013).

Latin America imported 11 billion USD worth of fertilizers on 2011 (FAOSTAT 2013). This is not necessarily positive, because when the application is not managed correctly, excess N and P can have negative environmental consequences.

Nitrogen fertilizers contribute to the already mentioned problems of nitrous oxide emission and soil acidification (Álvarez and Rubio 2009). Additionally, the leaching of nitrate to groundwater presents a toxic contamination hazard for the neighboring population.

Phosphorus is a limiting nutrient for plant growth and for that reason, when the excess of P reaches water bodies by runoff or leaching, the nutrient's extra input causes a quick proliferation of algae. As it grows, algae consume all of the nutrients also required by water plants and animals. When the population of algae reaches its maximum density, it starts to die and sink. At the bottom, microbial decomposition of the organic matter depletes the oxygen in the deeper water creating a "dead zone," a layer of anoxic conditions deadly to fish and other organisms. This is a highly damaging phenomenon known as eutrophication (Figueruelo and Dávila 2004).

In the past the need of fertilizers was determined by the soil's nutrient deficiency and plant requirement. Since the rise of genetic engineering and the development of new technologies to enhance plant productivity, the concept of deficiency has lost validity. Nowadays in order to achieve the potential yield of different plant species, the natural nutrients in the soil are not enough and require additional input. It is feasible to say that now the use of fertilizers is independent of either the availability or lack of a particular nutrient in the soil (Lavado 2012).

Pesticides. Farming has many hindrances that set back productivity. If we categorize according to the type of interaction with the species of interest we find competitors (take up growth resources), predators (feed from the product), and pathogens (produce diseases). These three external factors bring about shortcomings on the yield that translate in low revenue. In order to eradicate these disrupting agents there is an ongoing search to develop means of control. These controls are often in the form of chemical annihilators called pesticides. As in farming the bigger loss is caused by weeds on crops, the market is dominated by herbicides (Oerke 2006).

Until the early twentieth century, pesticides were natural and inorganic compounds. The end of the 1930s was the spawning of the synthetic organic pesticides era, with a movement towards the specifically targeted development of evermore selective substances. The environmental and health-related hazards many of these compounds pose overshadow the enhancement of productivity.

According to their chemical structure, pesticides are divided into organic and inorganic groups. A preponderance of inorganic pesticides belong to the group of toxic element compounds like copper and lead arsenic compounds, fluorinated insecticides, copper and sulfur fungicides, and inorganic herbicides. Organic pesticides have been largely developed. Classification of organic pesticides is by their chemical structure and composition, and they are chlorinated hydrocarbons, chlorophenoxy, organophosphates, carbamates, among others. Also, considered different groups are polychlorinated biphenyls (PCBs) and dioxins.

In-depth studies of chlorinated hydrocarbons are readily available, perhaps because of their extended use in warfare and their particular lingering persistence in the environment. The main representatives in this group are dichlorodiphenyltrichloroethane (DDT), methoxychlor, lindane (γ-hexachlorocyclohexane), 1,4-dichlorobenzene, hexachlorobenzene, aldrin, dieldrin, endrin, chlordane, heptachlor, mirex, kepone, and toxaphene. Their persistence ranges from 4 to 12 years and their high fat solubility drives to bioaccumulation on the trophic chain.

Chlorophenoxy are hormonal herbicides discovered in the 1940s. The leaves absorb the compound and translocate it to the meristems of the plant causing uncontrolled and unsustainable growth leading to plant death. The infamous 2,4-D (2,4-dichlorophenoxyacetic acid) and 2,4,5-T (2,4,5-trichlorophenoxyacetic acid) were the components of Agent Orange, used as a defoliant by the US military during the Vietnam War as part of a herbicidal warfare program. They are believed to be carcinogenic to humans (Alavanja et al. 2013; Bokán et al. 2013).

Organophosphates are organic derivate compounds of phosphoric and thiophosphoric acid. Although in the beginning they were developed for military purposes, their insecticide attributes were discovered during World War II. Some examples from this group are tetraethyl pyrophosphate (TEPP), parathion, dichlorvos (DDVP), malathion, and dipterex. Their mode of action is irreversible inactivation of acetylcholinesterase, which is essential to nerve function in insects, humans, and many other animals. Although their toxicity is much higher than chlorinated hydrocarbons, organophosphates are considered less dangerous because of their shorter life-span.

Carbamates are compounds derivatives from carbamic acid. Their use started on the late 1950s, when carbaril was synthesized. The insecticide carbamates act similarly to organophosphates, but other carbamates have nematicide, herbicide, and fungicide features. Carbaril is not as toxic or persistent as DDT, and a quick breakdown rate makes it environment friendly. Another noteworthy representative from this group is Baygon.

Polychlorinated biphenyls and dioxins are not pesticides, but they are found in agronomic practices. Aroclor (PCB as commercialized by Monsanto) serves as pesticide extender, which is a chemical that enhances the effectiveness or life-span of a pesticide. However 2,3,7,8-tetrachlorodibenzo-p-dioxin (TCDD) is a by-product dioxin from the synthesis of 2,4,5-T.

Although many benefits come from the use of pesticides, some also have serious drawbacks, such as toxicity risk to humans and nontarget species. According to the Stockholm Convention (2013) on persistent organic pollutants from 2001, 9 of the 12 worst offenders (the dirty dozen) belong to the chlorinated hydrocarbons group.

Heavy metals. Soils contaminated with potentially toxic elements (PTE) portray a hazardous situation. The term PTE is broad and encompasses heavy metals (atomic weight greater than 63 and density greater than 5–6 g/dm^3), other metals with lower atomic weights, and different nonmetals and metalloids. The commonality of the elements in this group is their toxicity at high concentrations. One manifestation of their toxicity is the production of free radicals that cause cellular oxidative stress and replace other metals in pigments or enzymes disrupting their proper functioning (Lavado 2009).

Anthropic contamination from agriculture is due to the use of a wide variety of production inputs. Fertilizers, gypsum and organic amendments, pesticides, and other agrochemical studies point out the importance of manure nowadays. Manipulated diets to enhance growth and increase feeding efficiency of poultry and other livestock add elements like arsenic (As), cobalt (Co), copper (Cu), iron (Fe), manganese (Mn), selenium (Se), and zinc (Zn) to the fodder. This raises the concentration of the mentioned PTEs in soils amended with manure (Lavado 2009; Khan et al. 2013).

Arsenic, as a component of arsenic pesticides or herbicides, still finds its way into agricultural soil (Huertas and Michán 2013). However, As-based herbicides are no longer used as weed killers, with the exception of monosodium methanearsonate (MSMA), a broadleaf weed herbicide for use on cotton (Bencko and Foong 2013). At present, irrigation excess dissolves regolith minerals that contain As and mobilize it by leaching it from natural sources (Lavado 2009).

1.6 Research on Remediation Strategies

By now it is known that dilution is not the solution to pollution. In fact, over the years research of several pollutants shows that in most of the cases, dilution exacerbates the negative impacts of contamination, because either the diluent body has limited buffer capacity and spreads the contamination, or because contaminants are immobilized and bioaugmented (Feldman 2013; Pani and Muwal 2013).

Bioremediation is a waste management technique. It involves the use of living organisms or their products to eliminate or neutralize contaminants from a specific site or environmental compartment. This elimination occurs by treating the pollutant at the site (in situ) or elsewhere (ex situ) by removing the contaminated material to be treated. Currently, there are numerous methods for bioremediation and most of them involve living plants and microorganisms (Niti et al. 2013).

Phytoremediation is a method where decontamination of the pollutant is via plants. It may occur in five ways: phytostabilization, phytoextraction, phytovolatilization, phytofiltration, and rhyzodegradation. It can also use hyperaccumulators, which are plants that extract heavy metals by accumulating them on their vegetable tissues (Atangana et al. 2014). This is the primary strategy for the bioremediation of heavy metals.

In Latin America, in spite of extensive mining of wealthy ore deposits and the resulting accruing-associated heavy metal pollution of those areas, there are few plants that identify as metal accumulators or metal tolerant. However, there is a slowly growing line of research that analyzes the tolerance and extraction-accumulation ability of some Latin American plants towards different pollutants in order to assess their potential use on mercury (Hg), As, Cu, Zn, cadmium (Cd), lead (Pb), and silver (Ag) bioremediation. Most authors have found roots to be the part of the plant with higher accumulative potential for the abovementioned metals (Bech et al. 2002; Ginocchio and Baker 2004; Benimeli et al. 2010), except for Hg which has been found in higher concentrations on leaf tissue (Romero Núñez et al. 2011). Data regarding plant's attributes can be put into a model to select species with the most appropriate extraction potential (Guala et al. 2011). The removal efficiency of Cu, molybdenum (Mo), chromium (Cr), Pb, titanium (Ti), Fe, lanthanum (La), Zn, cerium (Ce), and Mn by aquatic macrophyte studies shows that they act as good indicators of trace metals (Valitutto et al. 2007).

On a parallel path, research is being done to analyze the biotechnological potential of novel microorganisms for bioremediation applications. Most of the bioremediation techniques foresee microbial activity as an avenue to eliminate the pollution but they have different modes of action. Bioventing, biosparing, bioaugmentation, and biostimulation supply oxygen and nutrients to optimize the growth conditions of indigenous bacteria present in the soil to allow the efficient breakdown of the pollutants that are present. Generally these methods can be implemented without disturbing the local activities.

Bioleaching involves metal extraction from their ores and it could be thought as microbial mining. This constitutes a clean alternative to the use of cyanide from traditional leaching, but the inconvenient slowness of microorganisms to reach the levels of production of a conventional mining leaves the methodology momentarily relegated to contamination bioremediation. It involves ferrous iron and sulfur-oxidizing bacteria.

Because of the fast changes in agriculture, new chemical compounds are discharged in the environment every year. To alleviate the negative impacts from anthropic activities, research groups worldwide work on discovery and development of novel plants and microorganisms capable of bioremediation.

1.7 Concluding Remarks

Latin America's different crops and production models are not only driven by a ranging variety of weather and landscape. They also reflect the idiosyncrasy of their land owners from different cultures and the land policies enforced by each country. The production scale influences the potential environmental damage, but even more important is the knowledge and use of Good Agricultural Practices (GAPs).

When analyzing the different agronomic models used throughout Latin America it is possible to see at least three common factors conspiring strongly against sustainability. Intertwined among each other are the lack of education in GAPs, the low capital investment to see their implementation, and the bad governance. To reduce vulnerability in agriculture, the government has to ensure market stability and to supply small- and medium-scale farmers the capital to invest in the implementation and enhancement of GAPs.

Agricultural activity originates sociopolitical and environmental problems. Although the scientific community takes actions to remediate the latter, more effort should be put into strategies to prevent environmental and human damage. Both, industrial and ecological, technological paradigms are extreme and should be replaced by an integrated paradigm based on a preventive approach.

The discovery of innocuous agrochemicals and bio-control and bio-fertilizing agents paves the way to a cleaner production. Therefore, research on bioremediation along with clean production strategies has to develop simultaneously to ensure sustainability in agriculture.

References

Alavanja MCR, Ross MK, Bonner MR (2013) Increased cancer burden among pesticide applicators and others due to pesticide exposure. Cancer J Clin 63:120–142

Albanesi A, Anriquez A, Dominguez Nuñez JA et al (2013) Soil quality. Biological properties and evaluation in semiarid ecosystems. In: Albanesi A (ed) Microbiología agrícola. Un aporte de la investigación en Argentina, Magna, Tucumán, pp 1–26

Álvarez CR, Rubio G (2009) Acidez del suelo. In: Taboada MA, Lavado RS (eds) Alteraciones de la fertilidad de los suelos. El halomorfismo, la acidez, el hidromorfismo y las inundaciones. Editorial Facultad Agronomía, Buenos Aires, pp 59–76

Álvarez R, Steinbach HS (2012) Efecto del uso agrícola sobre el nivel de materia orgánica. In: Álvarez R, Rubio G, Álvarez CR, Lavado RS (eds) Fertilidad de suelos. Caracterización y manejo en la Región Pampeana. Editorial Facultad Agronomía, Buenos Aires, pp 201–230

Atangana A, Khasa D, Chang S et al (2014) Tropical agroforestry, Phytoremediation in tropical agroforestry. Springer, Amsterdam, pp 343–351

Beadle GW (1980) The ancestry of corn. Sci Am 242:112–119

Bech J, Poschenrieder C, Barceló J et al (2002) Plants from mine spoils in the South American area as potential sources of germplasm for phytoremediation technologies. Acta Biotechnol 22:5–11

Bencko V, Foong FYL (2013) The history, toxicity and adverse human health and environmental effects related to the use of agent orange. In: Simeonov LI, Macaev FZ, Simeonova BG (eds)

Environmental security assessment and management of obsolete pesticides in southeast Europe. Springer, Varna, pp 119–130
Benimeli CS, Medina A, Navarro CM et al (2010) Bioaccumulation of copper by *Zea mays*: impact on root, shoot and leaf growth. Water Air Soil Pollut 210:365–370
Benintende SM, Benintende MC (2013) Mineralizable N by anaerobic incubation used to establish fertilization needs. In: Albanesi A (ed) Microbiología agrícola. Un aporte de la investigación en Argentina, Magna, Tucumán, pp 111–122
Bokán K, Syberg K, Jensen K et al (2013) Genotoxic potential of two herbicides and their active ingredients assessed with comet assay on a fish cell line, Epithelioma Papillosum Cyprini (EPC). J Toxicol Environ Health 76:1129–1137
Borras SM Jr, Franco JC, Gómez S et al (2012) Land grabbing in Latin America and the Caribbean. J Peasant Stud 39:845–872
FAOSTAT (2013) Food and Agriculture Organization of the United Nations. http://faostat3.fao.org/faostat-gateway/go/to/home/E. Accessed Nov 2013
Feldman E (2013) A course in environmental sustainability. Paper presented at the 17th annual conference of the national collegiate inventors and innovators alliance (NCIIA), Washington, DC, 22–23 Mar 2013
Figueruelo JE, Dávila MM (2004) Química física del ambiente y de los procesos medioambientales. Reverté, Barcelona
Food and Agriculture Organization of the United Nations (FAO) (2012a) FAO statistical yearbook 2012. World Food and Agriculture, Rome
Food and Agriculture Organization of the United Nations (FAO) (2012b) The state of food and agriculture. Investing in agriculture for a better future. FAO, Rome
Ginocchio R, Baker AJM (2004) Metallophytes in Latin America: a remarkable biological and genetic resource scarcely known and studied in the region. Rev Chil Hist Nat 77:185–194
Guala SD, Vega FA, Covelo EF (2011) Development of a model to select plants with optimum metal phytoextraction potential. Environ Sci Pollut R 18:997–1003
Harlan JR (1971) Agricultural origins: centers and noncenters. Science 174:468–474
Harris DR (1990) Vavilov's concept of centers of origin of cultivated plants: its genesis and its influence on the study of agricultural origins. Biol J Linn Soc 39:7–16
Huertas MJ, Michán C (2013) Indispensable or toxic? The phosphate versus arsenate debate. Microb Biotechnol 6:209–211
Khan K, Lu Y, Khan H et al (2013) Heavy metals in agricultural soils and crops and their health risks in Swat District, northern Pakistan. Food Chem Toxicol 58:449–458
Kyoto Protocol (2013) United nations framework convention on climate change (UNFCCC). http://unfccc.int/2860.php. Accessed Nov 2013
Lavado RS (2009) Contaminación de suelos de la Región Pampeana con elementos potencialmente tóxicos. In: Taboada MA, Lavado RS (eds) Alteraciones de la fertilidad de los suelos. El halomorfismo, la acidez, el hidromorfismo y las inundaciones. Editorial Facultad Agronomía, Buenos Aires, pp 151–163
Lavado RS (2012) La fertilidad del suelo, sus características y la utilización de fertilizantes. In: Álvarez R, Rubio G, Álvarez CR, Lavado RS (eds) Fertilidad de suelos. Caracterización y manejo en la Región Pampeana. Editorial Facultad Agronomía, Buenos Aires, pp 15–22
Lindquist EJ, D'Annunzio R, Gerrand A et al (2012) Global forest land-use change 1990–2005, FAO forestry paper no 169. Food and Agriculture Organization of the United Nations and European Commission Joint Research Centre (FAO & JRC), Rome
Ma K, Conrad R, Lu Y (2013) Dry/wet cycles change the activity and population dynamics of methanotrophs in rice field soil. Appl Environ Microb. doi:10.1128/AEM.00850-13
Marlenko N (2003) Aplicaciones en uso de la tierra. In: Navone SM (ed) Sensores remotos aplicados al estudio de los recursos naturales. Editorial Facultad Agronomía, Buenos Aires, pp 119–144
McGrath JM, Lobell DB (2013) Regional disparities in the CO2 fertilization effect and implications for crop yields. Environ Res Lett 8:1–9

Navone SM, Maggi AE (2013) Características espectrales. In: Navone SM (ed) Sensores remotos aplicados al estudio de los recursos naturales. Editorial Facultad Agronomía, Buenos Aires, pp 101–116
Niti C, Sunita S, Kamlesh K et al (2013) Bioremediation: an emerging technology for remediation of pesticides. Res J Chem Environ 17:88–105
Oerke EC (2006) Crop losses to pests. J Agr Sci 144:31–43
Pani B, Muwal PK (2013) Alternative synthetic approaches for the industrial pollution control and prevention. J Acad Indus Res 8:511–513
Ravishankara AR, Daniel JS, Portmann RW (2009) Nitrous oxide (N2O): the dominant ozone-depleting substance emitted in the 21st century. Science 326:123–125
Romero Núñez SE, Marrugo Negrete JL, Arias Rios JE et al (2011) Hg, Cu, Pb, Cd, and Zn accumulation in macrophytes growing in tropical wetlands. Water Air Soil Pollut 216:361–373
Rosenberg M (1990) The mother of invention: evolutionary theory, territoriality, and the origins of agriculture. Am Anthropol 92:399–415
Roy RN, Finck A, Blair GJ et al (2006) Plant nutrition for food security. A guide for integrated nutrient management, FAO Fertilizer and plant nutrition bulletin No. 16. Food and Agriculture Organization of the United Nations (FAO), Rome
Serio LA (2013) Variabilidad y cambio climático. In: Murphy GM, Hurtado RH (eds) Agrometeorología. Editorial Facultad Agronomía, Buenos Aires, pp 149–164
Stockholm Convention (2013) United nations environment programme (UNEP). http://chm.pops.int/. Accessed Nov 2013
Thrall PH, Bever JD, Burdon JJ (2010) Evolutionary change in agriculture: the past, present and future. Evol Appl 3:405–408
United Nations Framework Convention on Climate Change (UNFCCC) (2007) Investment and financial flows to address climate change. UNFCCC, Bonn
Valitutto RS, Sella SM, Silva-Filho EV et al (2007) Accumulation of metals in macrophytes from water reservoirs of a power supply plant, Rio de Janeiro State, Brazil. Water Air Soil Poll 178:89–102
Vavilov NI (1992) Origin and geography of cultivated plants (trans: Löve D). Cambridge University Press, Cambridge
Yang S, Peng S, Hou H et al (2013) Controlled irrigation and drainage of a rice paddy field reduced global warming potential of its gas emissions. Arch Acker Pfl Boden. doi:10.1080/03650340.2013.772687

Chapter 2
Organochlorinated Contaminants in General Population of Argentina and Other Latin American Countries

Adriana S. Ridolfi, Gloria B. Álvarez, and María E. Rodríguez Girault

Abstract Organochlorinated contaminants integrate the persistent organic pollutants (POPs) group, according to the Stockholm Convention. Persistent organic pollutants are synthetic chemicals highly lipophilic that cause harmful effects on human health. The extensive use of organochlorine pesticides (OCPs) in agriculture and polychlorinated biphenyls (PCBs) on industry, in confluence with its resistance to metabolic degradation, determined its persistence in the environment.

Studies on population of Argentina and other Latin American countries show exposure to POPs, whose levels in adipose tissue, serum, and breast milk mainly depend on the age, sex, and place of residence. Most studies were based on researches of OCP in exposed and unexposed populations. Blood was the biological matrix mostly used. The most frequently found pesticides were *pp*′-dichlorodiphenyldichloroethane (*pp*′-DDE), *pp*′-dichlorodiphenyltrichloroethane (*pp*′-DDT), hexachlorobenzene (HCB), and *β*-hexachlorocyclohexane (*β*-HCH) in all matrices investigated. Some studies examined the presence of PCBs in human tissues predominating the most persistent congeners 138, 153, 180, and 170.

While applications of POPs have decreased in recent years, human exposure in some Latin American countries continues because of their persistence and reuse of banned OCPs. Consequently, monitoring programs of the POP concentrations in general population are needed to provide data to government authorities in order to implement policies of bioremediation and to minimize the risk of human exposure to these persistent compounds.

A.S. Ridolfi (✉) • G.B. Álvarez • M.E. Rodríuez Girault
Cátedra de Toxicología y Química Legal, Laboratorio de Asesoramiento Toxicológico Analítico (CENATOXA), Facultad de Farmacia y Bioquímica, Universidad de Buenos Aires, Junín 956 7piso, 1113 CABA, Buenos Aires, Argentina
e-mail: aridolfi@ffyb.uba.ar

A. Alvarez, M.A. Polti (eds.), *Bioremediation in Latin America: Current Research and Perspectives*, DOI 10.1007/978-3-319-05738-5_2

2.1 Introduction

Organochlorinated contaminants integrate the persistent organic pollutants (POPs) group. These compounds have been added to the POP Stockholm Convention list (Stockholm Convention 2001). POPs are synthetic chemicals highly lipophilic, resistant to degradation that accumulated in the environment with harmful effects on human health (Porta et al. 2008).

Organochlorinated compounds are stored in the fat of human tissues, particularly dichlorodiphenyldichloroethylene [DDE, the main product of degradation of dichlorodiphenyltrichloroethane (DDT), hexachlorobenzene (HCB), the hexachlorocyclohexanes (HCH), and polychlorinated biphenyls (PCBs)] (Porta et al. 2008). Humans can be exposed to these contaminants primarily through food intake by their working activities or environmental exposure. Many studies have shown that the main contributors to the total intake of POPs are meat, fish, and dairy (Gasull et al. 2011).

Although some POPs have been banned or restricted, others remain deeply rooted in our societies and human exposure continues even to compounds banned like DDT (Porta and Zumeta 2002).

Organochlorine environmental contaminants such as *pp*′-DDE, PCBs, and other related compounds may be causally related to decreased male reproductive capacity and infertility; also an increased incidence of breast cancer and neurodevelopmental deficits in children are based on environmental, laboratory, animal, and human data. Organochlorine pesticides (OCPs) and hydroxy-PCBs modulate endocrine responses in animal models and cell culture; these compounds are identified as potential endocrine disruptors (Safe 2001; López-Cervantes et al. 2004; McKinlay et al. 2008; Varayoud et al. 2008). Milesi et al. (2012) found that low doses of neonatal exposure to endosulfan affect uterine function in adulthood.

Although the human exposure to PCBs and persistent chlorinated pesticides has declined, there are evidences that the presence of organochlorines in food may be a risk factor for neurologic, hormonal, genotoxic, and immunologic effects in infants and children (ATSDR 2000; Glynn et al. 2003; Freire et al. 2012; Boccolini et al. 2013).

Organochlorine pesticides were widely used in Argentina from the 1940s to the 1970s, although different government resolutions were restricting and prohibiting their use (Villaamil Lepori et al. 2013). However, endosulfan is still the most widely used insecticide in agriculture both in Argentina and other Latin American countries (Souza Casadinho 2008). By their known toxic effects, the Argentine National Health Service and Food Quality (SENASA) enacted a law which banned the importation of endosulfan (from June 30, 2012, resolution 511/11) (Souza Casadinho 2008; SENASA 2013; Villaamil Lepori et al. 2013).

Polychlorinated biphenyls have been widely used as additives to oils in electrical equipment (transformers and capacitors), hydraulic machines, adhesives, textiles, printing, and sealants. Coplanar PCBs (non-*ortho* and mono-*ortho*) are the most toxic, producing effects like 2,3,7,8-tetrachlorodibenzodioxin (2,3,7,8-TCDD or dioxin) in experimental animals. Toxic equivalency factors (TEFs) have been established based on the aryl hydrocarbon receptor (AhR) affinity. Dioxin toxic equivalencies (TEQs) were used to evaluate the toxicological risk associated to the presence of PCBs in humans (Mariottini et al. 2000). Polychlorinated biphenyls were banned in Argentina from 2002 (law 25670/02), but even today continue the decontamination of transformers (Miglioranza et al. 2013).

The lipophilic properties of POPs promote transport via circulating lipids and lipoproteins and storage in tissues with high-fat content. Assessing levels of organic pollutants in these matrices as biomarkers provide invaluable information on the biological effects observed in individuals exposed to these substances (Díaz-Barriga et al. 2003; Waliszewski et al. 2013). Body concentrations of OCPs and PCBs in human biological samples have been analyzed in several studies and indicate that most humans store POPs, although interindividual differences in concentrations are substantial in most populations worldwide (Cerrillo et al. 2005; Porta et al. 2008; Gasull et al. 2011). Hereby, assessing human contamination requires monitoring POP levels in representative samples.

Argentina has reported several studies on the impact of organochlorinated contaminants in the environment, food, and wildlife. In the 1970s were investigated organochlorine pesticides in water of the Parana and Uruguay rivers. Hexachlorobenzene, α-, β-, and γ-HCH isomers, and *pp*′-DDT were detected more frequently. Higher concentrations corresponded to *pp*′-DDT (5.6 μg/L) (García Fernández et al. 1979).

Organochlorine pesticide residues were analyzed in 101 samples of infant formula and dairy products collected in the local market of Buenos Aires, Argentina. The groups of pesticides that were found mainly were heptachlor (57 %) and HCH (53 %). Heptachlor and its epoxide exceeded the acceptable daily intake (ADI) in dairy products which would imply an increased risk to health of infants and toddlers (Villaamil Lepori et al. 2003, 2006).

Colombo et al. (2011) evaluated the bioaccumulation and risk associated to consumption of lipid-rich detritivorous fish. Chlorinated pesticides and PCBs were analyzed in Sábalo fish (*Prochilodus lineatus*) collected in the polluted Metropolitan Buenos Aires coast and in migrating specimens. In this study, very high concentrations of organic pollutants were found in fatty fish muscles. These data correlate with that reported in fish from different river basins of Brazil by Torres et al. (2010).

In a recent study along the Río Negro basin (Argentinean Patagonia), Miglioranza et al. (2013) report the occurrence and distribution of OCPs and PCBs in soil, sediment, suspended particle matter (SPM), stream water, and macrophytes. They found a clear predominance of OCPs among all matrices indicating the impact of agriculture on the watershed. The highest levels were found for *pp*′-DDE in agricultural soils from the Upper Valley. The insecticide endosulfan was also found in all matrices. Levels of PCB (153, 138, 110, and 101 congeners) were directly related with the presence of hydroelectric power plant.

Researches on the impact of POPs in the environment, food, and nontarget species have also been conducted in other Latin American countries (Díaz-Barriga et al. 2003; Pérez-Maldonado et al. 2010; Torres et al. 2010; Avancini et al. 2013;

Benítez-Díaz and Miranda-Contreras 2013). Several studies suggested that banned OCPs were being used to protect illegal crops from pests. *pp*′-Dichlorodiphenyltrichloroethane is still used for vector control in several tropical and subtropical areas of South America and there is evidence of recent illegal use in agriculture (Varona et al. 2010; Mercado et al. 2013), hence the importance of monitoring these persist compounds in the environment and in humans.

In Argentina, Brazil, Mexico, and Bolivia, new studies have evaluated the genotoxic risk of occupational exposure to pesticides in agricultural workers that may be correlated with the exposure time (Simoniello et al. 2008; Poma et al. 2010; Da Silva et al. 2012; Benedetti et al. 2013; Gómez-Arroyo et al. 2013).

In many countries throughout Europe, Asia, and North America OCP and PCBs in humans (blood serum, breast milk, adipose tissue) have been reported, showing a reduction of human exposure in recent years due to the prohibitions and restrictions of these substances (Dua et al. 1996; James et al. 2002; Glynn et al. 2003; Jaraczewska et al. 2006; Petrik et al. 2006; Sudaryanto et al. 2006; Thomas et al. 2006; Lucena et al. 2007; Tanabe and Kunisue 2007; Kozul and Romanić 2010; Shen et al. 2010; Kalantzi et al. 2011; Mishra et al. 2011; Bräuner et al. 2012; Cok et al. 2012).

Although adipose tissue levels have been preferentially used as an indicator of chronic human exposure to organochlorine contaminants, serum levels have been adopted in epidemiological studies as a less invasive and more practical alternative (Díaz-Barriga et al. 2003).

In this document, we present a data collection of exposure to OCP and PCBs in adipose tissue, serum, and breast milk in Argentine general population and population from other Latin American countries, in order to evaluate the degree of exposure to POPs and compare the levels found in inhabitants of different regions.

2.2 Persistent Organic Pollutants and Polychlorinated Biphenyls: Human Exposure in Latin America

2.2.1 Adipose Tissue

Organochlorine pesticides and PCBs accumulate in fatty tissue due to its lipophilic properties. In human adipose tissue, the average life of most of these compounds is several years. Adipose tissue biopsy has been used in epidemiological studies to assess chronic exposure to organochlorine. This tissue is selected because it gives adequate results with regard to accumulation status (Botella et al. 2004; Waliszewski et al. 2011). Although the presence of organochlorine compounds in human adipose tissue has been reported throughout the world, in Argentina and Latin America few studies are found. Muñoz-de-Toro et al. (2006a) investigated the residues of OCPs and PCBs in mammary fat tissue from 76 women not occupationally exposed to organochlorines living in a littoral region in northeastern Argentina. Organochlorine compounds that appeared most frequently were *pp'*-DDE (100 %), HCB (87 %), and β-HCH (75 %). On the other hand, the incidence of PCBs congeners was very low. *pp'*-Dichlorodiphenyldichloroethane and β-HCH residues reached the highest levels (4,794 and 1,780 mg/g, respectively). Significant positive association was found between organochlorine levels, body mass index, and women age ($P<0.05$). The diet was a relevant source of exposure, particularly the consumption of animal fat and freshwater fish. Organochlorines accumulated in the mother could be transferred transplacentally to the developing fetus and from breast milk to the nursing infant. The frequencies and concentrations of organochlorine residues in adipose tissue reported in this study were higher than those reported in developed countries (Botella et al. 2004; Porta et al. 2008).

Another study by the same group of researchers in a subsample of 55 women diagnosed with invasive breast carcinoma from Santa Fe city area (a littoral region in Argentina) found a positive correlation between progesterone receptor (PR) expression (an estrogen-induced protein) in the neoplastic cells and organochlorine levels in adipose tissue. It was observed that when organochlorine levels in adipose tissue reached levels higher than 2,600 mg/g, the estrogen receptor alpha (ERα)-positive breast carcinomas from postmenopausal women exhibited high proliferation. The authors conclude that organochlorine residues in adipose tissue adjacent to breast carcinoma generate an estrogenic microenvironment that may influence the biological behavior of the tumor through ERα activation and ERα-dependent proliferation (Muñoz-de-Toro et al. 2006b).

Ridolfi et al. (2008) conducted a case-control study in order to assess the levels of 17 PCB congeners in breast adipose tissue from 88 women (aged between 30 and 85 years) with malignant and benign breast tumors from Buenos Aires city, Argentina. Concentrations of PCB congeners reported in this study were higher in malignant tumors compared to benign (Fig. 2.1). The sum of dioxin-like congeners adjusted for age, body mass index (BMI), residence, and occupation was higher in malignant tumors than in benign (200.1 vs. 61.1 ng/g lipid, respectively)

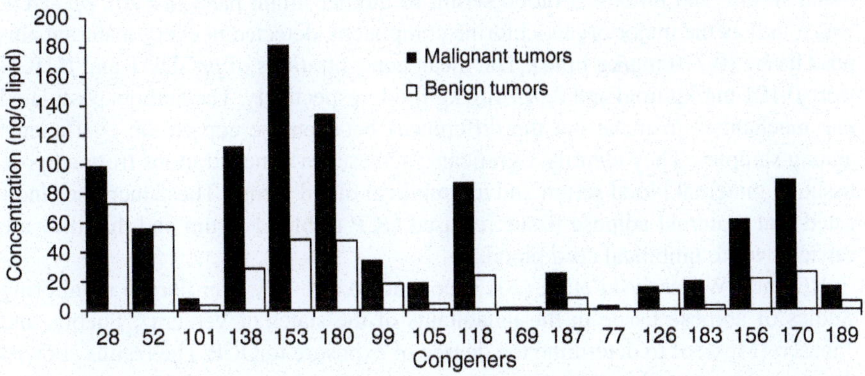

Fig. 2.1 Concentrations of congeners of polychlorinated biphenyls in malignant and benign breast tumors in Argentinean women

predominating the mono-*ortho* congeners 118 and 156. Association was found between toxicity equivalence to dioxins (TEQs) and breast cancer risk ($P<0.05$). The results suggest that exposure to PCBs dioxin-like will increase the risk of breast cancer in the studied population.

In Latin America, Mexico had reported the highest number of studies in human adipose tissue. Levels of DDT and its metabolites in human biological media from different communities of Mexico had been assessed by Díaz-Barriga et al. (2003). The authors found higher values of DDE in adipose tissue of exposed population compared to an urban non-exposed community (21.5 and 8.3 μg/g lipid, respectively). In this study the authors also observed higher concentrations of DDT in adipose tissue from workers of the malaria program compared with adipose tissue from general people living in malarious areas. It was clear that children and elderly people were identified as the most exposed when the concentrations of DDTs in adipose tissue were correlated by age. The elderly findings can be explained by chronic exposure to DDT, whereas the children concentration may be the result of exposure to multiple pathways (soil, household dust, air, water, food). The authors conclude that a DDT monitoring program in adipose tissue is needed in order to assess the body burden in Mexican population.

Waliszewski et al. (2011) monitored HCB (α-β-γ-HCH), *pp′*-DDE, *op′*-DDT, and *pp′*-DDT in 150 adipose tissue samples from inhabitants of Veracruz (Mexico). The following pesticides were detected: *pp′*-DDE (100 % of the samples; 1.643 mg/kg); *pp′*-DDT (99.3 % of the samples, 0.227 mg/kg); β-HCH (97.3 % of the samples, 0.063 mg/kg); and *op′*-DDT (93.3 % of the samples, 0.022 mg/kg). The samples grouped according to age showed a positively associated factor with OCP levels in adipose tissue of Veracruz inhabitants. Comparing OCP levels between 2008 and 2010, a decreased tendency for β-HCH, *pp′*-DDE, Σ-DDT (DDT sum), and *pp′*-DDE/*pp′*-DDT ratio levels was observed.

At the same place (Veracruz), Herrero-Mercado et al. (2011) conducted another study to determine levels and calculate ratios of copartition coefficients among *β-HCH*, *pp′*-DDE, *op′*-DDT, and *pp′*-DDT in maternal adipose tissue, maternal blood serum, and umbilical blood serum of mother-infant pairs ($n=70$). *pp′*-DDE was found as the major organochlorine component, detected in every maternal adipose tissue (0.770 mg/kg lipid). The mean concentrations of *pp′*-DDT and β-HCH were 0.101 mg/kg lipid and 0.027 mg/kg lipid, respectively. The authors postulated one mechanism to make out the differences between the copartition coefficients among samples. They identify significant increases in concentrations from adipose tissue to maternal blood serum and to umbilical blood serum. These increases indicated that maternal adipose tissue released OCP to blood serum and that they are carried over to umbilical cord blood.

Recently Waliszewski et al. (2013) conducted a review over human monitoring studies of adipose tissue in the inhabitants of the states of Veracruz, Puebla, and Tabasco (Mexico) to determine the degree of exposure to OCP. The results showed the presence of residues of organochlorine compound in general Mexican population; *pp′*-DDE was found in 100 % of samples and β-HCH, *pp′*-DDT, and *op′*-DDT predominated in Veracruz.

In Uruguay, the Poison Control Center (CIAT: Information and Toxicological Assessment Center), together with the Ministry of Agriculture, evaluated pesticide exposure through adipose tissue analysis of the Montevidean adult population. Both studies (1985 and 1996) on OCP in human fat show accumulation of these products in fat tissues (Mañay et al. 2004). Burger et al. (2000) evaluated a possible correlation between risk of breast cancer and residues of OCP in human body. Fifty-eight patients diagnosed with breast cancer and other 28 no-occupationally exposed with benign mammalian tumors were studied in Uruguay. There was a tendency toward higher levels of chlorinated pesticides in the problem cases than in the controls, but difference statistically significant was found only for β-HCH, being not possible to determine a consistent relationship between OCP residues and breast cancer. These results are consistent with studies in other regions (Safe 2001).

Another Latin American study compared human exposure to PCBs in adipose tissue samples collected from Concepción (Chile) and Siena (Italy). ΣPCBs were higher in Italian samples than those from Chile (493 and 53 ng/g, respectively). The congeners that prevailed in both groups were PCB 118, 138, 153, 170 180, and 187. Average concentrations of non-*ortho*-substituted coplanar congeners were below 1 pg/g. TEQ values were lower in Concepción, while in Sienese adipose tissue, toxic potential was much higher. According to the authors, the levels of PCBs found in Siena were similar to those reported by other industrialized countries of the northern hemisphere (Mariottini et al. 2000).

Table 2.1 presents the comparison of the mean concentrations of OCP most frequently found in human adipose tissue in different studies conducted in Argentina and other countries. Mexico had the highest level of DDT and metabolites, especially in malaria's areas. The concentrations found in Argentina and Uruguay were similar to those found in Spain by Botella et al. (2004), but higher than those reported in other European countries.

Table 2.1 Organochlorine pesticides (ng/g lipid) detected in human adipose tissue from Argentina and other countries (1998–2013)

Region	HCB	*β-HCH*	*pp'*-DDE	*pp'*-DDT	References
Argentina	68.3	367.4	918.8	7.9	Muñoz-de-Toro et al. (2006a)
Mexico (malarious zone)	ND	ND	28,900.0	8,100.0	Díaz-Barriga et al. (2003)
Mexico (urban zone)	ND	ND	6,000.0	1,100.0	Díaz-Barriga et al. (2003)
Mexico—Veracruz	ND	63.0	1,643.0	227.0	Waliszewski et al. (2011)
Mexico—Puebla	ND	73.0	916.0	83.0	Waliszewski et al. (2013)
Mexico—Tabasco	ND	49.0	1,034.0	116.0	Waliszewski et al. (2013)
Mexico—Veracruz	ND	27.0	770.0	101.0	Herrero-Mercado et al. (2011)
Uruguay (1996)	400.0	300.0	700.0	ND	Mañay et al. (2004)
USA—Long Island	19.7	22.2	546.7	17.0	Stellman et al. (1998)
Spain	ND	ND	508.0	61.0	Botella et al. (2004)

ND not determined

2.2.2 *Human Blood*

Persistent organic pollutant investigations in human blood from different regions of Argentina are scarce; however, its presence has been determined often in measurable concentrations (García Fernández et al. 1987; Lucero et al. 2008; Villaamil Lepori et al. 2013). Álvarez et al. (2006) investigated OCP in 100 samples of plasma from healthy volunteers not occupationally exposed of general population living in the metropolitan area of Buenos Aires (35 women and 65 men aged 18–82 were evaluated). Results showed that DDT group appeared most frequently (71 %) with a prevalence of *pp′*-DDE metabolite (57 %), followed by HCB (70 %), HCH group (57 %), and heptachlor and its epoxide (49 %). Maximum values of 9.87 and 8.05 μg/L were registered by *pp′*-DDD and *op′*-DDE, respectively (Table 2.2). OCP levels detected in this study were lower those reported by García Fernández et al. (1987) two decades earlier (Table 2.3). Results showed a significant decrease in the β- and γ-HCH isomers and *pp′*-DDT, banned in Argentina from the 1970s to 1980s. These results reinforce the concept that persistence of OCP metabolites is related with age and period of exposure (de Boer and Fiedler 2013; Waliszewski et al. 2013).

Another study conducted in Argentina evaluated the OCP levels in 649 blood samples analyzed during 2004–2012. People (aged between 1 and 85) came from different regions of the country. Results showed that the environment was the predominant etiology in children (97 %) and adults (85 %). Percentages of OCP were highest in children (0–15 years) [ΣDDT (53 %) and ΣHCH (41 %)] (Fig. 2.2). On the other hand, aldrin's group had a higher frequency in young adults [(16–30 and 31–50 years, 39 and 51 %, respectively)], followed by DDT (33 and 44 %) and HCH (31 and 47 %). Thirty percent of samples of older adults (51–85 years) presented HCH and others OCP. Concentration (μg/L) of pesticides (Fig. 2.3) showed different results with respect to the frequency of them (Fig. 2.2). Children had the lowest value of DDT (0.02 ± 0.63 μg/L), while the HCH group had the highest concentration (0.2012 ± 0.046 μg/L) exceeding the reference value for adults (0.09 ± 0.22 μg/L) reported by Álvarez et al. (2006). In young and adults DDT concentrations present low mean; the highest levels were found for HCHs and aldrin. In the older population (51–85 years) HCB had the highest mean concentration. The prevalence of OCP in this study was consistent with those reported in other countries (Fernandes Delgado et al. 2002; Petrik et al. 2006). However, endosulfan (a pesticide permitted until 2013 and used intensively in cultivating soybeans in Argentina) showed very low frequency of occurrence (4 %) and concentration (0.01–0.06 μg/L) which could be due to its shorter half-life compared to other chlorinated groups (Stockholm Convention 2001; Souza Casadinho 2008).

In the same study (between 2004 and 2012) the different chemical forms of DDT and its metabolites were compared between children and adults. Results revealed higher percentages of *pp′*-DDE and *op′*-DDD in children (Fig. 2.4a) and significantly lower concentrations in adults ($P < 0.05$) (Fig. 2.4b). On the other hand, the analysis of the HCH isomers showed a high percentage of adult samples containing β-HCH (Fig. 2.5a) while concentration of this isomer was low (Fig. 2.5b). In children,

Table 2.2 Concentration and frequency of organochlorine pesticides in plasma from people living in the metropolitan area of Buenos Aires, Argentina (2006)

Organochlorine pesticide	Mean concentration (μg/L)	Maximum concentration (μg/L)	Frequency (%)
HCB	0.19 ± 0.19	1.11	70
α-HCH	0.18 ± 0.32	1.78	38
β-HCH	0.08 ± 0.17	0.81	21
δ-HCH	0.01 ± 0.06	0.33	6
∑ HCH	0.09 ± 0.22	1.78	57
Lindane	0.03 ± 0.09	0.59	10
Heptachlor	0.32 ± 0.86	6.09	32
Heptachlor epoxide	0.04 ± 0.09	0.50	15
∑ Heptachlor	0.18 ± 0.63	6.09	49
Aldrin	0.20 ± 0.95	5.94	19
Dieldrin	0.02 ± 0.11	0.67	11
∑ Aldrin	0.11 ± 0.68	0.36	19
Endrin	0.03 ± 0.17	1.08	3
α-Chlordane	0.00 ± 0.00	0.00	4
γ-Chlordane	0.03 ± 0.09	0.60	6
∑ Chlordane	0.01 ± 0.06	0.67	8
α-Endosulfan	0.00 ± 0.00	0.00	4
β-Endosulfan	0.02 ± 0.08	0.67	6
∑ Endosulfan	0.01 ± 0.06	0.67	8
op′-DDE	0.10 ± 0.81	8.05	5
pp′-DDE	0.30 ± 0.44	2.46	57
op′-DDD	0.04 ± 0.20	1.80	6
op′-DDT	0.04 ± 0.20	1.57	2
pp′-DDD	0.11 ± 0.99	9.87	3
pp′-DDT	0.07 ± 0.64	6.35	71
∑ DDT	0.11 ± 0.63	10.28	71
Mirex	0.03 ± 0.16	1.29	6

Table 2.3 Comparison of organochlorine pesticides (μg/L) detected in 2 years (1987 vs. 2006) in blood samples from people living in the metropolitan area of Buenos Aires, Argentina

Organochlorine pesticides	1987	2006
α-HCH	3.30 ± 2.71	0.18 ± 0.32
β-HCH	9.41 ± 5.60	0.08 ± 0.17
δ-HCH (lindane)	2.81 ± 2.12	0.03 ± 0.09
Heptachlor	2.20 ± 1.60	0.32 ± 0.86
Heptachlor epoxide	0.81 ± 0.32	0.04 ± 0.09
pp′-DDE	8.10 ± 4.31	0.30 ± 0.44
pp′-DDT	7.70 ± 4.51	0.07 ± 0.64

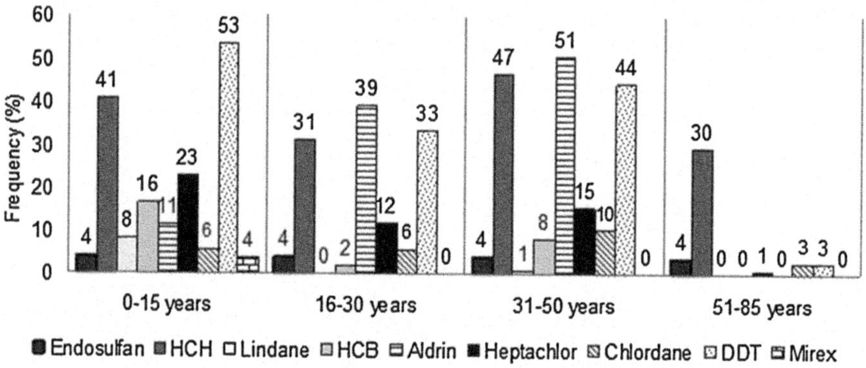

Fig. 2.2 Frequency of organochlorine pesticides by range of ages in Argentinean population

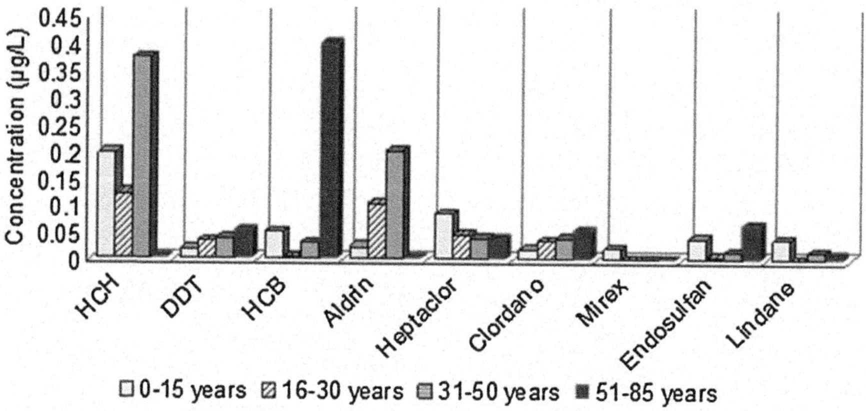

Fig. 2.3 Concentration of organochlorine pesticides by range of ages in Argentinean population

β-HCH showed higher concentration [(0.11 ± 0.43 µg/L), Fig. 2.5b]. It can be concluded that *β*-HCH isomer prevailed in the adults, which would result from their persistence, bioaccumulation, and prior use. In contrast, *pp′*-DDE and *op′*-DDD prevailed in children, probably because of its persistence in foods and in soils where DDT was applied. These compounds would also be transferred to the child through the placenta and breast milk. Concentrations of DDT were higher in adults, due to the property of biomagnifications (Lucena et al. 2007; Waliszewski et al. 2013).

During 2007, a descriptive and cross-sectional study of presence of OCPs in blood samples of healthy children was performed in Córdoba city (Argentina). Samples were collected in children (aged between 1 and 14 years) from two villages: Ituzaingó neighborhood [with high values and high frequency of OCPs registered in 2005 ($n = 142$)] and from children of the rest of the province [without exposure, reference population ($n = 62$)]. DDT group showed the highest frequency of occurrence in both groups (Ituzaingó: 65 %, reference: 27 %), followed by HCH isomers (Ituzaingó: 41 %, reference: 31 %) (Fig. 2.6a). Concentration of HCH was

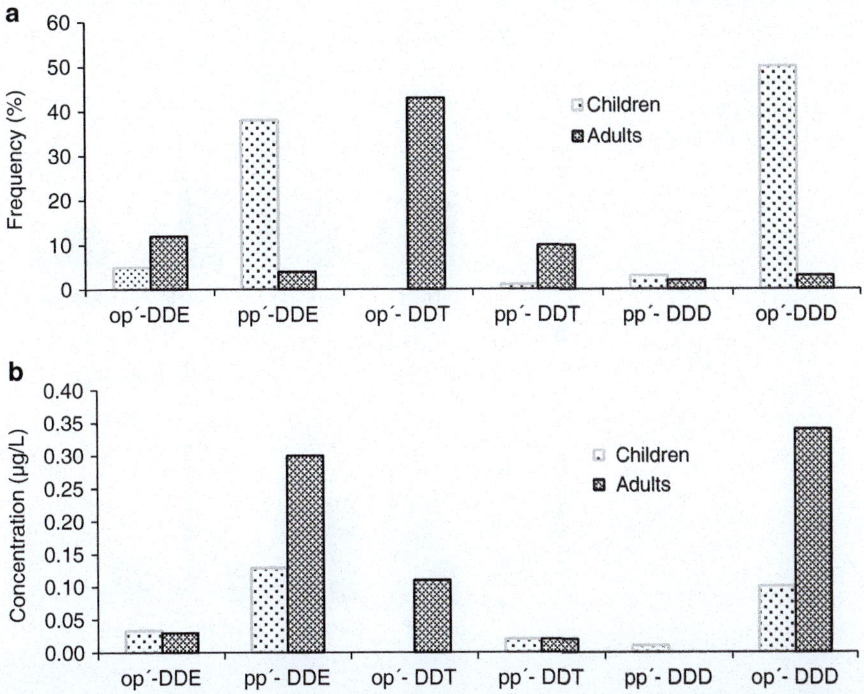

Fig. 2.4 Comparison of frequency (**a**) and concentration (**b**) of dichlorodiphenyltrichloroethane group in children and adults from Argentina

similar in both groups, but DDT was significantly lower in the reference, with respect to Ituzaingó population ($P<0.05$) (Fig. 2.6b) (Ridolfi et al. 2006, 2007; Rodríguez Girault et al. 2012).

In Argentina, as in many other countries, another side to the problematic POP pollution comes from waste management of both OCP and PCBs. It has been found that different provinces used pesticides expired or banned (Pechen de D´Angelo et al. 1998). In order to assess levels of OCP, Lucero et al. (2008) analyzed 167 blood samples from people living nearby undeclared deposits of pesticides (Cordoba province). The authors found mainly *pp′*-DDE, HCB, and *β*-HCH. The maximum concentration detected was 7.31 µg/L and corresponded to *pp′*-DDE. The *β*-HCH showed higher values than the other isomers of HCH. Good correlation between the concentration of pesticides *pp′*-DDE and age was observed. People of one of the studied sites showed *pp′*-DDT/*pp′*-DDE ratio of 0.76, according to DDT stored for over 20 years in their living place. This study provides information of the blood levels of OCP in a population with prolonged environmental exposure.

Occupational exposure is another relevant source of exposure to POPs where the income of these compounds occurs primarily through the skin and respiratory tract. Plasma levels of PCBs have been shown to be good biomarkers of recent exposure and have been proposed for monitoring exposed workers. In this context, analysis of plasma samples from 202 people working in the Electric Energy Argentina company

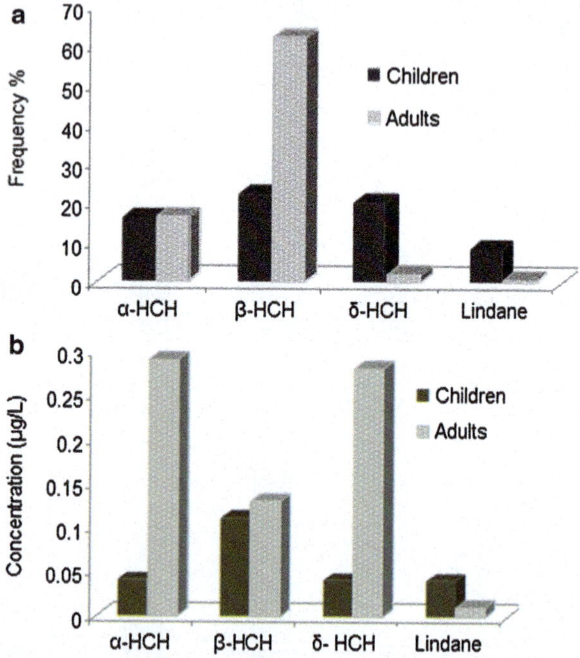

Fig. 2.5 Comparison of frequency (**a**) and concentration (**b**) of the hexachlorocyclohexane group in children and adults from Argentina

was performed (Rodríguez Girault et al. 2009). Samples were collected in 2007 after implementing preventive work practices. The values obtained were compared with values found in workers from the same company in 2005. At least one congener (among 28, 52, 101, 138, 153, and 180) was found in 92 % of the samples. The mean concentration of the sum of congeners was 0.86±0.96 µg/L. Congener 52 had the highest concentration (range: ND–3.6 µg/L) (Table 2.4). In contrast, congeners that occurred more frequently were the most persistent, 138 (65 %) and 153 (63 %), followed by 52 (54 %), 180 (49 %), 101 (21 %), and 28 (20 %). Levels of total PCBs found in this study (2007) were significantly lower compared with those obtained in 2005 by the same working group ($P<0.05$) (Fig. 2.7). The average level of the sum of PCB congeners found in both studies was below that reported by Turci et al. (2006) and (Fitzgerald et al. 2007) for the general population.

Polychlorinated biphenyls "dioxin-like" present a risk of encouraging the development of cancer and may act as endocrine disruptors. Due to these characteristics the presence of PCBs in the environment brings concerns (Safe 2001; James et al. 2002). In 2011 17 PCB congeners were investigated in adults and children from different regions of Argentina. Toxic equivalencies (TEQs) of dioxin-like congeners were calculated. Two hundred and twenty plasma samples [73 from adults from the metropolitan area of Buenos Aires and Mar del Plata city and 147 from children from two Argentinean towns, Palpalá (Jujuy province) and neighborhood Ituzaingó (Cordoba province)] were analyzed. The most prevalent congeners in adult and child population were 180, 153, 138, and 170. Of the PCBs "dioxin-like," the congener 118 had the highest frequency (Table 2.5). Concentration of total PCBs and TEQ was significantly lower in children ($P<0.05$) probably because of long

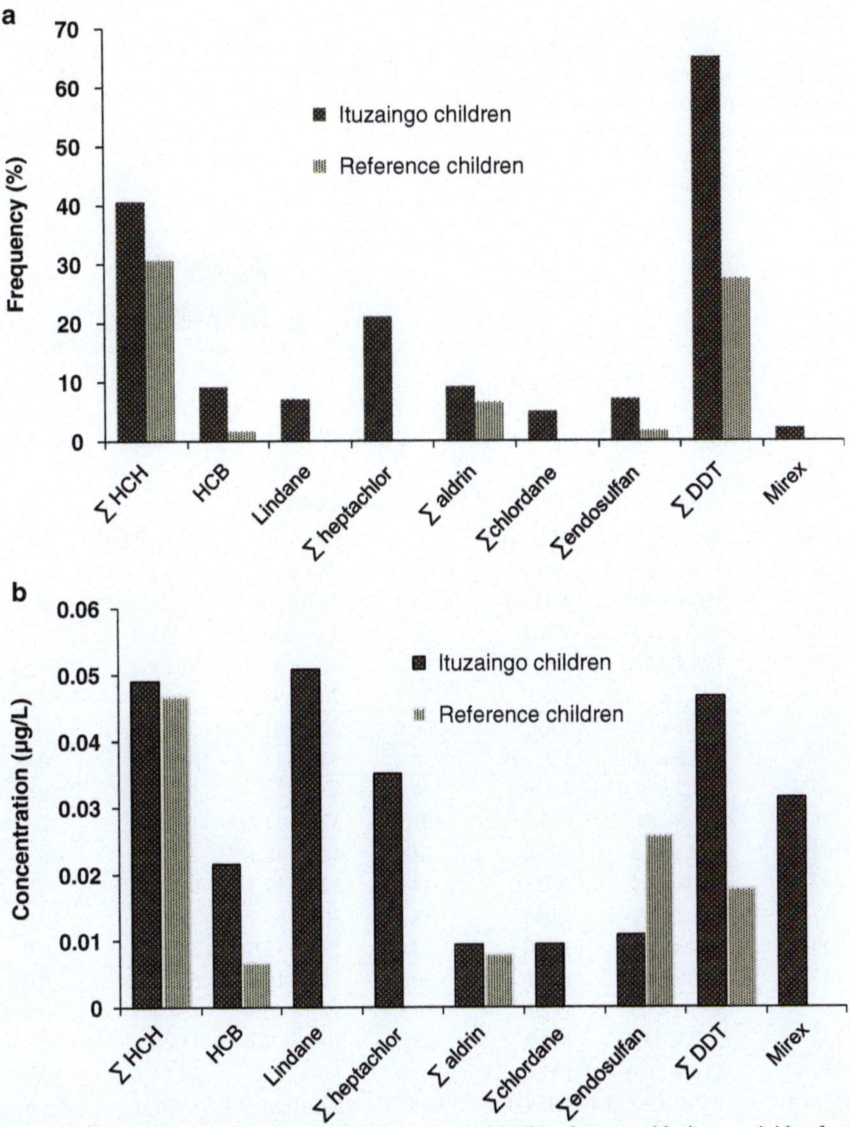

Fig. 2.6 Comparison of frequency (**a**) and concentration (**b**) of organochlorine pesticides from exposed children (Ituzaingó city) and not exposed (reference: Cordoba city)

Table 2.4 Polychlorinated biphenyl congeners (µg/L) determined in occupationally exposed population from Argentina

	Congeners						
	28	52	101	138	153	180	∑ Congeners
Concentration	0.1 ± 0.1	0.4 ± 0.7	0.1 ± 0.2	0.2 ± 0.2	0.1 ± 0.2	0.1 ± 0.1	1.0 ± 0.9
Range	ND–0.9	ND–3.6	ND–2.2	ND–2.1	ND–1.5	ND–1.0	ND–7.2

ND not detectable

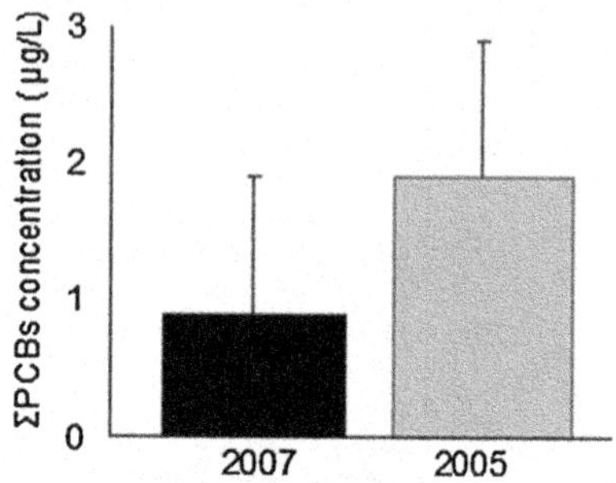

Fig. 2.7 Polychlorinated biphenyls in plasma obtained from occupational population exposed in 2 years (2005 and 2007) ($P<0.05$, Tukey test)

Table 2.5 Polychorinated biphenyl congeners (ng/g lipid) detected in adults and children from Argentina

	Adults			Children		
Congener	Mean	Maximum	Median	Mean	Maximum	Median
28	132.1 ± 142.3	795.1	92.0	28.1 ± 31.2	214.2	27.1
52	194.0 ± 306.2	1,410.0	84.1	87.0 ± 298.1	3,155.0	0.0
101	83.3 ± 110.0	720,2	58.2	73.4 ± 163.0	1,361.3	23.2
138	115.1 ± 57.4	391.1	109.4	46.1 ± 61.0	402.1	34.1
153	147.2 ± 174.0	835.0	75.1	70.2 ± 97.1	706.4	47.0
180	220.1 ± 154.1	876.2	200.0	42.1 ± 61.0	541.1	32.4
99	144.2 ± 261.0	1,212.0	33.1	54.2 ± 121.2	643.0	0.0
105	115.2 ± 89.0	3,359.2	114.3	8.0 ± 22.3	115.3	0.0
118	102.5 ± 84.2	615.1	101.1	32.4 ± 31.0	175.2	32.1
169	60.1 ± 82.3	302.0	24.0	40.1 ± 48.5	225.1	26.2
187	52.0 ± 58.1	270.3	44.2	12.1 ± 19.3	94.2	0.0
77	118.2 ± 186.2	1,044.2	18.1	30.0 ± 64.2	307.0	0.0
126	26.4 ± 26.1	110.0	22.0	19.4 ± 32.0	200.0	0.0
183	47.0 ± 53.3	194.1	30.3	11.2 ± 16.1	67.1	0.0
156	50.1 ± 45.0	172.3	50.1	12.3 ± 25.4	164.3	0.0
170	57.3 ± 33.2	198.2	55.2	30.1 ± 26.0	156.6	30.1
189	32.1 ± 53.0	195.1	0.0	1.2 ± 5.2	30.4	0.0
Σ PCB	1,696.0 ± 833.1	3,858.0	1,487.1	595.0 ± 465.1	3,997.0	475.4

exposure time in adults. These results were similar to those registered in other countries (Petrik et al. 2006; Shen et al. 2010).

Several studies have investigated levels of POPs in blood exposed occupationally and unexposed population in different countries of Latin America. In Mexico human exposure to DDT has been reported as a result of the presence of this insecticide in different environmental media. Women living in an area of malaria exposed to DDT (Veracruz city) had higher levels of DDT in blood than women who lived in San Luis Potosí, a control area not exposed to DDT (*pp′*-DDE: 14.5 μg/L and 1.8 μg/L, respectively). Correlation was also found between serum levels of DDE in mother's blood and umbilical cord blood, which shows the transplacental passage of this compound and the potential harmful effects on the fetus (Waliszewski et al. 1999).

Another study conducted in the Chiapas state (southwest of México) reported higher levels of DDTs in serum with respect to Veracruz. In Chiapas, children had higher levels of DDTs than adults and some senior workers occupationally exposed to DDT. Samples collected in San Luis Potosí and Chiapas showed high levels of DDE metilsulfonated ($MeSO_2$-DDE) that has been associated with toxic effects on the adrenal gland. The results obtained in this study identified three high-risk populations in Mexico: children, pregnant women, and workers (Díaz-Barriga et al. 2003).

Another study in Mexico assessed blood levels of PCBs and OCP in 229 children among 6–12 years who attended schools near contaminated sites of Mexico. Samples were collected in 2004 from nine sites with a history of contamination (agricultural areas with intensive use of pesticides, mining, and industrial) and urban or rural distribution. The results showed that all children had levels of *pp′*-DDE. The least persistent pesticide *pp′*-DDT was detected in 14 % of the study population. The community with higher concentrations of DDT and DDE was Puerto Madero, Chiapas, in southern Mexico, and this was due to the massive use of DDT in the malaria program. While the mean level of DDE for all the sites was approximately 2,000 ng/g lipid, the mean level was 11 times higher in Chiapas. In this study exposure to HCB was seen in 10 % of children. Three of the five sites with positive data of HCB were influenced from brick. Lindane was detected in a high percentage of children (85 %). This exposure could be due to the use of shampoos containing lindane that was still used to control scabies and lice. In this study PCB congener levels (52, 118, 138, 153, 170, and 180) were detected only in the community of Nicolas, and later it was identified that the source of PCB contamination was an oil used as fuel in brick kilns. As for health risks, the authors believe that while there is no health indicator clearly associated with DDT/DDE, previous studies have shown levels in exposed children correlated with apoptosis of immune cells, immunosuppression, and neurocognitive effects (Trejo-Acevedo et al. 2009).

In Uruguay, levels of OCP were compared in blood of exposed and not exposed population from 1979 to 1983; OCP residues were detected in all samples. The exposed population showed higher levels of *β*-HCH, *pp′*-DDE, *pp′*-DDT, dieldrin, and HCB compared to unexposed. Further studies (1985 and 1996) also assessed OCPs in the Montevidean adult population. A marked decrease in 1996 compared to 1979 was observed (Mañay et al. 2004).

In Rio de Janeiro city (Brazil), Fernandes Delgado et al. (2002) investigated levels of OCP and PCBs in blood samples from 33 volunteers (16 men, 17 women) aged between 19 and 63 years. Organochlorine residues of *op′*-DDT, *pp′*-DDT, *pp′*-DDD, *pp′*-DDE, aldrin, dieldrin, endrin, heptachlor, heptachlor-epoxide, *α-β-γ-HCH,* and HCB were detected in all samples. Polychlorinated biphenyl congeners (28, 52, 101, 138, 153, 180) were detected, as well. The *pp′*-DDE was found more frequently in a range of concentrations from 1.4 to 8.4 μg/L serum or from 0.200 to 3.452 μg/g lipids. Positive association between increased frequency and concentration of *pp′*-DDE with the age was found. This study suggests that the inhabitants of the urban area of Rio de Janeiro city would be exposed to relatively low levels of POPs.

In Colombia, Varona et al. (2010) explored exposure to OCP in blood of individuals living in a region of illegal crops. During 2005 and 2006, 99 serum samples were collected. Heptachlor (73 %), *pp′*-DDE (19 %), aldrin (15 %), *γ*-chlordane (12 %), dieldrin (11 %), α-chlordane (10 %), α-endosulfan (8 %), β-endosulfan (5 %), oxychlordane (3 %), *pp*′-DDT (3 %), and *op*′-DDT (2 %) were detected. Heptachlor presented a median of 8.7 ng/L and maximum of 43.8 ng/L. In this context, authors opined that prohibited OCPs were being used in some regions of Colombia.

Considering their persistence and toxicity, the Pan American Health Organization (PAHO) implemented a surveillance program in Mesoamerica in order to detect DDT residues in children's blood. This program was carried out in communities from Mexico, Guatemala, El Salvador, Honduras, Nicaragua, Costa Rica, and Panama. Results showed that in some areas the children had high concentrations of DDT, particularly in Mexico (mean level 50.2 ng/mL). The *pp*′-DDT/*pp*′-DDE quotient was higher than one in some communities reflecting recent exposure. Authors suggested that more efforts are needed to prevent the reintroduction of DDT into the region (Pérez-Maldonado et al. 2010).

In eastern Bolivia, human exposure to DDT in blood samples of farmers in three rural communities was investigated. *pp*′-DDE was found in 100 % of the samples, with a median concentration of 19.7 ng/mL (4,788.7 ng/g lipid), while *op*′-DDT was detected in three samples (4.3 %). The *pp*′-DDE serum concentrations were associated with length of residence in the study area, personal hygiene after work, and body mass index. The results revealed high levels of *pp*′-DDE in the population under study, which is due to high occupational exposure in the past and very polluted environment (Mercado et al. 2013). In Argentinean population, OCP levels are lower than other countries, with prevalence of HCH and DDT groups (Table 2.6).

Table 2.6 Organochlorine residues (μg/L) founded in human blood from different countries

Country/city	Period	*n*	Σ HCH	Σ DDT	References
Brazil	–	42	32.4	76.9	Minelli and Ribeiro (1996)
Germany	–	25	6.4	15.4	DeVoto et al. (1998)
Mexico	–	65	1.6	16.4	Waliszewski et al. (1999)
Sweden	–	790	51.6	836.1	Glynn et al. (2000)
Spain	1997–1999	141	1.1	2.2	Sala et al. (2001)
Japan	–	41	0.5	6.3	Hanaoka et al. (2002)
Belgium	–	251	–	8.2	Charlier and Plomteux (2002)
Portugal	2001–2002	203	13	93.5	Cruz et al. (2003)
Spain	1998–2000	102	13.8	49.4	Falcon et al. (2004)
Uruguay	1996	68	1.1	6.4	Mañay et al. (2004)
Spain	1997–1998	682	–	370.0	Zumbado et al. (2005)
Portugal	1997–2001	160	24.9	74.7	Lino and Silveira (2006)
Romania	2005	142	1,114.0	2,420.0	Dirtu et al. (2006)
Poland	2004	44	3.9	401.1	Jaraczewska et al. (2006)
UK	2003	154	15.0	100.2	Thomas et al. (2006)
Spain	1992–1995	405	1,291.4	4,895.8	Porta et al. (2008)
Delhi	–	112	23.5	1.7	Pathak et al. (2009)
Dibrugarh	2009–2010	169	348.1	417.0	Mishra et al. (2011)
Bolivia	2013	70	–	19.7	Mercado et al. (2013)
Argentina	2004–2012	649	0.3	0.4	Ridolfi (unpublished)

2.2.3 *Breast Milk*

Persistent organic pollutants are accumulated in the fat of the different tissues due to its lipophilic properties. These compounds are translocated and excreted through breast milk during lactation. Therefore it is an important route of exposure for infants during the first months of life. Early exposure to POPs during the neonatal period may affect child neurodevelopment and growth. The concentrations of these compounds in human milk are a nice indicator of environmental contamination (Díaz-Barriga et al. 2003; Der Parsehian 2008).

Der Parsehian (2008) studied in Argentina the presence of OCP in breast milk of 248 postpartum women from the Hospital Materno Infantil Ramon Sarda of Buenos Aires (from 2000 to 2001 and 2003 to 2004). Ninety-one percent of the samples showed OCP residues. Most frequently used pesticides were *pp'*-DDE (86.7 %), HCB (26.6 %), heptachlor epoxide (25.4 %), *β*-HCH (23.0 %), and chlordane (15.7 %). The highest concentration and the maximum value corresponded to *pp'*-DDE (8.98 ng/ml and 200.4 ng/ml, respectively). These results are consistent with other studies conducted in many countries.

Della Ceca et al. (2012) recently assessed exposure to OCP and PCBs in 59 breast milk samples from women living in Buenos Aires, Argentina, collected during 2010 and 2011. Concentrations of POPs found in breast milk decreased in the following order: DDTs ≈ PCBs > HCHs > heptachlor. PCB values ranged from 22 to 258 ng/g lipid, with a clear predominance of the most persistent congeners: 118 (13.0 ± 9.0 %), 138 (21.0 ± 9.0 %), 153 (27.0 ± 8.0 %), and 180 (20.0 ± 5.9 %). This reflects the persistence of higher molecular weight PCBs and their levels were comparable to previously reported in the literature for other areas (Kalantzi et al. 2004; Tanabe and Kunisue 2007; Polder et al. 2009; Kozul and Romanić 2010; Cok et al. 2012). Dichlorodiphenyltrichloroethane levels ranged from 7.7 to 500 ng/g lipid (76.0 ± 91.0), prevailing in all samples the main metabolite *pp'*-DDE (90.0 ± 17.0). This is consistent with increased lipophilicity and therefore higher bioaccumulation in fat tissue of DDE in comparison to DDT. Both compounds showed very low levels compared with values reported in other countries (Kalantzi et al. 2004; Jaraczewska et al. 2006; Azeredo et al. 2008; Polder et al. 2009). Hexachlorocyclohexane levels were higher (range: 5.8–197, mean: 33.0 ± 37.0 ng/g lipid) than those reported in literature (Sudaryanto et al. 2006; Polder et al. 2009). In both studies reported in Argentina, it was concluded that in maternal milk of the studied populations, the degradation products (DDE, heptachlor epoxide) and more persistent congeners and isomers (*β*-HCH, PCBs penta-hexa 180, 138, 118, and 153) predominated. This reflects the greater stability of these compounds and their high bioaccumulation potential.

In Latin America, data reported belong preferentially to levels of DDT and its metabolites, due to intensive use of this pesticide in areas of malaria.

Studies in Mexican population by Díaz-Barriga et al. (2003) reported that concentrations of DDT and metabolites (DDT, DDD, and DDE) in human milk samples collected between 1994 and 1995 were higher in agricultural communities exposed to this pesticide and in samples from areas of malaria with intensive use of DDT with respect to urban areas, where DDT was never applied (Mexico city). Following the banning of DDT since 2000, a decline in DDT levels in later years is expected. The estimated infant's daily intake in this study, considering the average concentrations of DDT and its body weight, was three times higher than that established by the World Health Organization's Acceptable Daily Intake (ADI) of 20 μg/kg/day.

In order to assess environmental exposure to OCP in Colombian population, Rojas-Squella et al. (2013) conducted a study in 32 milk samples from Colombian mothers. The results obtained corresponded to the most frequent *pp'*-DDE with a mean value of 203 ng/g lipid, and a range of <17 and 14,948 ng/g lipid. According to the authors, Colombia ranks fourth from bottom to top in terms of *pp'*-DDE average concentrations based on the results obtained from the POP Global Monitoring Plan report of 2009 of the Stockholm Convention.

Dichlorodiphenyltrichloroethane and its metabolites *pp'*-DDE and *pp'*-DDD were detected in 69 samples of maternal milk of inhabiting of Rio Madeira (Brazil). The range of concentrations found were *pp'*-DDE (10.7–7,271.5 ng/g lipid), *pp'*-DDD (ND–400.7 ng/g lipid), and *pp'*-DDT (3.0–2,534.1 ng/g lipid). Whereas the sum of all the average DDT concentration was 369.6 ng/g lipid and ranged from 25.4 to 9,361.9 and 8.7 % of the estimated daily intake (EDI), based on total DDT, it was higher than the acceptable daily intake proposed by the WHO. Although the studied population belonged to an area of numerous cases of malaria in the past, and intensive use of DDT until 1990, the concentrations found were lower than those reported in another study in Rio de Janeiro, in 2000, prior to the banned use of DDT for agriculture programs and malaria control (Azeredo et al. 2008).

Figure 2.8 shows the levels of DDTs and PCBs found in breast milk in Latin American countries. Argentina's population showed low levels compared with values reported in other areas.

The decreased concentrations of OCP and particularly DDT in different works reported in recent years are probably due to the prohibition of their use for agriculture in various Latin American countries, although their metabolites are still detected in the biological samples, due to their high persistence and bioaccumulation.

2.3 Concluding Remarks

The extensive use of OCP in agriculture and PCB on industry, in confluence with its lipophilicity and resistance to environmental and metabolic degradation, determined its persistence over the environment. These characteristics have defined the ability of POP bioaccumulation, increased by the generation of toxic metabolites of similar chemical characteristics which has developed current scenario of chronic exposure to residues of these compounds.

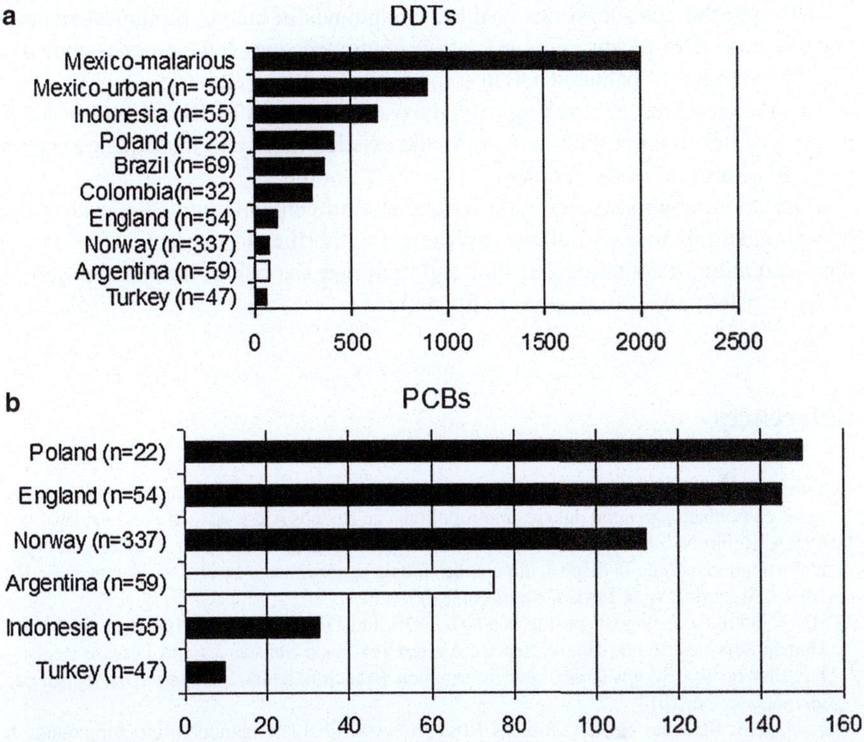

Fig. 2.8 Concentration (ng/g lipids) of dichlorodiphenyltrichloroethane (**a**) and polychlorinated biphenyls (**b**) in breast milk from different countries

In Argentina and Latin America there is little information on levels of POPs in human samples. Most studies are based on the investigation of OCP in exposed and unexposed populations. Blood was the most commonly used biological matrix. HCB, *pp'*-DDE, *pp'*-DDT, and *β*-HCH were the OCPs that were seen most frequently.

Some studies examined the presence of PCBs in serum, breast milk, and adipose tissue, predominantly the most persistent congeners 138, 153, 180, and 170.

The concept of toxicity due to POPs in child population is related to environmental long-term exposure to low concentrations, and increased susceptibility. Studies in Argentina show the presence of OCP in children even after decades of prohibition; for this reason children should be evaluated through epidemiological monitoring because they are not occupationally exposed, but reflect environmental exposure to a greater extent than adults.

The results of studies in Latin American populations demonstrate exposure to POPs, whose levels depend on age, sex, and place of residence.

Although the concentrations of these compounds in human biological samples have decreased in recent years, exposure continued due to their persistence and evidence of reuse of banned OCP in some Latin American countries.

Health consequences resulting from the use of OCP and PCBs are mainly due to the lack of information, failure to apply the existing laws, and inadequate supervision and awareness of the problem.

Population monitoring programs in conjunction with environmental evaluations will provide data that will allow government authorities to implement policies to bioremediation of contaminated sites and minimize the risk of human exposure to these persistent toxic compounds in the region.

References

Álvarez G, Rodríguez Girault ME, Bardoni N et al (2006) Valores guía de plaguicidas organoclorados en población general del área metropolitana de Buenos Aires. Acta Toxicol Argent 14:50

Ángelo A, Rubio NC, Kirs V et al (1998) Análisis del riesgo potencial para la salud y el medio ambiente derivado de la disposición clandestina de agroquímicos en el Cuy, Provincia de Río Negro, Argentina. Acta Toxicol Argent 6:28–33

ATSDR (2000) Toxicological profile of DDT, DDE and DDD. U.S. Department of Health and Human Services, Public Health Service Agency for Toxic Substances and Disease Registry, Division of Toxicology/Toxicology Information Branch, Atlanta, GA. http://www.atsdr.cdc.gov. Access Oct 2013

Avancini RM, Schettert Silva I, Simões Rosa AC et al (2013) Organochlorine compounds in bovine milk from the state of Mato Grosso do Sul-Brazil. Chemosphere 90:2408–2413

Azeredo A, Torres JP, de Freitas Fonseca M et al (2008) DDT and its metabolites in breast milk from the Madeira River basin in the Amazon, Brazil. Chemosphere 73:246–251

Benedetti D, Nunesa E, Sarmento M et al (2013) Genetic damage in soybean workers exposed to pesticides: evaluation with the comet and buccal micronucleus cytome assays. Mutat Res 752:28–33

Benítez-Díaz P, Miranda-Contreras M (2013) Contaminación de aguas superficiales por residuos de plaguicidas en Venezuela y otros países de Latinoamérica. Rev In Contam Ambie 29:7–23

Boccolini PMM, Boccolini CS, Meyer A et al (2013) Pesticide exposure and low birth weight prevalence in Brazil. Int J Hyg Envir Heal 216:290–294

Botella B, Crespo J, Rivas A et al (2004) Exposure of women to organochlorine pesticides in Southern Spain. Environ Res 96:34–40

Bräuner EV, Sørensen M, Gaudreau E et al (2012) A prospective study of organochlorines in adipose tissue and risk of non-Hodgkin lymphoma. Environ Health Perspect 120:105–111

Burger M, Mate M, Laviña R et al (2000) Role of the organochlorine pesticides in breast cancer. Rev Toxicol 17:79–82

Cerrillo I, Granada A, López-Espinosa MJ et al (2005) Endosulfan and its metabolites in fertile women, placenta, cord blood, and human milk. Environ Res 98:233–239

Charlier CJ, Plomteux GJ (2002) Determination of organochlorine pesticides residues in the blood of healthy individuals. Clin Chem Lab Med 40:361–364

Cok I, Mazmanci B, Mazmanci MA et al (2012) Analysis of human milk to assess exposure to PAHs, PCBs and organochlorine pesticides in the vicinity Mediterranean city Mersin, Turkey. Environ Int 40:63–69

Colombo JC, Cappelletti N, Williamson M et al (2011) Risk ranking of multiple-POPs in detritivorous fish from the Río de la Plata. Chemosphere 83:882–889

Cruz S, Lino C, Silviera MI (2003) Evaluation of organochlorine pesticide residues in human serum from an urban and two rural populations in Portugal. Sci Total Environ 317:23–35

Da Silva FR, Da Silva J, Allgayer MC et al (2012) Genotoxic biomonitoring of tobacco farmers: biomarkers of exposure, of early biological effects and of susceptibility. J Hazard Mater 225–226:81–90

de Boer J, Fiedler H (2013) Persistent organic pollutants. Trends Anal Chem 46:70–71

Della Ceca LS, Migoya C, Capelletti N et al (2012) Contaminantes orgánicos persistentes en leche materna de centros urbanos de la provincia de Buenos Aires. Augmdomus 4:92–102, http://revistas.unlp.edu.ar/domus/article/view/495. Accessed Oct 2013

Der Parsehian S (2008) Plaguicidas Organoclorados en leche materna. Rev Hosp Mat Inf Ramón Sardá 27:70–78, http://www.sarda.org.ar/Profesionales/Publicaciones/RevistaSarda/2008. Accessed Oct 2013

DeVoto E, Kohlmeier L, Heeschen W (1998) Some dietary predictors of plasma organochlorine concentrations in an elderly German population. Arch Environ Health 53:147–155

Díaz-Barriga F, Borja-Aburto V, Waliszewski S et al (2003) DDT in México. In: Fiedler H (ed) The handbook of environmental chemistry vol 3, Part O. Persistent organic pollutants. Springer, Berlin, p 372

Dirtu AC, Cernat R, Dragan D et al (2006) Organohalogenated pollutants in human serum from Iassy, Romania and their relation with age and gender. Environ Int 32:797–803

Dua VK, Pant CS, Sharma VP (1996) Determination of level of HCH and DDT in soil, water, and whole blood from bioenvironmental and insecticide sprayed areas of malaria control. Indian J Malariol 33:7–15

Falcon M, Oliva J, Osuna E et al (2004) HCH and DDT residues in human placentas in Murcia (Spain). Toxicology 195:203–208

Fernandes Delgado I, Barretto HHC, Kussumi TA et al (2002) Serum levels of organochlorine pesticides and polychlorinated biphenyls among inhabitants of Greater Metropolitan Rio de Janeiro, Brazil. Cad Saúde Pública (Rio de Janeiro) 18:519–524

Fitzgerald EF, Belanger EE, Gomez MI et al (2007) Environmental exposures to polychlorinated biphenyls (PCBs) among older residents of upper Hudson River communities. Environ Res 104:352–360

Freire C, Koifman RJ, Sarcinelli P et al (2012) Long term exposure to organochlorine pesticides and thyroid function in children from Cidade dos Meninos, Rio de Janeiro, Brazil. Environ Res 117:68–74

García Fernández JC, Marzi AA, Casabella AN et al (1979) Plaguicidas organoclorados en agua de los ríos Paraná y Uruguay. Ecotoxicología 1:51–78

García Fernández JC, Villamil EC, Chechi AL et al (1987) Niveles plasmáticos de plaguicidas organoclorados en la población general. Acta Bioquímica Clínica Latinoamericana 3:345–349

Gasull M, Bosch de Basea M, Puigdomènech E et al (2011) Empirical analyses of the influence of diet on human concentrations of persistent organic pollutants: a systematic review of all studies conducted in Spain. Environ Int 37:1226–1235

Glynn AW, Wolk A, Aune M et al (2000) Serum concentrations of organochlorines in men: a search for markers of exposure. Sci Total Environ 263:197–208

Glynn AW, Granath F, Aune M et al (2003) Organochlorines in Swedish women: determinants of serum concentrations. Environ Health Perspect 111:349–355

Gómez-Arroyo S, Martínez-Valenzuela C, Carbajal-López Y et al (2013) Riesgo genotóxico por la exposición ocupacional a plaguicidas en América Latina. Rev Int Contam Ambie 29:159–180

Hanaoka T, Takahashi Y, Kobayashi M et al (2002) Residuals of beta-hexachlorocyclohexane, dichlorodiphenyltrichloroethane, and relations with consumptions of dietary components in rural residents in Japan. Sci Total Environ 286:119–127

Herrero-Mercado M, Waliszewski SM, Caba M et al (2011) Organochlorine pesticide gradient levels among maternal adipose tissue, maternal blood serum and umbilical blood serum. B Environ Contam Tox 86:289–293

James RA, Hertz-Picciotto I, Willman E et al (2002) Determinants of serum polychlorinated biphenyls and organochlorine pesticides measured in women from the child health and development study cohort, 1963–1967. Environ Health Perspect 110:617–624

Jaraczewska K, Lulek J, Covaci A et al (2006) Distribution of polychlorinated biphenyls, organochlorine pesticides and polybrominated diphenyl ethers in human umbilical cord serum, maternal serum and milk from Wielkopolska region, Poland. Sci Total Environ 372:20–31

Kalantzi OI, Martin FL, Thomas GO (2004) Different levels of polybrominated diphenyl ethers (PBDEs) and chlorinated compounds in breast milk from two U.K. regions. Environ Health Perspect 112:1085–1091

Kalantzi OI, Geens T, Covaci A et al (2011) Distribution of polybrominated diphenyl ethers (PBDEs) and other persistent organic pollutants in human serum from Greece. Environ Int 37:349–353

Kozul D, Romanić SH (2010) Levels and distribution of PCBs and organochlorine pesticides in the air, pine needles, and human milk. Arh Hig Rada Toksikol 61:339–356

Lino CM, Silveira MIN (2006) Evaluation of organochlorine pesticides in serum from students in Coimbra, Portugal: 1997–2001. Environ Res 102:339–351

López-Cervantes M, Torres-Sánchez L, Tobías A et al (2004) Dichlorodiphenyldichloroethane burden and breast cancer risk: a meta-analysis of the epidemiologic evidence. Environ Health Perspect 112:207–214

Lucena RA, Allam MF, Jiménez SS et al (2007) A review of environmental exposure to persistent organochlorine residuals during the last fifty years. Curr Drug Saf 2:163–172

Lucero P, Nassetta M, De Romedi A (2008) Evaluación de la exposición ambiental a plaguicidas orgánicos persistentes en dos barrios de la provincia de Córdoba. Acta Toxicol Argent 16:41–46

Mañay N, Rampoldi O, Alvarez C et al (2004) Pesticides in Uruguay. Rev Environ Contam Toxicol 181:111–138

Mariottini M, Aurigi S, Focardi S (2000) Congener profile and toxicity assessment of polychlorinated biphenyls in human adipose tissue of Italians and Chileans. Microchem J 67:63–71

McKinlay R, Plant JA, Bell JNB et al (2008) Endocrine disrupting pesticides: implications for risk assessment. Environ Int 34:168–183

Mercado LA, Freille SM, Vaca-Pereira JS et al (2013) Serum concentrations of p, p′-dichlorodiphenyltrichloroethane (p, p′-DDE) in a sample of agricultural workers from Bolivia. Chemosphere 91:1381–1385

Miglioranza KSB, Gonzalez M, Ondarza PM et al (2013) Assessment of Argentinean Patagonia pollution: PBDEs, OCPs and PCBs in different matrices from the Río Negro basin. Sci Total Environ 452–453:275–285

Milesi MM, Varayoud J, Bosquiazzo VL et al (2012) Neonatal exposure to low doses of endosulfan disrupts the expression of proteins regulating uterine development and differentiation. Reprod Toxicol 33:85–93

Minelli EV, Ribeiro M (1996) DDT and HCH residues in the blood serum of malaria control sprayers. Bull Environ Contam Toxicol 57:691–696

Mishra K, Sharma RC, Kumar S (2011) Organochlorine pollutants in human blood and their relation with age, gender and habitat from North-east India. Chemosphere 85:454–464

Muñoz-de-Toro M, Beldoménico HR, García SR et al (2006a) Organochlorine levels in adipose tissue of women from a littoral region of Argentina. Environ Res 102:107–112

Muñoz-de-Toro M, Durando M, Beldoménico PM et al (2006b) Estrogenic microenvironment generated by organochlorine residues in adipose mammary tissue modulates biomarker expression in ERα-positive breast carcinomas. Breast Cancer Res 8:R47, http://breast-cancer-research.com/content/8/4/R47. Accessed Sept 2013

Pathak R, Ahmed RS, Tripathi AK et al (2009) Maternal and cord blood levels of organochlorine pesticides: association with preterm labor. Clin Biochem 42:746–749

Pérez-Maldonado IN, Trejo A, Ruepert C et al (2010) Assessment of DDT levels in selected environmental media and biological samples from Mexico and Central America. Chemosphere 78:1244–1249

Stockholm Convention on Persistent Organic Pollutants (2001) Stockholm, Sweden. www.chem.unep.ch. Accessed Oct 2013

Petrik J, Drobna B, Pavuk M et al (2006) Serum PCBs and organochlorine pesticides in Slovakia: age, gender, and residence as determinants of organochlorine concentrations. Chemosphere 65:410–418

Polder A, Skaare JU, Skjerve E et al (2009) Levels of chlorinated pesticides and polychlorinated biphenyls in Norwegian breast milk (2002–2006), and factors that may predict the level of contamination. Sci Total Environ 407:4584–4590

Poma ML, Bustillos NT, Ascarrunz ME (2010) Genotoxic damage caused by exposure to pesticides in farmers Luribay township. Biofarbo 18:31–43

Porta M, Zumeta E (2002) Implementing the Stockholm treaty on POPs. Occup Environ Med 59:651–652

Porta M, Puigdomènech E, Ballester F (2008) Monitoring concentrations of persistent organic pollutants in the general population: the international experience. Environ Int 34:546–561

Ridolfi A, Fernández R, Contartese C et al (2006) Determinación de plaguicidas organoclorados (OC) en niños del barrio Ituzaingó de Córdoba. Acta Toxicol Argent 14:68–69

Ridolfi A, Fernández R, Contartese C et al (2007) Evolución de residuos de plaguicidas organoclorados (OC) en niños de barrio Ituzaingó de Córdoba. Acta Toxicol Argent 15:45

Ridolfi A, Burlando S, Olivera M et al (2008) Riesgo de cáncer de mama asociado a congéneres de bifenilos policlorados (PCBs) en tejido adiposo mamario. Medicina 68:172–173

Rodríguez Girault ME, Álvarez G, Carretero M et al (2009) Bifenilos policlorados (PCBs): niveles en plasma en población expuesta ocupacionalmente. Acta Toxicol Argent 17:2–3

Rojas-Squella X, Santos L, Baumann W et al (2013) Presence of organochlorine pesticides in breast milk samples from Colombian women. Chemosphere 91:733–739

Safe S (2001) Hydroxylated polychlorinated biphenyls (PCBs) and organochlorine pesticides as potential endocrine disruptors. In: Metzler M (ed) The handbook of environmental chemistry vol 3. Part L. Springer, Berlin, p 156

Sala M, Ribas-Fito N, Cardo E et al (2001) Levels of hexachlorobenzene and other organochlorine compounds in cord blood: exposure across placenta. Chemosphere 43:895–901

SENASA (2013) Resol 511/11 del Servicio Nacional de Sanidad y Calidad Agroalimentaria, Ministerio de Agricultura, Ganadería y Pesca de la Nación Argentina. http://www.senasa.gov.ar/contenido.php. Accessed Oct 2013

Shen H, Ding G, Han G et al (2010) Distribution of PCDD/Fs, PCBs, PBDEs and organochlorine residues in children's blood from Zhejiang, China. Chemosphere 80:170–175

Simoniello MF, Kleinsorge EC, Scagnetti JA et al (2008) DNA damage in workers occupationally exposed to pesticide mixtures. J ApplToxicol 28:957–965

Souza Casadinho J (2008) Alternativas al endosulfán en la soja: El caso de Argentina. Oficina para América Latina y el Caribe del Pesticide Action Network (PAN) Internacional. http://www.rapaluruguay.org/endosulfan/Alternativas_endosulfan pdf. Accessed Oct 2013

Stellman SD, Djordjevic MV, Muscat JE et al (1998) Relative abundance of organochlorine pesticides and polychlorinated biphenyls in adipose tissue and serum of women in Long Island, New York. Cancer Epidemiol Biomarkers Prev 7:489–496

Sudaryanto A, Kunisue T, Kajiwara N et al (2006) Specific accumulation of organochlorines in human breast milk from Indonesia: levels, distribution, accumulation kinetics and infant health risk. Environ Pollut 139:107–117

Tanabe S, Kunisue T (2007) Persistent organic pollutants in human breast milk from Asian countries. Environ Pollut 146:400–413

Thomas GO, Wilkinson M, Hodson S et al (2006) Organohalogen chemicals in human blood from the United Kingdom. Environ Pollut 141:30–41

Torres JPM, Azevedo e Silva CE, Meire R et al (2010) POPs (PCBs and Organochlorine Pesticides) in fat from tropical detritivorous fish (Prochilodus sp) from Brazil the "fish-watch" approach. Organohalogen Compd 72:116–171

Trejo-Acevedo A, Díaz-Barriga F, Carrizales L et al (2009) Exposure assessment of persistent organic pollutants and metals in Mexican children. Chemosphere 74:974–980

Turci R, Finozzi E, Catenacci G et al (2006) Reference values of coplanar and non-coplanar PCBs in serum samples from two Italian population groups. Toxicol Lett 162:250–255
Varayoud J, Monje L, Bernhardt T et al (2008) Endosulfan modulates estrogen-dependent genes like a non-uterotrophic dose of 17β-estradiol. Reprod Toxicol 26:138–145
Varona ME, Díaz-Criollo SM, Lancheros-Bernal AR et al (2010) Organochlorine pesticide exposure among agricultural workers in Colombian regions with illegal crops: an exploration in a hidden and dangerous world. Int J Environ Health Res 20:407–414
Villaamil Lepori EC, Rodríguez Girault ME, Ridolfi A et al (2003) Residuos de plaguicidas organoclorados en productos lácteos infantiles y su aporte a la ingesta diaria admisible. Acta Toxicol Argent 11:81–82
Villaamil Lepori E, Ridolfi A, Álvarez G et al (2006) Residuos de plaguicidas organoclorados en leches infantiles y productos lácteos y su evaluación del riesgo. Acta Toxicol Argent 14:55–59
Villaamil Lepori EC, Bovi Mitre G, Nassetta M (2013) Situación actual de la contaminación por plaguicidas en Argentina. Rev Int Contam Ambie 29:25–43
Waliszewski SM, Aguirre AA, Infazon RM et al (1999) Levels of organochlorine pesticide in blood serum and umbilical blood serum of mothers living of Veracruz, México. Fresenius Environ Bull 8:171–178
Waliszewski SM, Caba M, Herrero-Mercado M et al (2011) Monitoring of organochlorine pesticide residue levels in adipose tissue of Veracruz, México inhabitants. Bull Environ Contam Toxicol 87:539–544
Waliszewski SM, Caba M, Gomez-Arroyo S et al (2013) Niveles de plaguicidas organoclorados en habitantes de México. Rev Int Contam Ambie 29:121–131
Zumbado M, Goethals M, Alvarez-Leon EE et al (2005) Inadvertent exposure to organochlorine pesticides DDT and derivatives in people from the Canary Islands (Spain). Sci Total Environ 339:49–62

Chapter 3
Strategies to Ameliorate Soils Contaminated with Boron Compounds

Norma B. Moraga, María Julia Amoroso, and Verónica B. Rajal

Abstract Boron derivatives are widely used in everyday modern lifestyle as fertilizer in agriculture, in ceramic and glass products, as melting products, in metallurgic furnaces, and others. The North West of Argentina, where Salta province is situated, is one of the most important boron mineral reserves in the world. This province is ranked as the first Latin American borate producer, the first world producer of hydroboracite (69,025 tons in 2012), and the third world producer of borates (179,358 tons in 2012). This implies an intensive exploitation of boron minerals, which brings out major environmental pollution. There are many places contaminated with boron compounds in the province and many strategies have been developed to purge contaminated soils, from physicochemical to biological technologies all around the world. Biological remediation using microorganisms implies various interesting mechanisms, such as biomineralization and exopolysaccharide formation. This chapter summarizes the most common technologies used in remediation of boron-contaminated soils, particularly in Salta, and trends in bioremediation strategies using native actinobacteria.

N.B. Moraga • V.B. Rajal (✉)
Instituto de Investigaciones para la Industria Química (INIQUI),
Consejo Nacional de Investigaciones Científicas y Técnicas (CONICET);
Universidad Nacional de Salta (UNSa),
Av. Bolivia 5150, 4400 Salta, Argentina
e-mail: vbrajal@gmail.com

M.J. Amoroso
Planta Piloto de Procesos Industriales Microbiológicos (PROIMI),
CONICET, Av. Belgrano y Pasaje Caseros, 4000 Tucumán, Argentina

Facultad de Bioquímica, Química y Farmacia,
Universidad Nacional de Tucumán, 4000 Tucumán, Argentina

Universidad del Norte Santo Tomás de Aquino (UNSTA),
4000 Tucumán, Argentina

A. Alvarez, M.A. Polti (eds.), *Bioremediation in Latin America: Current Research and Perspectives*, DOI 10.1007/978-3-319-05738-5_3

3.1 Introduction

Salta province—in the North West of Argentina—is part of the third most important boron mineral reserve in the world. Boron compound mining activity has great impact on the province economy not only for the foreign currency income but also for both direct and indirect employment it brings. However, this intense exploitation does have an environmental impact despite the methods used to avoid or ameliorate contamination in the production and industrialization processes. There are many places contaminated with boron (B) compounds in the province, but particularly one of them, Baradero (located in the south of the city of Salta), is extremely dangerous because of the high B content detected in soil and water and the proximity to inhabited neighborhoods. Many studies have been conducted there, first to determine boron levels in soil and groundwater (PNUMA 2004; INTA 1997, 2000, 2007; Suarez et al. 1997; Romero 2003) and second to ameliorate contamination (Albarracín et al. 2007; Albarracín 2008; Moraga et al. 2013).

3.1.1 Remediation Strategies

The treatments currently being performed to ameliorate environmental contamination produced by B are oriented to clean production, in situ residue disposal, recycling process streams, waste reuse, process and operation reengineering, and reduction of toxic discharges (Flores 2004). As to the refuse, numerous physical, chemical, and biological amelioration techniques have been developed for their treatment, the first two being the more expensive.

3.1.1.1 Physicochemical Methods

Thus, different methods are employed on the one hand to diminish contamination and, on the other, for effluent treatment (ionic exchange, removal with organic solvents, adsorption of metallic hydroxides, alkaline precipitate, processes with membranes and others) in order to avoid noxious effects in current and future exploitations (Flores 2004). Other physicochemical methods have also been developed to reduce environmental contamination with B, such as adsorption-coagulation (Chong et al. 2009; Kavak 2009), electrocoagulation (Yilmaz et al. 2008), inverse osmosis (Öztürk et al. 2008), and precipitation (Itakura et al. 2005), none of them applied in Salta province.

For many years sanitary dumping (sanitary refilling) has been a reasonably profitable process. But it is not a remediation process because it just moves the passive somewhere else without removing the toxic matter with all the problems involved, such as scarcity of available places and high transportation costs. On the other hand, incineration and storage generate environmental problems such as contaminant emission into the air, toxic matter accumulation, and lixiviation, respectively.

Reactivity reduction is widely used in heavy metals as it means changing their oxidation state, since this makes them less available, hence less toxic. This is not possible in the case of B as it has only one oxidation state. However, some recovery technologies for soils with high B content have been developed, such as water lixiviation and compound addition.

Water lixiviation. Water application to plant in excess of the required amount is known as lixiviation (Hoffman 1990). When employed in soils where contamination is not high enough to prevent plant growth, it may be efficient; the washing fraction must be high enough to eliminate B excess but low enough to prevent the loss of soil nutrients essential for plant growth, especially in sandy and acid soils. The amount of water required to lixiviate B at a given depth varies widely. If a low fraction of lixiviation is used, B resulting concentration near the soil surface will be close to that of the usual watering process whereas its concentration near the lower part of the root will presumably be much higher. Excessive watering to eliminate B excess is an old practice still applied in farming for soil recovery as it allows moderately satisfactory B levels in soils (Shennan et al. 1995). If B level reduction is attempted by water lixiviation in soils with low B level, a fraction of B is absorbed making its elimination much slower than that of unabsorbed elements. Furthermore, soluble B reduction may not be permanent after lixiviation since B may be regenerated through the mineralization of the organic matter in the soil, or by means of the meteorization processes of the soil minerals (Peryea et al. 1985). Anyway, soil B regeneration diminishes with every lixiviation, which points to a finite B concentration.

Compound addition. As B soil adsorption depends on the pH (Keren and Bingham 1985), a common practice is that of adding lime to increase pH and thus promote B adsorption from the soil as a short-term solution (Bartlett and Picarelli 1973). In soils with sodium content B toxicity may be ameliorated with the addition of gypsum, which improves water infiltration and changes the easily soluble sodium metaborate into the less soluble calcium metaborate (Nable et al. 1997). Applications with $Ca(H_2PO_4)_2$ also reduce B availability for plants, especially in acid soils.

Prather (1977) reported that sulphuric acid can efficiently help in recovering soils with high B concentration. Despite the pH decrease in the soil brought about by sulphuric acid addition, soluble B concentration is increased in the soil water and thus, as time goes by B is reabsorbed.

Combining soils with low B concentration with materials with excess B may limit B upward and downward migration. Such combination depends on what the soil is used for, its water depth penetration, water pattern movements in it, its topography, organic matter, and physicochemical features. Boron tends to migrate to the soil surface as water evaporates.

3.1.1.2 Biological Methods

Biological remediation techniques apply natural processes to eliminate harmful chemicals from the environment by means of two methods, phytoremediation and bioremediation.

Phytoremediation. Consists of using plants, plant microbial systems, and agronomic techniques to restrain, immobilize, remove, stabilize, and/or degrade environmental contaminant compounds (EPA 2000; Cunningham and Ow 1996). Sowing B-tolerant plant species may be a means to recover soils containing high contamination levels through plant absorption of soluble B. Under some circumstances, B toxicity may be ameliorated in plants adding zinc (Zn) in the soil. Swietlik (1995) demonstrated in barley (*Hordeum vulgare*) and orange (*Citrus aurantium*) with Zn deficiency solutions that B can be accumulated by plants and in excess it may become toxic but when Zn is added, B excess accumulated by plants does not become toxic. But as low Zn amount and high B concentration are found in some alkaline soils in semiarid regions, the likelihood of reducing B toxicity by adding Zn should be taken into account.

Sasmaz and Obek (2009) and Del-Campo Marín and Oron (2007) conducted some studies about the use of biological materials such as water lentils (*Lemna gibba*) to treat water contaminated with B, and they found that these species could uptake till 40 % of the initial concentration B in the first 2 days, this value being the maximum of saturation. On the other hand, Taştan et al. (2012) reported that *Chlorella* sp. microalgae are B accumulators and can be used to remove the element from contaminated waters.

Laboratory experiments were performed to prove B tolerance in native species (rapid growth herbs and trees such as *Lolium multiflorum* and *Aspioderma* sp. (white quebracho) in substrata contaminated with different B concentrations to pilot field experiments later in the Baradero site (Albarracín et al. 2007, 2008). However, despite phytoremediation advantages as compared to other techniques, the proposal was discontinued considering the long term required for obtaining results.

Bioremediation. Consists in using microorganisms and manipulating their metabolic activities to eliminate contaminants or, at least, to turn them into less aggressive chemical species, minimizing the environment's predicament and facilitating the biodegrading enzymatic processes responsible for its self-cleansing activity (EPA 2004).

There are many conclusive studies about bioremediation of organochlorine compounds (Benimelli et al. 2008; Fuentes et al. 2010, 2011, 2013a, b; Cuozzo et al. 2012; Alvarez et al. 2012), hydrocarbons (Canals 2005; Gentili et al. 2006), and heavy metals (Schippers 2007; Cabrera et al. 2007; Plaza et al. 2011; Albarracín et al. 2008; Polti et al. 2009, 2011a, b). Numerous bacterial groups have been studied to this purpose, but so far there have been very few microorganisms mentioned in published literature which may grow in natural environments with high B concentrations, none of them in Salta. Until now only in Cyanobacteria (Mateo et al. 1986) and in *Bacillus boroniphilus* (Ahmed et al. 2007a) has B as boric acid been singled out as an essential element for their growth. In *Azobacter* sp., B stimulates nitrogen fixation mechanisms and in sea bacteria such as *Vibrio harveyi* it intervenes in Quorum Sensing mediation process mechanisms (Chen et al. 2002). Also in *Arthrobacter nicotinovorans*, phenyl boric acid (Negrete-Raymond et al. 2003) is eliminated mainly as orthoboric acid [$B(OH)_3$]. Some bacteria were isolated from soils naturally contaminated with B minerals in Hisrcik, Turkey, such as *Bacillus boroniphilus* sp. nov. (Ahmed et al. 2007a), *Gracilibacillus boraciitolerans* sp. nov. (Ahmed et al. 2007b), Chimaereicella boritolerans sp. nov. (Ahmed et al. 2007c) *Lysinibacillus boronitolerans* (Ahmed et al. 2007d), and *Lysinibacillus parviboronicapiens* (Miwa et al. 2009) which need B for their growth and are able to tolerate concentrations over 450, 450, 300, and 150 mM boric acid, respectively. Miwa et al. (2008) published the existence of *Variovorax boronicumulans* sp. nov., a Gram-negative bacterium able to accumulate B. This mechanism enables contaminating B removal from the soil through its accumulation inside the cells.

Even though actinobacteria belong to the most abundant bacterial group present in all soils, showing an important biodegrading activity through the secretion of a great number of extracellular enzymes which enables them to metabolize recalcitrant compounds (Keiser et al. 2000), there has been no report so far about the action of this bacterial group on B compounds. Soil actinobacteria use different strategies to keep their population stable ranging from fast proliferation cycles to sporulation and slow growth in periods of stress or lack of nutrients. These features and the fact that they are infrequent pathogens give them great potential in bioremediation processes (Ravel et al. 1998). There are no precedents of this type of remediation strategies in Salta province soils, which makes the development of this fast economic environment friendly type of technology allowing the repair of previous contaminations an ideal remediation system.

3.2 Bioremediation Mechanisms

There are several mechanisms to develop metal-resistant microorganisms. Majzlik et al. (2011) proposed the following: through intra- and extracellular mechanisms; excreting the metals through transport systems; sequestering compounds through cytosol agents which may bind to the metal and detoxicate the interior of cells; excreting chelator compounds to the extracellular medium to bind metals, thus freezing or paralyzing them, and binding the metals to the cell membrane since the latter is prone to bonding them and thus preventing their entrance. Other authors also proposed organic acid excretion to make metals soluble (Gadd 1999), and developing cytoplasmatic protection mechanisms through inclusion bodies which retain a great number of metal cations (González and Jensen 1998). Two other well-known mechanisms are of great importance to the authors: biomineralization and exopolysaccharide formation.

Biomineralization. Biomineralization is the process by which organisms (from prokaryotes to eukaryotes, including human) produce minerals. Microorganisms are the second group in importance in this respect and can produce a great variety of different minerals depending on the mechanism of the mineralization process: biologically controlled mineralization (BCM) and biologically induced mineralization

(BIM) (Pérez-González et al. 2011). In the latter (BIM), biomineral formation takes place as a consequence of the changes in the system oversaturation triggered by the capture or the excretion of different metabolites (active mechanism; Lowenstam 1981) and by the contribution of crystallization nuclei such as cell components (cell wall, membrane, organic, and lysis cell remnants) favoring precipitation (passive mechanism; Lowenstam 1981). Thus, microorganisms modify their nearest extracellular environment generating the conditions required for the mineral to precipitate (Gadd 2010).

Minerals formed by BIM are almost always deposited outside the inducing organism. Therefore, mineralization takes place in an open environment, and not inside an intracellular space limited to this purpose, so there is neither cell nor macromolecular machinery specialized in a specific role in the biomineralization process. In general, biominerals count on calcium as majority cation, iron being the second most common. Phosphates, oxides, and carbonates are the most numerous anions (Pérez-González et al. 2011).

There is evidence of numerous bacterial genii forming biominerals, such as *Pseudomonas* sp. (Baskar et al. 2006), *Bacillus* sp. (Castanier et al. 2000), *Vibrio* sp. (Rivadeneyra et al. 1993), reducing sulphate bacteria and cyanobacteria (Wright 1999), *Myxobacteria* sp. (González-Muñoz et al. 1996), *Halobacillus* sp. (Rivadeneyra et al. 2004), *Shewanella* sp. (Pérez-González et al. 2010), as well as *Streptomyces* sp. (Cañaveras et al. 1999; Schütze et al. 2013).

Exopolysaccharide formation. Polysaccharides are carbohydrate polymers whose structural diversity (from linear repetitions of a monomer to large branching-out structures of different sugars) reflects their functionalities. There exist two types of polysaccharides, the reserve ones (such as glycogen) and the structural ones (such as cellulose). Exopolysaccharides (EPS) belong to the latter group (Bernal and Llamas 2012). They may be produced by algae, plants, animals, and bacteria and are responsible for interactions among cells and on the cell surface. Published research states that B intervenes in quorum sensing mediation mechanisms in sea bacteria *Vibrio harveyi* (Chen et al. 2002), this mechanism being one of the main ones involved in extracellular polysaccharides or EPS generation.

Exopolysaccharides may have different compositions even if generated by the same species, such as *Pseudomonas aeruginosa* which produces three different EPS (alginato, Pel, and Psl) (Ryder et al. 2007). Their composition not only varies with the microorganism but also with the ambient conditions where they are able to develop (Mayer et al. 1999). These include oxygen and nutrient availability (Vu et al. 2009), temperature, pH, and the presence of other stressors such as heavy metals (Sheng et al. 2005) and high salt concentrations, among others. Different strategies related to microorganism use have been developed to detoxify natural environments because of all the ecologic and economic advantages they have (Singh et al. 2006). The importance of bacterial EPS lies in their potential application in remediating mining or industrial contaminated environments (Rawlings and Johnson 2007). About the use of EPS in the removal of heavy metals, it is based on the possibility for uniting these polymeric matrices or complex metals in solution and flocculating them (Pal and Paul 2008). Within the group of Actinobacteria, Amycolatopsis sp. AB0 stands out because it generates an EPS in the presence of copper (Albarracín et al. 2008).

Native actinobacteria. Numerous microorganisms were isolated from B-contaminated soils in Salta province. Some of these soils were naturally contaminated (Tincalayu, an exploitation mine of B minerals in the West of the province, and Animaná where pollution comes from natural leaching of bedrock in the Calchaquíes Valley) and one with anthropogenic pollution (Baradero, where B minerals were processed in Salta Capital) (Moraga et al. 2013).

Eight strains were sequenced and genetically identified as actinobacteria, seven of the genus *Streptomyces* and the other one of *Lentzea* genus. Seven strains had >99 % homology with some species already identified. Moreover, these organisms were able to tolerate up to 5 % w/v NaCl, so they were classified as halotolerant according to the criteria of Zahran (1997).

Resistance and tolerance to boric acid of the eight strains were analyzed, taking as reference the maximum and minimum B concentrations found in soils. Those strains that were most tolerant were studied on their growth in liquid medium with concentrations of 20 and 40 mM and subsequently evaluated for their potential use as agents of indigenous soil remediation.

These studies allowed us to find four strains whose B adaptation mechanisms were different and very promising, such as the formation of a biomineral through biologically induced mineralization mechanism. Other two revealed the formation of EPS observed growth in liquid medium. Marguesin and Schiner (2001) suggested that the halotolerant microorganisms can survive in saline environments through various mechanisms, including the production of biosurfactants and EPS stands, many of which use ions or solutes in composition.

Whereas bioremediation refers to the application of biological systems for cleaning organic and inorganic contamination with microorganisms like bacteria and fungi, detoxification through the immobilization of contaminants included in the formation of biominerals or EPS would be a way to remediate. Furthermore, the products thus formed by the microorganisms, containing the element included, can have catalytic properties and novel features as nanoparticles, so their study is relevant for the development of new biomaterials with structural, technological, and environmental implications (Lloyd et al. 2008; Theng and Yuan 2008; Petkov et al. 2009; Hennebel et al. 2009).

3.3 Concluding Remarks

Although there is an intense development of remediation strategies for B-contaminated soil all over the world, phytoremediation and bioremediation technologies in Argentina are still emerging and experimental. This is probably because the success of these treatments depends not only on the species (either plants or bacteria), but also on the integrated ecosystem where they are immerse. New trends in bioremediation technologies require an interdisciplinary approach. There is no doubt that increasing our understanding of microbiology, mineralogy, geomicrobiology, and biotechnology, among other disciplines will lead us to useful and ecological applications in remediating B mining or industrial contaminated environments.

References

Ahmed I, Yokota A, Fujiwara T (2007a) A novel highly boron tolerant bacterium, *Bacillus boroniphilus* sp. nov., isolated from soil, that requires boron for its growth. Extremophiles 11:217–224

Ahmed I, Yokota A, Fujiwara T (2007b) *Gracibacillus boraciitolerans* sp. nov, a highly boron-tolerant and moderately halotolerant bacterium isolated from soil. Int J Syst Evol Microbiol 57:796–802

Ahmed I, Yokota A, Fujiwara T (2007c) *Chimaereicella boritolerans* sp. nov, a boron-tolerant and alkaliphilic bacterium of the family Flavobacteriaceae isolated from soil. Int J Syst Evol Microbiol 57:986–992

Ahmed I, Yokota A, Yamazoe A et al (2007d) Proposal of *Lysinibacillus boronitolerans* gen. nov. sp. nov., and transfer of *Bacillus fusiformis* to *Lysinibacillus fusiformis* comb. nov. and *Bacillus sphaericus* to *Lysinibacillus sphaericus* comb. nov. Int J Syst Evol Microbiol 57:1117–1125

Albarracín FS (2008) Fitorremediación: evaluación del potencial de especies nativas en la remediación de suelos contaminados con boro. Graduate thesis, Universidad Nacional de Salta, Argentina

Albarracín FS, de Viana M, Flores H (2007) Germinación y supervivencia de dos especies vegetales en altas concentraciones de boro. Avances en Energías Renovables y Medio Ambiente 11:49–56

Albarracín V, Winik B, Kothe E et al (2008) Copper bioaccumulation by actinobacteria *Amycolatopsis* sp. AB0. J Basic Microbiol 48:323–330

Alvarez A, Benimeli C, Sáez J, Fuentes M, Cuozzo S, Polti M, Amoroso M (2012) Bacterial Bio-Resources for Remediation of Hexachlorocyclohexane. Int J Mol Sci 13:15086–15106

Bartlett RJ, Picarelli CJ (1973) Availability of boron and phosphorus as affected by liming an acid potato soil. Soil Sci 116:77–80

Baskar S, Baskar R, Mauclaire L et al (2006) Microbially induced calcite precipitation in culture experiments: possible origin for stalactites in Sahastradhara caves, Dehradun, India. Curr Sci India 90:58–64

Benimelli CS, Fuentes MS, Abate CM et al (2008) Bioremediation of lindane-contaminated soil by *Streptomyces* sp. M7 and its effects on *Zea mays* growth. Int Biodeter Biodegr 61:233–239

Bernal P, Llamas M (2012) Promising biotechnological applications of antibiofilm exopolysaccharides. Microbial Biotechnol 5:670–673

Cabrera G, Viera M, Gómez JM et al (2007) Bacterial removal of chromium (VI) and (III) in a continuous system. Biodegradation 18:505–513

Canals MV (2005) Biorremediación de suelos contaminados por hidrocarburos: caracterización microbiológica, química y ecotoxicológica. Ph.D. thesis. Universidad de Barcelona, España

Cañaveras JC, Hoyos M, Sánchez-Moral S et al (1999) Microbial communities associated with hydromagnesite and needle-fiber aragonite deposits in a Karstic Cave (Altamira, Northern Spain). Geomicrobiol J 16:9–25

Castanier S, Le Métayer-Levrel G, Perthuisot JP (2000) Bacterial roles in the precipitation of carbonate minerals. In: Riding RE, Awramik SM (eds) Microbial sediments. Springer, Berlin, pp 32–39

Chen X, Schauder S, Potier N et al (2002) Structural identification of a bacterial quorum-sensing signal containing boron. Nature 415:545–549

Chong MF, Lee KP, Chieng HJ et al (2009) Removal of boron from ceramic industry wastewater by adsorption-flocculation mechanisms using palm oil mil boiler (POMB) bottom ash and polymer. Water Res 43:3326–3334

Cunningham S, Ow D (1996) Promises and prospects of phytoremediation. Plant Physiol 110:715–719

Cuozzo SA, Fuentes MS, Bourguignon N et al (2012) Chlordane biodegradation under aerobic conditions by indigenous *Streptomyces* strains. Int Biodeter Biodegr 66:19–24

Del-Campo Marín CM, Oron G (2007) Boron removal by duckweed *Lemna gibba*: a potential method for the remediation of boron-polluted waters. Water Res 41:4579–4584

EPA (Environmental Protection Agency) (2000) Introduction to phytoremediation. EPA-report EPA/600/R-99/107. http://cluin.org/search/default.cfm?search_term=600%2FR-99%2F107&t=all&advlit=0. Accessed Jan 2014

EPA (Environmental Protection Agency) (2004) Introduction to phytoremediation. EPA-report. http://cluin.org/techfocus/default.focus/sec/Bioremediation/cat/Overview/. Accessed Jan 2014

Flores HR (2004) El Beneficio de los Boratos. Historia, Minerales, Yacimientos, Usos, Tratamiento, Refinación, Propiedades, Contaminación, Análisis Químico, Salta, Argentina

Fuentes MS, Benimeli CS, Cuozzo SA et al (2010) Isolation of pesticide-degrading actinomycetes from a contaminated site: bacterial growth, removal and dechlorination of organochlorine pesticides. Int Biodeter Biodegr 64:434–441

Fuentes MS, Sáez JM, Benimeli CS et al (2011) Lindane biodegradation by defined consortia of indigenous *Streptomyces* strains. Water Air Soil Pollut 222:217–231

Fuentes MS, Alvarez A, Sáez JM et al (2013a) Methoxychlor bioremediation by defined consortium of environmental *Streptomyces* strains. Int J Environ Sci Technol. doi:10.1007/s13762-013-0314-0

Fuentes MS, Briceño GE, Sáez JM et al (2013b) Enhanced removal of a pesticides mixture by single cultures and consortia of free and immobilized *Streptomyces* strains. BioMed Res Int. doi:10.1155/2013/392573

Gadd GM (1999) Fungal production of citric and oxalic acid: importance in metal speciation, physiology and biogeochemical processes. Adv Microb Physiol (Lond) 41:4748

Gadd GM (2010) Metals, minerals and microbes: geomicrobiology and bioremediation. Microbiology 156:609–643

Gentili AR, Cubitto MA, Ferrero M, et al (2006) Bioremediation of crude oil polluted seawater by a hydrocarbon-degrading bacterial strain immobilized on chitin and chitosan flakes. Int Biodeter Biodegr 57:222–228

González-Muñoz MT, Ben Omar N, Martínez-Cañamero M et al (1996) Struvite and calcite crystallization induced by cellular membranes of Myxococcus xanthus. J Cryst Growth 163(4): 434–439

González H, Jensen T (1998) Nickel sequestering by polyphosphate bodies in *Staphylococcus aureus*. Microbios 93:179–185

Hennebel T, Gusseme BD, Verstraete W (2009) Biogenic metals in advanced water treatment. Trends Biotechnol 27:90–98

Hoffman GJ (1990) Leaching fraction and root zone salinity control. In: Tanji KK (ed) Agricultural salinity assessment and management. American Society of Civil Engineers, New York, pp 238–261

INTA, Instituto Nacional de Tecnología Agropecuaria. Secretaría de Agricultura, Ganadería, Pesca y Alimentación (1997) Informe de muestras en agua y suelos. Centro Regional NOA, Salta, Argentina

INTA, Instituto Nacional de Tecnología Agropecuaria. Secretaría de Agricultura, Ganadería, Pesca y Alimentación (2000) Informe de análisis de suelos. Centro Regional NOA, Salta, Argentina

INTA, Instituto Nacional de Tecnología Agropecuaria. Secretaría de Agricultura, Ganadería, Pesca y Alimentación (2007) Informe de análisis de suelos. Centro Regional NOA. Salta, Argentina

Itakura T, Sasai R, Itoh H (2005) Precipitation recovery of boron from wastewater by hydrothermal mineralization. Water Res 39:2543–2548

Kavak D (2009) Removal of boron from aqueous solutions by batch adsorption on calcined alunite using experimental design. J Hazard Mater 163:308–314

Keiser T, Viv MJ, Buttner MJ et al (2000) Preparation and analysis of genomic plasmid DNA. In: Keiser T, Viv MJ, Buttner MJ, Chater KF, Hopwood DA (eds) Practical streptomyces genetics. The John Innes Foundation, Norwich, pp 161–210

Keren R, Bingham FT (1985) Boron in water, soils, and plants. Adv Soil Sci 1:230–276

Lloyd JR, Pearce CI, Coker VS et al (2008) Biomineralization: linking the fossil record to the production of high value functional materials. Geobiology 6:285–297

Lowenstam HA (1981) Minerals formed by organisms. Science 211:1126–1131

Majzlik P, Strasky A, Adam V et al (2011) Influence of zinc (II) and copper (II) ions on *Streptomyces* bacteria revealed by electrochemistry. Int J Electrochem Sci 6:2171–2191

Marguesin R, Schiner F (2001) Potential of halotolerant and halophilic microorganisms for biotechnology. Extremophiles 5:73–83

Mateo P, Bonilla I, Fernández-Valiente E et al (1986) Essentiality of boron for dinitrogen fixation in Anabaena sp. PCC 7119. Plant Physiol 94:1154–1560

Mayer C, Moritz R, Kirschner C et al (1999) The role of intermolecular interactions: studies on model systems for bacterial biofilms. Int J Biol Macromol 26:3–16

Miwa H, Ahmed I, Yoon J et al (2008) *Variovorax boronicumulans* sp. nov., a boron-accumulating bacterium isolated from soil. Int J Syst Evol Microbiol 58:286–289

Miwa H, Ahmed I, Yokota A et al (2009) *Lysinibacillus parviboronicapiens* sp. nov., a low boron-containing bacterium isolated from soil. Int J Syst Evol Microbiol 59:1427–1432

Moraga NB, Poma R, Amoroso MJ et al (2013) Isolation and characterization of indigenous *Streptomyces* and *Lentzea* strains from soils containing boron compounds in Argentina. J Basic Microb. doi:10.1002/jobm.201200714

Nable RO, Bañuelos GS, Paull JG (1997) Boron toxicity. Plant Soil 193:181–198

Negrete-Raymond A, Weder B, Wackett L (2003) Catabolism of arylboronic acids by *Arthrobacter nicotinovorans* strain PBA. Appl Environ Microbiol 69:4263–4267

Oztürk N, Navak D, Köse TE (2008) Boron removal from aqueous solution by reverse osmosis. Desalinization 223:1–9

Pal A, Paul A (2008) Microbial extracellular polymeric substances: central elements in heavy metal bioremediation. Indian J Microbiol 48:49–64

Pérez-González T, Jiménez-López C, Neal AL et al (2010) Magnetite biomineralization induced by *Shewanella oneidensis*. Geochim Cosmochim Acta 74:967–979

Pérez-González T, Valverde-Tercedor C, Jiménez-López C (2011) Biomineralización bacteriana de magnetita y aplicaciones. http://www.ehu.es/sem/seminario_pdf/SEMINARIO_SEM_7_058.pdf. Accessed Nov 2013

Peryea FJ, Bingham FT, Rhoades JD (1985) Mechanisms for boron regeneration. Soil Sci Soc Am J 49:840–843

Petkov V, Ren Y, Saratovsky I et al (2009) Atomic-scale structure of biogenic materials by total X-ray diffraction: a study of bacterial and fungal MnOx. ACS Nano 3:441–445

Plaza J, Viera M, Donati E et al (2011) Biosorption of mercury by Macrocystis pyrifera and Undaria pinnatifida: Influence of zinc, cadmium and nickel. J Environ Sci 23(11):1778–1786

PNUMA, Programa de las Naciones Unidas para el Medio Ambiente. Oficina Regional para América Latina y el Caribe (2004) México DF. http://www.rolac.unep.mx/centinf/esp/. Accessed Jan 2014

Polti MA, García R, Amoroso MJ et al (2009) Bioremediation of chromium (VI) contaminated soil by *Streptomyces* sp. MC1. J Basic Microbiol 49:285–292

Polti MA, Atjián MC, Amoroso MJ et al (2011a) Soil chromium bioremediation: synergic activity of actinobacteria and plants. Int Biodeter Biodegr 65:1175–1181

Polti MA, Amoroso MJ, Abate CM (2011b) Intracellular chromium accumulation by *Streptomyces* sp. MC1. Water Air Soil Pollut 214:49–57

Prather RJ (1977) Sulphuric acid as an amendment for reclaiming soils high in boron. Soil Sci Soc Am Proc 41:1098–1101

Ravel J, Amoroso MJ, Colwell RR et al (1998) Mercury resistant actinomycetes form Chesapeake Bay. FEMS Microbiol Lett 162:177–184

Rawlings DE, Johnson DB (2007) The microbiology of biomining: development and optimization of mineral-oxidizing microbial consortia. Microbiology 153:315–324

Rivadeneyra MA, Delgado R, Del Moral A et al (1993) Precipitation of calcium carbonate by *Vibrio* sp. from an inland saltern. FEMS Microbiol Ecol 13:197–204

Rivadeneyra MA, Párraga J, Delgado R et al (2004) Biomineralization of carbonates by Halobacillus trueperi in solid and liquid media with different salinities. FEMS Microbiol Ecol 48 (1):39–46

Romero O (2003) Proyecto de valoración económica del impacto ambiental producido por la industria del boro sobre los acuíferos del cuadrante sur de la ciudad de Salta. Graduate thesis, Universidad Nacional de Salta, Argentina

Ryder C, Byrd M, Wozniak DJ (2007) Role of polysaccharides in *Pseudomonas aeruginosa* biofilm development. Curr Opin Microbiol 10:644–648

Sasmaz A, Obek E (2009) The accumulation of arsenic, uranium, and boron in *Lemma gibba* exposed to secondary effluents. Ecol Eng 35:1564–1567

Schippers A (2007) Microorganisms involved in bioleaching and nucleic acid-based molecular methods for their identification and quantification. In: Donati ER, Sand WW (eds) Microbial processing of metal sulfides. Springer, Amsterdam, pp 3–33

Schütze E, Weist A, Klose M et al (2013) Taking nature into lab: biomineralization by heavy metal resistant *Streptomycetes* in soil. Biogeosci Discuss 10:2345–2375

Sheng G, Yu H, Yue Z (2005) Production of extracellular polymeric substances from *Rhodopseudomonas acidophila* in presence of toxic substances. Appl Microbiol Biotechnol 69:216–222

Shennan C, Grattan SR, May DM et al (1995) Feasibility of cyclic reuse of saline drainage in a tomato cotton rotation. J Environ Qual 24:476–486

Singh R, Paul D, Jain RK (2006) Biofilms: implications in bioremediation. Trends Microbiol 14:389–397

Suarez O, Camacho C, Visuara M (1997) Estudio de impacto ambiental del predio de la ex-planta de fabricación de ácido bórico de la empresa. Industrias Químicas Baradero, Salta

Swietlik D (1995) Interaction between zinc deficiency and boron toxicity on growth and mineral nutrition of sour orange seedlings. J Plant Nutr 18:1191–1207

Taştan BE, Duygu E, Dönmez G (2012) Boron bioremoval by a newly isolated *Chlorella* sp. and its stimulation by growth stimulators. Water Res 46:167–75

Theng BKG, Yuan G (2008) Nanoparticles in the soil environment. Elements 4:395–399

Vu B, Chen M, Crawford RP et al (2009) Bacterial extracellular polysaccharides involved in biofilm formation. Molecules 14:2535–2554

Wright DT (1999) The role of sulphate-reducing bacteria and cyanobacteria in dolomite formation in distal ephemeral lakes of the Coorong region, South Australia. Sediment Geol 126:147–157

Yilmaz AE, Boncukcuoğlu R, Kocakerim MM et al (2008) Boron removal from geothermal waters by electrocoagulation. J Hazard Mat 153:146–151

Zahran HH (1997) Diversity, adaptation and activity of the bacterial flora in saline environments. Biol Fertil Soil 25:211–223

Chapter 4
Advances in Chile for the Treatment of Pesticide Residues: Biobeds Technology

Gabriela Briceño, Gonzalo Tortella, Olga Rubilar, Graciela Palma, and M. Cristina Diez

Abstract The widespread use of pesticides for agricultural and for nonagricultural purposes worldwide has resulted in the presence of pesticide residues in various environmental matrices. The occurrence of pesticide residues in surface waters, groundwater, and large volumes of soil is mainly due to the inadequate management of these compounds. In this context, a biobed system was developed in Sweden in response to the need for a simple and effective way to minimize environmental contamination from pesticide manipulation, particularly when filling the spraying equipment, a typical point source of contamination. Biobeds are based on the adsorption and degradation potential of organic biomixtures composed of top soil, peat, and straw that fills a deep hole in the ground and a grass layer that covers the surface. Recently, the use of biobeds has expanded to other countries in Europe and Latin America. In Chile, four biobeds similar to the European ones have been installed, making this country a pioneer in this type of decontamination system. This chapter gives a general overview of biobeds technology and the advances in research at laboratory scale related to the treatment of pesticide residues in a biobed system in Chile.

G. Briceño
Departamento de Ingeniería Química, Universidad de La Frontera, Casilla 54-D, Temuco, Chile

Departamento de Ciencias Químicas y Recursos Naturales, Universidad de La Frontera, Casilla 54-D, Temuco, Chile

Núcleo Científico y Tecnológico en Biorecursos BIOREN-UFRO, Universidad de La Frontera, Casilla 54-D, Temuco, Chile

G. Tortella • O. Rubilar • M.C. Diez (✉)
Departamento de Ingeniería Química, Universidad de La Frontera, Casilla 54-D, Temuco, Chile

Núcleo Científico y Tecnológico en Biorecursos BIOREN-UFRO, Universidad de La Frontera, Casilla 54-D, Temuco, Chile
e-mail: cristina.diez@ufrontera.cl

G. Palma
Departamento de Ciencias Químicas y Recursos Naturales, Universidad de La Frontera, Casilla 54-D, Temuco, Chile

Núcleo Científico y Tecnológico en Biorecursos BIOREN-UFRO, Universidad de La Frontera, Casilla 54-D, Temuco, Chile

A. Alvarez, M.A. Polti (eds.), *Bioremediation in Latin America: Current Research and Perspectives*, DOI 10.1007/978-3-319-05738-5_4

4.1 Introduction

Biobeds are biological biopurification systems developed in Sweden in response to the need for a simple and effective method to minimize environmental contamination from pesticide manipulation, particularly when filling the spraying equipment, a typical point source of contamination (Torstensson and Castillo 1997; Castillo et al. 2008). Biobeds are a low-cost alternative for treating pesticide waste and washings, providing a matrix to absorb the pesticides and facilitate their biodegradation. The biologically active matrix, usually called a biomixture, is composed of straw, peat, and soil in a volumetric proportion of 2:1:1 (respectively) that fills a deep hole (60 cm) in the ground and a grass layer that covers the surface (Torstensson and Castillo 1997; Castillo et al. 2008) (Fig. 4.1).

Each component of a biobed system has a specific function. The biomixture (Fig. 4.1a) is the most important, although each component plays a role in its composition. For example, the straw stimulates the growth of ligninolytic microorganisms such as white-rot fungi and therefore the production of extracellular ligninolytic enzymes, such as peroxidases and phenoloxidases (Tortella et al. 2013a). The peat contributes to high sorption capacity and regulates the humidity of the system, and the soil provides sorption capacity in the biobed and pesticide-degrading microorganisms, including actinobacteria that can act synergistically with the fungi (Briceño et al. 2013a). The grass layer (Fig. 4.1b) that covers the biobed system increases the biobed efficiency, retaining the pesticides on top of biobeds and then controlling the leaching, helps keep the system humid, and promotes evapotranspiration and further pesticide degradation at root level. The gravel layer (Fig. 4.1c) acts as a filter to prevent organic residues from passing out of the biomixture. The waterproofing system (Fig. 4.1d) consists of the lining of the walls and bottom of the bed that prevent the contact of pesticides with adjacent soil. The recirculation system (Fig. 4.1e) consists of a concrete well connected to the biobed and its main functions are receiving the percolate pesticide residues from the biomixture and recirculating them to the biobed. The ceiling (Fig. 4.1f) prevents the entry of precipitation and finally, the support system for application equipment consists of a metallic structure or another material that supports the tractor and application equipment.

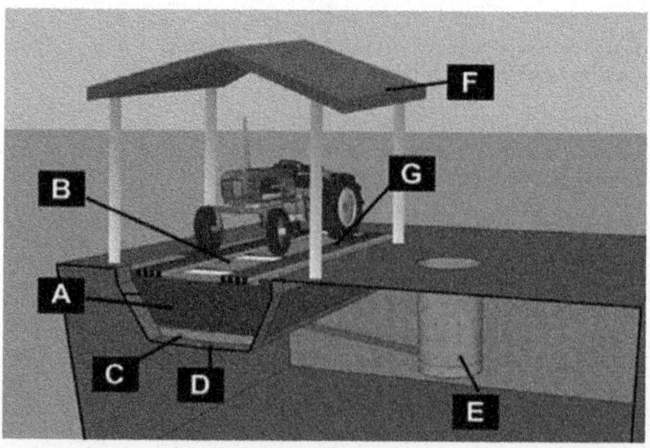

Fig. 4.1 General representation of a biobed system. Biomixture (*A*), grass layer (*B*), gravel layer (*C*), waterproofing system (*D*), recirculation system (*E*), ceiling (*F*), and support system for application equipment (*G*)

Biobed systems can be modified in several different ways according to the amount of pesticide residues to be treated, land available for installation, environmental conditions, geographic location, etc. Therefore, countries incorporating biobed technology have adapted it to local conditions, in some cases modifying the biomixture, construction, and the names.

4.2 Biobeds in the World

The first biobed was built in Sweden in 1993. Today, in Sweden and other parts of the world large numbers of biobeds are functioning on farms and have proven to be efficient at reducing pesticide water-body contamination (Castillo et al. 2008; Vischetti et al. 2008; De Wilde et al. 2010a). Biobeds as an on-farm biopurification systems or "BPS" as described in the literature (Karanasios et al. 2012; Verhagen et al. 2013) have attracted attention in several countries, where work is being conducted to adapt them to local conditions and applications (Antonious 2012).

The biobed system has been evaluated in several countries including the UK, Italy, Belgium, France, Greece, and the USA, where their implementation has led to modifications to the original biobed (Castillo et al. 2008; Antonious 2012; Marinozzi et al. 2013). For example, the depth was modified in the UK to increase the retention time of the pesticide in the bed. In Italy, this technology is known as a biomass bed and utilizes biomixtures as filters through which pesticide-contaminated water is circulated and decontaminated. Because peat is not readily found in Italy, organic materials, such as urban and garden compost, peach stones, and citrus peel, are being tested as replacements (Castillo et al. 2008).

It is estimated that there were about 2,800 bioremediation systems such as biobeds in the world and this technology is becoming more. For example, in the UK the number of biobeds grew from 75 in 2007 to 150 in 2010 (Husby 2010). The use of biobeds has expanded to Latin America, in countries such as Peru, Guatemala, and Ecuador, where some pilot/field-scale studies are being developed. Recently, this system has also been developed in Chile, although with significant modifications, and biobeds are currently being built and used at field scale (Diez et al. 2013a; Tortella et al. 2013a).

4.2.1 Evidence of Pesticide Removal in Biobeds

Several studies have demonstrated that biobeds can effectively retain and degrade a wide range of pesticides, either alone or in mixtures (Torstensson and Castillo 1997; Castillo et al. 2007; Fogg et al. 2003, 2004; Vischetti et al. 2004; Castillo and Tortensson 2007; Vischetti et al. 2008). For example, studies with mecoprop and isoproturon have shown that these pesticides can be degraded in biobeds (Henriksen et al. 2003). Niels et al. (2006) evaluated the degradation and leaching of 21 pesticides. They determined that no traces of 10 out of 21 applied pesticides were detected in the percolate. Fogg et al. (2003) evaluated the ability of biobeds to degrade pesticide mixtures (isoproturon and chlorothalonil) and the concentration effect. They found that, with the exception of isoproturon at concentrations greater than 11 mg/kg, degradation was more rapid in the biomix than in topsoil. The degradation of either isoproturon or chlorothalonil was unaffected by the presence of the other pesticides.

Many factors affecting the performance of biobeds as well as adaptation or modification of the original Swedish biobed have been studied by different authors (Castillo et al. 2008; Karanasios et al. 2012). In terms of the biomix composition, laboratory-based studies showed that mixtures of soil-organic waste may be able to degrade high concentrations and complex mixtures of pesticides (Fogg et al. 2004). Castillo and Tortensson (2007) observed that a straw, peat, and soil ratio of 50:25:25 % v/v (respectively) is recommended for the organic mixture composition, because such a mixture favors a low pH, convenient for lignin-degrading fungi and phenoloxidase production and activity. Karanasios et al. (2010) focused their research on identifying various by-products of the local agricultural practice (either raw or composted), which could be used as alternatives to peat or even straw. They provide the first evidence that straw can be substituted in biomixtures by other lignocellulosic materials readily available in southern Europe.

Vischetti et al. (2004) compared the behavior of chlorpyrifos in two biobed systems: a Swedish biobed and a modified Italian biobed system. They reported that chlorpyrifos half-lives were similar in both biomixtures assessed, but the microbial biomass content was reduced by 25 % and 50 % with 10 mg/kg and 50 mg/kg of chlorpyrifos in the Italian biomix, respectively. Coppola et al. (2007) and Vischetti et al. (2007) studied the biodegradation of chlorpyrifos in a biobed system adapted to Italian conditions. They found that the Italian biomix showed several differences compared to the Swedish biomix in the chlorpyrifos degradation. Vischetti et al. (2008) studied the effect of initial concentration, co-application and repeated applications on chlorpyrifos, and metalaxyl degradation in a biobed. They concluded that both pesticides were degraded relatively quickly due to the presence of the varied microbial community capable of degrading both pesticides.

Spliid and Husby (2010) presented a new closed biobed with recirculation and evaporation for use in colder climates like Denmark. They reported that biobeds used under these climate conditions require special precautions to avoid problems with surplus water. The water is collected from the bottom and recirculated to the biobed. In Belgium, De Wilde et al. (2010a, b, c) studied sorption and degradation processes on an increasing spatial scale (micro- and macroscale). Their main conclusions were the following: sorption coefficients determined in batch sorption experiments are often not suitable for describing solute transport at the column or field scale; matrix composition had no significant influence on pesticide leaching and degradation; however, the addition of cow manure enhanced the degradation of some pesticides; the use of pesticide-primed material significantly enhanced degradation of metalaxyl; an increasing flux had a pernicious influence on sorption and degradation of most of the pesticides studied.

Otherwise, considering that little is known regarding the interactions between biomixture microflora and pesticides, Marinozzi et al. (2013) recently studied the dissipation of the fungicides azoxystrobin, fludioxonil, and penconazole, commonly used in vineyards, in a biomixture composed of pruning residues and straw used in vineyard biopurification systems. The study also examined the impact of fungicides on the microbial community, and the main results showed that fungicides affected the microbial community differently, with penconazole being the compound with the highest adverse effect on both the size and the activity of the biomixture microflora. By contrast, there was a significant change in the structure of the microbial community for penconazole and fludioxonil. High biodegradation and high mineralization capacity are desirable in biobeds. Therefore, attempts have been made to boost their biodegradation capacity via inoculation with pesticide-degrading microorganisms (Karanasios et al. 2012). In this context, Verhagen et al. (2013) demonstrated that bioaugmentation with a mixed degrading enrichment culture can vastly improve the functionality of an on-farm biopurification system. The authors went on to suggest that the use of both plastic carriers and biomix substrata can harbor a stable microbial community that can effectively degrade chlorpropham and 3-chloroaniline.

4.3 Biobeds in Chile

The growth in fruit sector exports has increased concern regarding the suitable and safe use of pesticides. At the same time, there is little knowledge or education regarding the effects and use of pesticides in urban or rural populations. Therefore, research and education programs are needed to encourage actions and effective policies that regulate pesticide use.

The point source pollution by pesticides, a result of accidental spillages or inadequate residue management, has been rigorously investigated in the last decade in Europe, finding that it plays a key role in soil and water pollution (Gregoire et al. 2009). In Chile, point source pollution by pesticides has not been evaluated; nevertheless, it is expected that the situation could be similar to Europe, particularly in basins with greater pressure from pesticide use, as in fruit production. Therefore, suitable and efficient management of pesticide residue is a topic that requires further discussion.

As a way to reduce pesticide contamination, the Fund of Scientific and Technological Development (FONDEF) financed the project D09R1006 entitled, "Proper Handling of Pesticide Residues in Fruit Production through the Implementation and Diffusion of Biobeds." Through this project has developed biobed technology, which Chile is pioneering in Latin America with four units at field scale in the La Araucanía region (38° 44′ 24″ S, 72° 35′ 25″ W) (Diez et al. 2013a). These systems were installed in the INIA-Carillanca Experimental Station, the Maquehue Experimental Station at the Universidad de La Frontera, on Santa Olga farms owned by Agrícola San Clemente, and San José Farms, with these last two being fruit production companies (Fig. 4.2). It should be mentioned that the

Fig. 4.2 Biobeds installed at the INIA-Carillanca experimental station, Santa Olga Farm owned by Agricola San Clemente, and experimental station Maquehue of the Universidad de La Frontera (from *left* to *right*)

biobeds were built like Sweden biobeds with modifications, but the biomixture of straw, peat, and soil in the proportion 2:1:1 (respectively) was maintained. For further details see www.lechosbiologicos.cl.

The biobeds installed in at the INIA-Carillanca experimental station were monitored continuously for such parameters as temperature, humidity content, pH modifications, and enzymatic activity in the biomixture. Pesticide degradation was also studied. The evaluated compounds were atrazine, azinphos-methyl, captan, chlorothalonil, chlorpyrifos, diazinon, isoproturon, and methidathion, which were applied at a concentration of 32 mg active ingredient (a.i.)/kg. After 120 days of pesticide application, pesticide residue analysis showed that about 97 % of the captan, chlorothalonil, chlorpyrifos, diazinon, isoproturon, and methidathion were removed, whereas about 89 % of the atrazine and azinphos-methyl were removed (Diez et al. 2013a).

The biomixture is a principal element controlling the degradation efficacy of the biobed (Karanasios et al. 2010). Therefore, during the recent implementation of biobeds in Chile, several studies at laboratory scale were performed to optimize the functionality of this pesticide biopurification system. In this way, the effect of operating conditions, biomixture composition, and stabilization time of the biomixture on pesticide removal was evaluated (Fernández-Alberti et al. 2012; Urrutia et al. 2013). An additional innovation was to use other lignocellulosic residues such as sawdust, barley husks, and oat husk in place of straw and biochar in place of peat (Diez et al. 2013b, c). On the other hand, bearing in mind that the biological decomposition of pesticides is the most important and effective way to remove these compounds from the environment (Dabrowska et al. 2004), biomixture-pesticide-microorganism interactions have been evaluated (Tortella et al. 2013b).

4.3.1 *Pesticide Degradation in Biobeds: Studies at Laboratory Scale*

The following topic describes the main results obtained from the study at laboratory scale performed by Chilean researchers about pesticide removal in biobed system (Table 4.1). In general, the laboratory studies were performed using 30 cm

Table 4.1 Pesticide removal in biomixtures with different composition

Pesticide	Removal (%)	Biomixture composition	Reference
Organophosphorus			
Chlorpyrifos	18/65	(Clay soil 25 %; straw 50 %; peat 5 %; biochar 20 %)/(sandy soil 25 %; straw 50 %; peat 20 %; biochar 5 %)	Diez et al. (2013c)
	50/70	(Straw 50 %; soil 25 %; peat 25 %, 80 water holding capacity and 0 day of preincubation)/(straw 50 %; soil 25 %; peat 25 %, 60 % of water holding capacity and 15 days of preincubation)	Fernández-Alberti et al. (2012)
	60–75	Straw 50 %; soil 25 %; peat 25 %	Diez et al. (2013b)
	70	Straw 50 %; soil 25 %; peat 25 % + NPK fertilization	Tortella et al. (2010)
	70 and 85	Straw 50 %; soil 25 %; peat 25 % with 0 and 15 days of preincubation, respectively	Tortella et al. (2012)
Diazinon	28/70	(Clay soil 25 %; straw 50 %; peat 5 %; biochar 20 %)/(trumao soil 25 %; straw 50 %; peat 10 %; biochar 15 %)	Diez et al. (2013c)
	60–80	Soil 25 %; peat 25 %; straw 25 %; barley husks 25 %	Diez et al. (2013b)
Triazine			
Atrazine	78–96	Straw 50 %; soil 25 %; peat 25 %	Tortella et al. (2013c)
	90	Straw 50 %; soil 25 %; peat 25 % + terpenes	Tortella et al. (2013d)
	38/70	(Clay soil 25 %; straw 50 %; peat 5 %; biochar 20 %)/(trumao soil 25 %; straw 50 %; peat 10 %; biochar 15 %)	Diez et al. (2013c)
	80–95	Soil 25 %; peat 25 %; straw 25 %; barley husks 25 %	Diez et al. (2013b)
	80/90	(Straw 50 %; soil 25 %; peat 25 %)/(straw 50 %; soil 25 %; peat 25 % + orange peel)	Tortella et al. (2013e)

(continued)

Table 4.1 (continued)

Pesticide	Removal (%)	Biomixture composition	Reference
Benzimidazole			
Carbendazim	87–96	Straw 50 %; soil 25 %; peat 25 %	Tortella et al. (2013b)
	28/70	(Clay soil 25 %; straw 50 %; peat 5 %; biochar 20 %)/ (trumao soil 25 %; straw 50 %; peat 10 %; biochar 15 %)	Diez et al. (2013c)
	70–85	Soil 25 %; peat 25 %; straw 25 %; barley husks 25 %	Diez et al. (2013b)
Urea			
Isoproturon	30/70	(Clay soil 25 %; straw 50 %; peat 5 %; biochar 20 %)/ (clay soil 25 %; straw 50 %; peat 20 %; biochar 5 %)	Diez et al. (2013c)
	50–75	Straw 50 %; soil 25 %; peat 25 %	Diez et al. (2013b)
Dicarboximide			
Iprodione	16/63	(Clay soil 25 %; straw 50 %; peat 15 %; biochar 10 %)/ (sandy soil 25 %; straw 50 %; peat 5 %; biochar 20 %)	Diez et al. (2013c)
	20–30	Straw 50 %; soil 25 %; peat 25 %	Diez et al. (2013b)

Fig. 4.3 Biobed system at laboratory scale used to evaluate pesticide degradation

width × 20 cm height × 50 cm length biobeds which were incubated at room temperature and dark condition. Figure 4.3 shows the biobed model used to evaluate pesticide removal.

Effect of preincubation and water-holding capacity. Apart from the composition of the biomixture, an important factor for biobed efficacy is the age or maturity of the biomixture prior to its use in pesticide degradation. The progressive biodegradation of the biomixture component generates a series of microbial communities and enzymatic activities that enables the efficient dissipation of pesticides in the biobed system and avoids metabolite accumulation (Castillo et al. 2008). The moisture level in the biobed is also a relevant parameter to promote different microbial environments that can influence the oxygen level, the microbial activity, and the amount of pesticide in the solution (Fernández-Alberti et al. 2012). In this light, Fernández-Alberti et al. (2012) evaluated the degradation and adsorption of chlorpyrifos (160 mg a.i./kg) and formation 3,5,6-trichloro-2-pyrinidol (TCP) in a biomixture prepared with an Andisol, peat, and straw in a volumetric proportion of 1:1:2 at different preincubation times (0, 15, and 30 days) and with different moisture contents (40, 60, and 80 % of water-holding capacity). Moreover, ligninolytic enzyme activity and microbial respiration in the biomixture were periodically analyzed. The main results of this study showed that the biomix had a greater capacity to retain chlorpyrifos than topsoil. Moreover, the preincubation period, water-holding capacity, and concentration of the chlorpyrifos in the biomix influenced degradation of the contaminant and TCP formation as well as the biological activities in the biomix. Finally, the author concluded that a biomixture with an Andisol, peat, and straw (1:1:2), preincubated for 15 days and incubated with 60 % of water-holding capacity, is capable of degrading chlorpyrifos efficiently. In another study, Tortella et al. (2012) evaluated the effect of using a typical composition of Swedish biomixture at different maturity stages on the degradation of chlorpyrifos. The study was conducted using a biomixture at three maturity stages: 0, 15, and 30 days, where chlorpyrifos was added to the biobeds at a final concentration of 200, 320, and 480 mg/kg. Chlorpyrifos degradation in the biomixture as well as formation of TCP and hydrolytic and phenoloxidase activities were measured. The results showed that the biomixture efficiently degraded chlorpyrifos (degradation efficiency >50 %) in all the maturity stages evaluated. However, chlorpyrifos degradation decreased as the pesticide concentrations increased. 3,5,6-trichloro-2-pyrinidol formation occurred in all biomixtures, but a major accumulation was observed in the biomixture with 30 days of preincubation. Moreover, significant differences were found in both phenoloxidase and hydrolytic activities in the three maturity stages evaluated. These two biological activities were also affected by the increase in pesticide concentration. As a conclusion the authors reported that chlorpyrifos can be degraded efficiently in all the maturity stages evaluated.

Effect of biomixture composition. The composition of the Swedish biomixture has been efficient in degrading several pesticides (Vischetti et al. 2004; Castillo and Tortensson 2007). However, the biomixture had to be adapted due to the greater availability of other lignocellulosic wastes in some countries. Urrutia et al. (2013) evaluated the potential use of readily available wastes as barley husk, sawdust, and oat husk, as total or partial substitutes for straw in a biomixture for pesticide degradation studies. The results showed that a biomixture composed of oat husk was highly efficient in pesticide degradation, with half-life ($t_{1/2}$) values of 28, 58, and 26 days for atrazine, chlorpyrifos, and isoproturon, respectively. On the other hand, comparable for degrading capacities with the straw based biomixture were obtained with sawdust and barley husk, but only as partial replacement. By contrast, high $t_{1/2}$ values (more than 100 days) were obtained in biomixtures with total substitution of straw with sawdust or barley husk. Moreover, high and stable biological activity was observed in the biomixtures composed of oat husk. Therefore, the authors reported that straw can be partially or totally replaced by oat husk, thereby permitting an efficient degradation of pesticide mixture, and that straw can be only partially replaced by barley husk and sawdust in the biomixture to allow efficient pesticide degradation. Recently, Diez et al. (2013b) assessed two alternate lignocellulosic materials (barley husks and pine sawdust) as partial substitutes for straw in a biomixture on the degradation of a repeatedly applied mixture of six pesticides (atrazine, isoproturon, iprodione, chlorpyrifos, diazinon, and carbendazim). The results showed that the highest degradation efficiency was found in mixtures containing straw and barley husks. Each biomixture tested achieved a high degradation (50–90 %) of all the pesticides used except iprodione. Moreover, repeated applications of pesticides resulted in a slowing of the degradation rate of all pesticide types in all biomixtures. The study concluded that the straw in the traditional biomixture can be partially replaced by other lignocellulosic materials to efficiently degrade a mixture of pesticides even when the pesticides are added in successive applications and high concentrations. Finally, in the study by Diez et al. (2013c) biochar was evaluated as a partial replacement of peat in pesticide-degrading biomixtures formulated with different soil types. Each biomixture was prepared with one type of soil (clay, trumao, and sandy), straw, peat, and biochar in different volumetric proportions. In each biomixture, the residual pesticide (atrazine, carbendazim, chlorpyrifos, isoproturon, iprodione, and diazinon) concentrations were measured at 0 day and after 40 days. The results showed that at the end of the pesticide degradation assay, changes were observed in the biomixtures that demonstrated differences among their pesticide degradation abilities. In general, pesticide degradation was higher in the control biomixtures (without biochar) than in biomixtures prepared with biochar. One exception was iprodione, which presented higher degradation efficiency when biochar was included in the biomixture. The author indicated that although the use of biochar to replace peat in the biomixtures did not significantly improve pesticide degradation, a decrease in the initial residue concentration of the pesticides was observed. Therefore, biochar may represent an interesting material to replace peat in biomixtures designed to degrade and/or adsorb pesticides.

Effect on biobed microbial community. The current literature suggests that microbial communities in a pesticide-contaminated biomixture are adversely affected, though recovery is normally observed over time (Tortella et al. 2013c). In this context, and to gain a better understanding of the pesticide-biomixture-microorganism interaction, Tortella et al. (2013b, c) recently investigated carbendazim and atrazine dissipation, and the effect on the microbial community. In the first study, the impact of repeated carbendazim applications on the extent of carbendazim dissipation, microbial diversity, community-level physiological profile, and enzymatic activity within the biomixture was evaluated. After three successive carbendazim applications, the post-application carbendazim dissipation was 87 %, 94 %, and 96 %, respectively. Although microbial enzymatic activity was affected significantly by carbendazim application, it was able to recover after each carbendazim pulse. Likewise, the numbers of culturable bacteria, fungi, and actinobacteria were slightly affected by the addition of the compound, but the inhibitory effect of the pesticide application was temporary. Denaturing gradient gel electrophoresis (DGG) and Biolog Ecoplate™ assays showed that the microbial populations remained relatively stable over time compared to the control. With these results the authors demonstrated the high dissipation capacity of this biomixture and highlighted the microbiological robustness of this biological system. In the second study, the effects of repeated atrazine application (40 mg a.i./kg) on its degradation, microbial communities, and enzyme activities were studied in a peat-based biomixture composed of straw, soil, and peat in the volumetric proportions of 2:1:1 to be used in an on-farm biopurification system. The results showed that atrazine removal efficiency was high (96, 78, and 96 %) after each atrazine application and did not show a lag phase. Microbial enzyme activities were significantly reduced with atrazine application but rapidly recovered. On the other hand, the microbial diversity obtained by Biolog Ecoplate™ was similar after the first and second atrazine applications; however, an inhibitory effect was observed after the third application. After each atrazine application, culturable fungi were reduced, but rapidly recovered with no significant changes in culturable bacteria and actinobacteria compared to the control. Analysis through DGGE patterns revealed that the microbial community structure remained relatively stable over time compared to the controls. The authors concluded that after successive atrazine applications, the peat-based biomixture had a good degradation capacity. Moreover, microbiological assays demonstrated the robustness of the peat-based biomixture from a microbiological point of view to support pesticide degradation (Tortella et al. 2013c).

Effect of biomixture biostimulation. As has been observed, pesticide degradation in biobeds can be limited or improved by several factors. Biostimulation of the indigenous microorganisms through the addition of nutrients is an important aspect to consider because the enrichment of the indigenous microbial populations is the most widely used tool in a bioremediation procedure (Tortella et al. 2010). In this context and in order to ascertain the effect of biomixture stimulation, Tortella et al. (2010) evaluated the degradation of chlorpyrifos (160 a.i. mg/kg) using a biomixture biostimulated with inorganic fertilizer as nitrogen (N), phosphorus (P), and potasio (K) at three concentrations (0.1, 0.5, and 1.0 % w/w). Chlorpyrifos degradation, TCP accumulation, and biological activity of the biomix were evaluated. The results showed that the chlorpyrifos was dissipated efficiently (>75 %) after 40 days of incubation and no additional dissipation was obtained by increasing the NPK concentration after 20 days of incubation. 3,5,6-Trichloro-2-pyrinidol accumulation occurred in all the NPK concentrations evaluated and its concentration increased with the increase of NPK addition, raising the probability of leaching of this compound. Finally, the biological activity in the biomixture increased due to the presence of NPK in all the evaluated concentrations. In conclusion, the results demonstrated that the biomix prepared with an Andisol and biostimulated with NPK nutrient can be recommended in biobeds as a viable alternative to chlorpyrifos dissipation, thereby avoiding the likelihood of soil and water contamination (Tortella et al. 2010). By contrast, and taking into account that biostimulation of organic-pollutant-degrading microorganisms by adding volatile organic compounds such as terpenes has been used to increase pollutant biodegradation in contaminated soils (Bento et al. 2005; Tyagi et al. 2011; Dudášová et al. 2012), Tortella et al. (2013d) studied the effect of the terpenes α-pinene, eucalyptol, and limonene, individually and as mixtures, on atrazine biodegradation and on biological activity in a biobed biomixture. The results showed that terpenes added individually at relatively low concentrations (50 μg/kg) significantly enhanced atrazine degradation and biological activity during the first days of incubation. No significant effect on atrazine degradation was found from adding the terpene mixture, and, interestingly, an inhibitory effect on phenoloxidase activity was found during the first 20 days of incubation when mixed terpenes were present at 100 μg/kg. With this study it was concluded that successive applications of terpenes or the addition of materials that slowly release terpenes could sustain the atrazine degradation enhancement. However a contrary response was observed when natural wastes rich in terpenes as pine needles, eucalyptus leaves, and orange peels are added to the biomixture, where an enhancement of atrazine dissipation can be observed (Tortella et al. 2013e).

Effect of biomixture bioaugmentation. Inoculation of microorganisms into biobeds has not been a frequent practice, but the few international studies related to fungal inoculation are promising. In Chile the first approach related to biomixture bioaugmentation was performed by Diez and Tortella (2008), where inoculation with *Anthracophyllum discolors* Sp4 immobilized in lignocellulose material increased the degradation of pentachlorophenol in two biological systems: biobeds and fixed-bed columns. Recently, Elgueta et al. (2013) studied the formulation of different supports based on agro-forestry waste for the immobilization of *Anthracophyllum discolors* to be inoculated in a biobed system to increase atrazine degradation. A biomixture composed of an Andisol soil, peat, and straw was contaminated with 60 mg/kg of atrazine and inoculated with 10 % (w/w) of fungal *Anthracophyllum discolors*. The main results showed that $t_{1/2}$ of atrazine decreased from 14.5 days in the non-inoculated biomixture to 6 days in inoculated biomixture. On the other hand, inoculation with *A. discolor* in the biomixture contaminated with atrazine produced a stimulation in the fungal communities at the end of the experiment. With these results the author concluded that the bioaugmentation using *A. discolor* improved the atrazine degradation in a biobed system biomixture.

The presence of peat in the biomixture produces a low pH that enhances the growth of lignin-degrading fungi on the straw, which results in the production of ligninolytic enzymes and the subsequent degradation of pesticides. However, in some countries, the peat is replaced by compost for economic and/or environmental reasons (Fogg et al. 2004; Vischetti et al. 2004; Coppola et al. 2007), and the pH in the system can increase to values that are not favorable to the growth of certain fungi (Rousk et al. 2009). Under these conditions, bacterial activity plays a more important role. Briceño et al. (2013a) discussed the feasibility of using actinobacteria as inocula in biobed systems. Two actinobacteria isolated from agricultural soil and characterized as degrading organophosphorus pesticides (Briceño et al. 2012, 2013b) were used in the bioaugmentation of a biomixture contaminated with 25 mg/kg of chlorpyrifos and diazinon. The results showed that inoculation of actinobacteria *Streptomyces* sp. AC5 improved the enzymatic activity and microbial respiration in the biomixture. In addition, after 45 days of incubation, 48 % and 36 % of residual chlorpyrifos was found in the non-inoculated and inoculated biomixtures, respectively. However, when both strains, *Streptomyces* sp. AC5 and AC16, were inoculated in the biomixture, no effects on pesticide degradation were observed. Consequently, the authors indicated that future assays are needed to clarify the effect on organophosphorus pesticide degradation in bioaugmented biomixtures with an actinobacteria consortium (Briceño et al. 2013c).

4.4 Concluding Remarks

There is general awareness of the detrimental effects of improper pesticide use on the environment. Soil and water contamination from various sources can be prevented to a large extent by good farming practices, but additional measures are required for point-source contamination from the incorrect handling of pesticides in agriculture. Biobeds are a feasible biotechnological tool proven effective in pesticide removal. Therefore, biobeds have been implemented in several European countries, resulting in a reduction of environmental pollution caused by pesticide residues. Latin America is no stranger to the implementation of biobeds, and recently, four of them were installed in Chile (La Araucanía region), making the country a pioneer in the implementation of this technology in South America. Given the increasing adoption of this technique in different areas and for different purposes, adjustments are needed and consequently a considerable amount of research has been conducted at the laboratory and field scales. In Chile, the investigations to date have validated the efficient removal of pesticides in a biobed system, studying operational parameters to optimize the functioning of the biobeds, to understand the interactions between pesticide degradation, microorganisms, biomixtures, and others. The results obtained over the years have been used to create the "Manual of Construction and Operation of Biobeds," which has been used to disseminate biobed technology in the diverse public and private agricultural sector. The main goal has been to transmit biobed technology in Chile, giving special attention to the adequate pesticide manipulation to protect the environment and natural resources.

Biobed technology is appropriate for nationwide implementation; however, it must be adapted to local conditions where in many cases the traditional biomixture must be modified by adding different residues according to its availability. In this situation, many questions remain to be answered and research needs to be conducted to obtain an adequate and efficient functioning of biobeds as pesticide biopurification systems. Having knowledge of the efficiency of biobed technology for removing pesticide residues, it is hoped that it can be utilized in various parts of Chile and throughout Latin America.

Acknowledgments The authors gratefully acknowledge the support of the Center of Research and Development for Organic Waste Management (CIDGRO) Cod. 09FC02-6021, Chile.

References

Antonious G (2012) On-farm bioremediation of dimethazone and trifluralin residues in runoff water from an agricultural field. J Environ Sci Heal B 47:608–621

Bento FM, Camargo FAO, Okeke BC et al (2005) Comparative bioremediation of soils contaminated with diesel oil by natural attenuation, biostimulation and bioaugmentation. Bioresour Technol 96:1049–1055

Briceño G, Fuentes MS, Palma G et al (2012) Chlorpyrifos biodegradation and 3,5,6-trichloro-2-pyridinol production by actinobacterias isolated from soil. Int Biodet Biodeg 73:1–7

Briceño G, Pizzul L, Diez MC (2013a) Biodegradation of Pesticides by Actinobacteria and their Possible Application in Biobed Systems. In: Amoroso MJ, Benimeli CS, Cuozzo SA (eds) Actinobacteria application in bioremediation and production of industrial enzymes. Taylor & Francis, Boca Raton, FL, pp 245–255

Briceño G, Fuentes MS, Rubilar O et al (2013b) Removal of insecticide diazinon from liquid media by free and immobilized *Streptomyces* sp. isolated from agricultural soil. J Basic Microbiol 53:1–10

Briceño G, Rubilar O, Tortella G et al (2013c) Bioaumentación de una biomezcla con actinobacterias degradadores de residuos de plaguicidas organofosforados. Workshop Internacional y Taller Nacional Valorización de Residuos, oportunidad para la innovación. Universidad de la Frontera, Pucón, Chile

Castillo MP, Tortensson L (2007) Effect of biobed composition, moisture, and temperature on the degradation of pesticides. J Agric Food Chem 55:5725–5733

Castillo MP, Andersson A, Ander P et al (2001) Establishment of the white rot fungus Phanerochaete chrysosporium on unsterile straw of solid substrate fermentation systems intended for degradation of pesticides. World J Microbiol Biotechnol 17:627–633

Castillo MP, Torstensson L, Stenström J (2008) Biobeds for environmental protection from pesticide uses a review. J Agric Food Chem 56:6206–6219

Coppola L, Castillo MP, Monaci E et al (2007) Adaptation of the biobed composition for chlorpyrifos degradation to southern Europe conditions. J Agr Food Chem 55:396–401

Dabrowska D, Kot-Wasik A, Namieoenik J (2004) The Importance of degradation in the fate of selected organic compounds in the environment. Part II. Photodegradation and biodegradation. Pol J Environ Stud 13:617–626

De Wilde T, Spanoghe P, Sniegowksi K et al (2010a) Transport and degradation of metalaxyl and isoproturon in biopurification columns inoculated with pesticide-primed material. Chemosphere 78:56–60

De Wilde T, Spanoghe P, Ryckeboer J et al (2010b) Transport and degradation of pesticides in a biopurification system under variable flux, part I: a microcosm study. Environ Pollut 158: 3309–3316

De Wilde T, Spanoghe P, Ryckeboer J et al (2010c) Transport and degradation of pesticides in a biopurification system under variable flux, part II: a macrocosm study. Environ Pollut 158: 3317–3322

Diez MC, Tortella GR (2008) Pentachlorophenol degradation in two biological systems: biobed and fixed-bed column, inoculated with the fungus Anthracophyllum discolor. The proceeding of the ISMOM, Chile

Diez MC, Palma G, Altamirano C et al (2013a) Manual de construcción y operación de lechos biológicos. Universidad de La Frontera, Temuco

Diez MC, Tortella G, Briceño G et al (2013b) Influence of novel lignocellulosic residues in a biobed biopurification system on the degradation of pesticides applied in repeatedly high doses. Electron J Biotechnol 16:1–11

Diez MC, Levio M, Briceño G et al (2013c) Biochar as a partial replacement of peat in pesticide-degrading biomixtures formulated with different soil types. J Biobased Mater Bioenergy 7:1–7

Dudášová H, Lukáčová L, Murínová S et al (2012) Effects of plant terpenes on biodegradation of polychlorinated biphenyls (PCBs). Int Biodet Biodeg 69:23–27

Elgueta S, Campo M, Diez MC (2013) Formulación de inóculos fúngicos a partir de desechos agroforestales para la degradación de atrazina en un sistema de biopurificación. Workshop Internacional y Taller Nacional Valorización de Residuos, oportunidad para la innovación. Universidad de la Frontera, Pucón, Chile

Fernández-Alberti S, Rubilar O, Tortella G et al (2012) Chlorpyrifos degradation in a Biomix: effect of pre-incubation and water holding capacity. J Soil Sci Plant Nutr 12:785–799

Fogg P, Boxall A, Walker A (2003) Degradation of pesticides in biobeds: the effect of concentration and pesticide mixtures. J Agr Food Chem 51:5344–5349

Fogg P, Boxall A, Walker A et al (2004) Degradation and leaching potential of pesticides in biobed systems. Pest Manag Sci 60:645–654

Gregoire C, Elsaesser D, Huguenot D et al (2009) Mitigation of agricultural nonpoint-source pesticide pollution in artificial wetland ecosystems. Environ Chem Lett 7:205–231

Henriksen VV, Helweg A, Spliid NH et al (2003) Capacity of model biobeds to retain and degrade mecoprop and isoproturon. Pes Manag Sci 59:1076–1082

Husby J (2010) Introduction of the workshop. 3rd European biobed workshop, Piacenza, Italy

Karanasios E, Nikolaos G, Tsiropoulos A et al (2010) Novel biomixtures based on local Mediterranean lignocellulosic materials: evaluation for use in biobed systems. Chemosphere 80:914–921

Karanasios E, Tsiropoulos NG, Karpouzas DG (2012) On-farm biopurification systems for the depuration of pesticide wastewaters: recent biotechnological advances and future perspectives. Biodegradation 23:787–802

Marinozzi M, Laura Coppola L, Elga Monaci E et al (2013) The dissipation of three fungicides in a biobed organic substrate and their impact on the structure and activity of the microbial community. Environ Sci Pollut Res 20:2546–2555

Niels H, Helweg A, Heinrichson K (2006) Leaching and degradation of 21 pesticides in full-scale model biobeds. Chemosphere 65:2223–2232

Rousk J, Brookes P, Bååth E (2009) Contrasting soil pH effects on fungal and bacterial growth suggest functional redundancy in carbon mineralization. Appl Environ Microbiol 6:1589–1596
Spliid NN, Husby J (2010) New closed biobed with recirculation and evaporation for use under colder climates. 3rd European biobed workshop, Piacenza, Italy
Torstensson L, Castillo MP (1997) Use of biobeds in Sweden to minimize environmental spillages from agricultural spraying equipment. Pesticide Outlook 8:24–27
Tortella G, Rubilar O, Cea M et al (2010) Biostimulation of agricultural biobeds with NPK fertilizer on chlorpyrifos degradation to avoid soil and water contamination. J Soil Sci Plant Nutr 10:464–475
Tortella G, Rubilar O, Castillo MP et al (2012) Chlorpyrifos degradation in a biomixture of biobed at different maturity stages. Chemosphere 88:224–228
Tortella GR, Durán N, Rubilar O et al (2013a) Are white-rot fungi a real biotechnological option for the improvement of environmental health? Crit Rev Biotechnol. doi:10.3109/07388551.2013.823597
Tortella GR, Mella-Herrera R, Sousa D et al (2013b) Carbendazim dissipation in the biomixture of on-farm biopurification systems and its effect on microbial communities. Chemosphere 93: 1084–1093
Tortella GR, Mella-Herrera T, Sousa D et al (2013c) Atrazine dissipation and its impact on the microbial communities and community level physiological profiles in a microcosm simulating the biomixture of on-farm biopurification system. J Hazard Mater 260:459–467
Tortella GR, Rubilar O, Stenström J et al (2013d) Using volatile organic compounds to enhance atrazine biodegradation in a biobed system. Biodegradation 24:711–720
Tortella GR, Rubilar O, Cea M et al (2013e) Natural wastes rich in terpenes and their relevance in the matrix of an on-farm biopurification system for the biodegradation of atrazine. Int Biodeter Biodegr 85:8–15
Tyagi M, da Fonseca MMR, de Carvalho CCCR (2011) Bioaugmentation and biostimulation strategies to improve the effectiveness of bioremediation processes. Biodegradation 22:231–241
Urrutia C, Rubilar O, Tortella G et al (2013) Degradation of pesticide mixture on modified matrix of a biopurification system with alternatives lignocellulosic wastes. Chemosphere 92:1361–1366
Verhagen P, De Gelder L, Boon N (2013) Inoculation with a mixed degrading culture improves the pesticide removal of an on-farm biopurification system. Curr Microbiol 67:466–471
Vischetti C, Capri E, Trevisan M et al (2004) Biomassbed: a biological system to reduce pesticide point contamination at farm level. Chemosphere 55:823–828
Vischetti C, Coppola L, Monaci E et al (2007) Microbial impact of the pesticide chlorpyrifos in Swedish and Italian biobeds. Agron Sustain Dev 27:267–272
Vischetti C, Monaci E, Cardinali A et al (2008) The effect of initial concentration, co-application and repeated applications on pesticide degradation in a biobed mixture. Chemosphere 72: 1739–1743

Chapter 5
Bioremediation of Soils Contaminated with Pesticides: Experiences in Mexico

M. Laura Ortiz-Hernández, Alexis Rodríguez, Enrique Sánchez-Salinas, and M. Luisa Castrejón-Godínez

Abstract The worldwide use of pesticides for pest control in agriculture and some industrial processes has contributed to improve the food and goods production. However, their intensive use has resulted in the release of a wide range of xenobiotic compounds to the environment, widespread among air, water, and soil. The existence of contaminated sites is an important environmental and health concern today. For the treatment of pesticide-contaminated soil, several strategies involving biological, physicochemical, and thermal processes have been developed to remediate them, being the bioremediation approaches, among the more successful because they are environmental friendly, economic, and versatile. The pesticide soil pollution on Mexico, as around the world, is a serious concern, so that different research groups had developed biological strategies for the assessment of pesticide biodegradation, and bioattenuation, biostimulation, bioaugmentation, and composting schemes for the treatment, remediation, and detoxification of pesticide-contaminated sites. In this chapter we present information about involved processes in bioremediation of soils contaminated with pesticides and particularize on the status of pesticides in Mexico as well as the efforts undertaken in the biodegradation of pesticides and the potential for their application in contaminated soil bioremediation.

M.L. Ortiz-Hernández (✉)
Centro de Investigación en Biotecnología, Universidad Autónoma del Estado de Morelos, Av. Universidad 1001, Colonia Chamilpa, Cuernavaca, Morelos 62209, Mexico

Programa de Gestión Ambiental Universitario, Universidad Autónoma del Estado de Morelos, Av. Universidad 1001, Colonia Chamilpa, Cuernavaca, Morelos 62209, Mexico
e-mail: ortizhl@uaem.mx

A. Rodríguez • E. Sánchez-Salinas
Centro de Investigación en Biotecnología, Universidad Autónoma del Estado de Morelos, Av. Universidad 1001, Colonia Chamilpa, Cuernavaca, Morelos 62209, Mexico

M.L. Castrejón-Godínez
Programa de Gestión Ambiental Universitario, Universidad Autónoma del Estado de Morelos, Av. Universidad 1001, Colonia Chamilpa, Cuernavaca, Morelos 62209, Mexico

A. Alvarez, M.A. Polti (eds.), *Bioremediation in Latin America: Current Research and Perspectives*, DOI 10.1007/978-3-319-05738-5_5

5.1 Introduction

Pesticides are important tools in agriculture as they help to minimize economic losses caused by weeds, insects, and diseases. The use of pesticides has benefitted modern society by improving the quantity and quality of the world's production while keeping the cost of that food supply reasonable. Unsurprisingly, pesticide use has become an integral part of modern agricultural systems. Because of continuous pest problems, their usage possibly cannot be discontinued in the near future (Mosleh et al. 2003; Anjum et al. 2012). Extensive and improper use of these chemicals has already caused considerable environmental pollution and their continuous exposure leads to greater health risk, to nontarget organisms, as plants, animals, and the human population, since these substances can bioaccumulate and biomagnify in living organisms or even arrive by diffusion and/or advection at different trophic levels (Islas-Pelcastre et al. 2013). Residues of pesticides in animal products and other food items ultimately get accumulated in human beings especially in the adipose tissue, blood, and lymphoid organs and can cause immunopathological effects which acquire immunodeficiency, autoimmunity, and hypersensitivity reactions like eczema, dermatitis, and allergic respiratory diseases. Many pesticides are also known to cause mutations in chromosomes of man and animals, thereby leading to carcinoma of liver and lungs (Niti et al. 2013).

The rapid increase in demand and development of industrial chemicals, fertilizers, pesticides, and pharmaceuticals to sustain and improve quality of life worldwide has resulted in the contamination and high prevalence of these chemicals in air, water, and soils, posing a potential threat to the environment and a health concern in many communities, making their removal and detoxification a more urgent undertaking (Anjum et al. 2012).

Long after their use, the pesticides remain in soils and sediments where they can enter the food chain directly or percolate down to the water table. Once in the groundwater, these compounds can enter drinking water wells and cause health problems. These chemicals are also subject to long-range atmospheric transport. One of the primary concerns is the ability of these chemicals to accumulate within the adipose tissue of animals. Indirect accumulation or biomagnification in higher trophic level organism, such as mammals, may cause health problems over time because of the increasing levels of toxic compounds within the body (Kaufman 1983; Csizer 2002; Mörner et al. 2002). In soil, the accumulation of pesticides may alter the biota at several levels; they can exert toxic effect over the soil fauna as nematodes, microarthropods, and earthworms, a useful organism for evaluating contamination of the soil environment with toxic chemicals as pesticides, herbicides, among others (Edwards and Bohlen 1992; Martikainen et al. 1998). Pesticides can also affect the microbial activities and have an impact on microbial population (Ibekwe et al. 2001; Pandey and Singh 2004; Klose et al. 2006; Tejeda et al. 2011).

Soil pollution is a consequence of the accumulation of a wide range of chemical compounds generated either by natural or industrial processes. The existence of contaminated sites is an important environmental problem today (Bustamante et al. 2012).

Several strategies involving biological, physicochemical, and thermal processes have been developed to remediate contaminated sites (Bollag and Bollag 1995; Vidali 2001; Rubilar et al. 2011). Methods such as incineration, excavation, landfilling, and storage are expensive, and sometimes difficult to execute (Vidali 2001; Jain et al. 2005), inefficient, and often exchange one problem for another (Bollag and Bollag 1995). Alternatively, biological processes offer several advantages over conventional technologies for soil decontamination, because they are often more environmentally friendly, economic, and versatile, and they can reduce the concentration and toxicity of a large number of contaminants (Vidali 2001; Jain et al. 2005). However, these processes are limited by the low water solubility of the contaminants, limiting their availability to microorganisms (Bollag and Bollag 1995; Volkering et al. 1998).

The toxicity or the contamination of pesticides can be reduced by the bioremediation process that exploits the natural ability of plants and microbes for decontamination; organic pollutants in theory can be completely mineralized into water and carbon dioxide (Rayu et al. 2012). Bioremediation technologies have been successfully employed in the field and are gaining more and more importance with increased acceptance of eco-friendly remediation solutions. Owing to complex nature of pesticides, more versatile and robust techniques need to be developed which can produce the desired result in a very cost-effective manner (Singh and Walker 2006).

This chapter provides general information about the principles of bioremediation and examples of the experiences of the soil bioremediation process application in Mexico are described.

5.2 Principles of Bioremediation

Bioremediation, which is the use of microorganism consortia or microbial processes to degrade and detoxify environmental contaminants, is a fast-growing promising remediation technique increasingly being studied and applied in practical use for pollutant cleanup (Juwarkar et al. 2010). Bioremediation is the intentional use of biodegradation or contaminant accumulation processes to eliminate environmental pollutants from sites where they have been released. Bioremediation technology uses the physiological potential of microbes and plants for the degradation of pollutants (Odukkathil and Vasudevan 2013). The bioremediation employs living organisms, most often microorganisms, plants, or both, or the products generated from metabolism of these living organisms to degrade, detoxify, or sequester toxic chemicals present in natural waters and soils and can be adapted for use to treat contaminated soils, sediments, sludge, water, or even air (Juwarkar et al. 2008; Crawford and Rosenberg 2013). Though bioremediation has been used to varying degrees for more than 60 years, for example in petroleum land farming, it historically has been implemented as a *black box* engineering solution where amendments are added and the pollutants are degraded. This approach is often successful but sometimes the results are less than desirable; that is, no degradation of the contaminant is observed or even production of more toxic-derived products.

The key to successful bioremediation is to harness the naturally occurring catabolic capability of organisms to catalyze transformations of environmental pollutants (Vidali 2001; Chakraborty et al. 2012).

The bioremediation of pesticide-polluted soils is important due to the characteristics of these compounds, such as their ubiquitous distribution, their low bioavailability, their high persistence in soil, and their potential effect on human health. To date, all the efforts invested in research on bioremediation have been applied to optimizing microbial activity by adding nutrients, or to bioaugmentation. Nevertheless, success has not completely been achieved, which has been attributed to the high adherence to soil particles and the low water solubility of these compounds that limits their availability to microorganisms, which is a potential problem for bioremediation of contaminated sites. Due to their lipophilic character, organochlorine pesticides are less water soluble, followed by organophosphate compounds and carbamates with medium solubilities (Cameotra and Makkar 2010; Zheng and Wong 2010).

The feasibility and success of the soil bioremediation process are mainly related to the nature and chemical properties of the pesticides (for principles of pesticide biodegradation, see below). Some environmental conditions are required for the soil remediation, as pH, moisture content, nutritional state, microbial diversity of the site, temperature, and oxidation-reduction potential (redox-potential) are also required. So, the selection of appropriate technology among the wide range of bioremediation strategies developed to treat contaminants depends on three basic principles: (1) the amenability of the pollutant to biological transformation (biochemistry), (2) the accessibility of the contaminant to microorganisms (bioavailability), and (3) the opportunity for optimization of biological activity (bioactivity) (Dua et al. 2002).

Below we provide information about the interaction of pesticides in soil, the basic principles of pesticide biodegradation, and the technologies that have been developed for the bioremediation of contaminated soils. This will help to understand the variables involved in soil bioremediation and it will help in the success of this process.

5.2.1 *Interactions of Pesticides with Soil*

The study of pesticide behavior in the soil is really interesting, because it is a heterogeneous, complex, and dynamic system, in which different reactions (chemical and biochemical) take place, and also it plays a role as receptor of polluting substances. Pesticides can be directly incorporated into the soil by surface application on the crops, injection, or inadequate aspersion techniques or indirectly through plant leaves. When the pesticide enters in contact with the soil, sorption is the first process, including adsorption/desorption phenomena. The first one permits fixation of the compounds to the soil particles; the last one releases the pesticide into soil solutions. The sorption process is related with the persistence and pesticide degradation, because the physicochemical and biological characteristics of soils play a key role (Madrigal-Monárrez et al. 2008). The behavior of pesticides in soils, the efficiency, persistence, and potential as environmental contaminants, depends on their retention and degradation on soil constituents (Worrall et al. 2001).

Sorption is one of the major processes affecting the interactions among pesticides, soil, water, and the immobile and mobile solid phases of a soil (McCarthy and Zachara 1989). The rates of pesticide degradation, volatilization, hydrolysis, and photolysis are directly dependent upon the sorption process as it ultimately determines the pesticide concentration in the soil solution. Key factors governing pesticide sorption include soil properties as size, shape, configuration, molecular structure, chemical functions, solubility, polarity, polarizability, and charge distribution of interacting species, organic matter and clay contents, pH, water content, and temperature but also the structure of the pesticide (Odukkathil and Vasudevan 2013). Structural factors influencing the sorption process are the pesticide's molecular size, hydrophobicity, molecular charge, hydrogen bonding, arrangement, and interactions of molecular fragments and its coordination (Suyala et al. 2013). The pesticide sorption process incorporates a wide range of different chemical mechanisms including ion exchange, cation bridging, ion-dipole interactions, ligand exchange, charge transfer, hydrogen bonding, and van der Waals forces (Bailey and White 1970; Gevao et al. 2000).

Bound pesticide residues can be defined as the fraction of pesticides which cannot be readily extracted from the soil without altering the chemical structure of the original pesticides or its metabolites (Gevao et al. 2000). A variety of agricultural factors are capable of influencing the fate and binding of pesticides in soil which includes concentration of pesticide applied, rate and mode of application, repeated application, ageing, and use of organic and inorganic soil amendments. The environmental significance of a bound residue, however, depends not on its non-extractability under laboratory test conditions, but on its bioavailability (Odukkathil and Vasudevan 2013).

5.2.2 Principles of Pesticide Biodegradation

There is a big amount of reports on biotic degradation of pesticides in the environment, since microorganisms are able to degrade different substrates than the natural carbon sources due to their large genetic plasticity, quick generation times, and good ability to survive in different media. An important characteristic of microorganisms is the capability of adaptation through mutation. This is in order to develop the ability of degrading toxic or complex compounds, probably because of the evolution of more adequate transport systems or through the cell wall (Alexander 1994; Banat et al. 2000; Van Hamme et al. 2003; Yeomans et al. 2004; Haritash and Kaushik 2009; Chino-Flores et al. 2011; Abdel-Razek et al. 2013). Pesticides are transformed by metabolic reactions leading to changes in their chemical structure through diverse reactions, which give place to inorganic compounds such as CO_2,

H_2O, halides, ammonium, and phosphates, among others. In this case, the process is known as degradation (Zhang et al. 2003).

Biodegradation involves the breakdown of pesticides or any other organic compounds, usually by microorganisms, to less complex compounds and ultimately to water and CO_2 and other elements from the original compound. The complete breakdown of pesticides into inorganic components is termed as biomineralization. In some cases the degradation leads to formation of less complex and less toxic organic compounds, referred to as partial biodegradation. The pesticide thus transformed or degraded by the microorganism is used as a carbon source, nitrogen source, any other mineral source, or a final electron acceptor in respiratory chain (Odukkathil and Vasudevan 2013).

Knowledge about the metabolic principles of biodegradation, including the enzymatic systems, may be used to establish bioremediation strategies for pesticide-polluted soils (Pongrac et al. 2007). Soil microorganisms have the ability to degrade pesticides and convert them into simpler nontoxic compounds. This process is known as biodegradation. There is significant role of metabolic activities of bacteria, fungi, actinobacteria, and plants in the degradation process (Nawaz et al. 2011). Most of the pesticides reaching to the soil are biodegradable but there are certain pesticides that show complete resistance to biodegradation. These are called recalcitrant (Aislabie and Lloyd-Jones 1995; Richins et al. 1997; Mulchandani et al. 1999).

Microorganism's biodegradation ability depends on the physical, chemical, and microbiological characteristics of the soil and the chemical properties of the pollutant (Leahy and Colwell 1990; Banat et al. 2000; Van Hamme et al. 2003; Haritash and Kaushik 2000). This degradation takes place through the reactions in which heterotrophic microorganisms nourish themselves and gain the right amount of energy to satisfy their needs by taking pesticides as their main source of carbon (Yeomans et al. 2004).

Pesticide degradability decreases as molecular weight and degree of branching increase. Pesticide metabolism includes a three-phase process. In phase I, the original compound is transformed through oxidation, reduction, or hydrolysis reactions. Thus, more water-soluble and less toxic products than the original one are normally generated. The second phase includes the conjugation of the pesticide or its metabolites to sugars and amino acids, which also results in an increased water solubility and a decreased toxicity. Phase III involves the conversion of metabolites from phase II into less toxic secondary conjugates. In these processes, bacteria and fungi generate intra- or extracellular enzymes such as hydrolases, peroxidases, oxygenases, and others (Abraham et al. 2002; Li et al. 2007; Ortiz-Hernández et al. 2011).

Pesticide degradation follows different metabolic pathways depending on the nature of pesticide, environmental conditions, and nature of microbes. This involves (1) oxidative transformation mediated by oxidative enzymes (cytochrome p450, peroxidases, and polyphenol oxidases). (2) Hydrolytic transformation mediated by hydrolytic (hydrolases) which cleaves bonds of the substrate by adding hydrogen or hydroxyl group from water molecules. (3) Reductive transformation mediated by reductive enzymes (nitroreductase) by which removal of anion occurs by reduction. (4) Conjugation reaction by which exogenous or endogenous natural compound is added to pesticide facilitating mineralization. This occurs by using existing enzymes and hence it is a co-metabolic process. This process includes xyloxylation, alkylation, acylation, and nitrosylation. This type of biotransformation occurs in fungal biodegradation of pesticides. (5) Reductive dehalogenation is mediated by reductive dehydrohalogenase enzyme. During the process organohalide acts as a terminal electron acceptor for ATP production (Odukkathil and Vasudevan 2013).

Under aerobic conditions, pesticides with aromatic structures are transformed by mono- and dioxygenases producing dihydroxylated derivatives such as metabolites. Under anaerobic conditions, the transformation follows reducing routes, which makes the pesticide lose its aromaticity before the ring breaks. Following the breaking of the aromatic ring, the aromatic compounds dehalogenate, although some bacteria are able to dehalogenate without the previous rupture of the ring. Until now, there are three known classes of dehalogenation reactions in aromatic compounds: the oxidative reactions, where the halogen is removed from the ring during oxygenation; hydrolytic reactions, where halogen is replaced by a hydroxyl group (it may occur in aerobic conditions); and reductive reactions, where halogen is replaced by hydrogen (it occurs only in sulfate-reductive and metanogenic conditions) (Commandeur and Parsons 1994).

5.2.3 Bioavailability of Pesticide for Microorganisms

During bioremediation of pesticide-contaminated soil, bioavailability is one of the main constraints. Many studies have suggested that soil-bound pesticides are unavailable for microbial degradation. Less bioavailability of pesticide is one of the reasons for the persistence of many of the pesticides which are in use now. Major reasons for the less bioavailability are the unequal spatial distribution of microorganisms and pesticides and the retardation of substrate diffusion by soil matrix (Harms and Bosma 1997).

Bioavailability is the amount of a chemical in soil able to interact with organisms inhabiting the soil environment. Bioavailability is affected by many factors, including properties of pesticides and soils, aging time in soil, climate, and the organisms of concern. A *bioavailable compound* is freely available to cross an organism's cellular membrane from the medium the organism inhabits at a given time. Once transfer across the membrane has occurred, storage, transformation, assimilation, or degradation can take place within the organism; however, these processes are obviously distinct from the transfer between the medium (e.g., soil) and the organism. A *bioaccessible compound* is available to cross an organism's cellular membrane from the environment, if the organism has access to the chemical (Figs. 5.1 and 5.2) (Semple et al. 2004).

Bioavailability of pesticide is a major constraint in the bioremediation of pesticide-contaminated soil. Bioavailability in this context of bioremediation can be defined as the amount of pesticides that can be readily taken up by microbes. According to this definition, bioavailability of the pesticides to the microbes affects

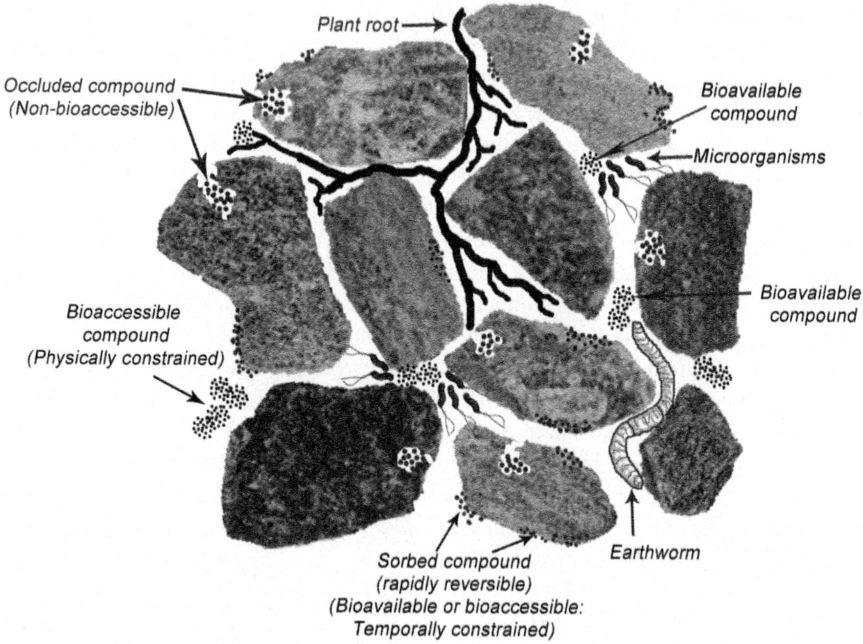

Fig. 5.1 Bioavailable and bioaccessible fractions of the pesticides in soil as defined by physical location. Also present is the relationship of soil-associated contaminant molecules in relation to bioaccessible fraction (modified from Semple et al. 2004)

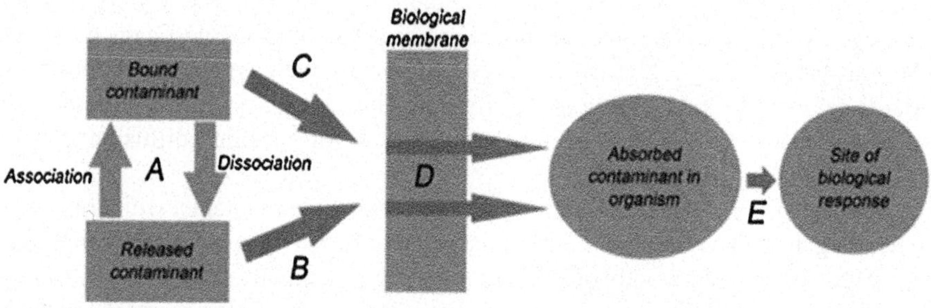

Fig. 5.2 Bioavailability processes (*A–D*). *A*, Release of a bound or recalcitrant chemical to a more accessible form, *B* and *C* describe the transport of chemicals to a cellular membrane, and *D* represents the uptake of a chemical across a cellular membrane. Bioavailability addresses process *D*, whereas bioaccessibility encompasses processes *A–D* (modified from Semple et al. 2004)

a bioremediation process in many ways as follows: (1) At low pesticide concentration microorganisms fail to produce some energy which induces the catabolic gene systems involved in biodegradation that slows down bioremediation. (2) At low contaminant concentration in a less nutrient environment, microbial cell may degrade the pollutant but the low nutrients in the environment reduce their growth rate which eventually leads to decrease in the uptake of pesticides by the microorganisms. This slows down the bioremediation process (Odukkathil and Vasudevan 2013).

The process of bioremediation depends on the metabolic potential of microorganisms to detoxify or transform the pollutant molecule, which is dependent on both accessibility and bioavailability (Antizar-Ladislao 2010). There is a considerable debate in the literature on what constitutes the bioavailable fraction and the methods of its measurements (Alexander 2000; Vasseur et al. 2008). Following entry into the soil environment, pollutants rapidly bind to the mineral and organic matter (solid phases) via a combination of physical and chemical processes. Sorption, complexation, and precipitation constitute the pollutant-soil interaction. The ability of soils to release (desorb) pollutants determines its susceptibility to microbial degradation, thereby influencing effectiveness of the bioremediation process. In soil aggregates which are the smallest composite units in the heterogeneous soil environment, bioavailability is limited by transport of the pollutant molecule to a microbial cell, i.e., diffusion of pesticide out of a soil aggregate to the cell attached to the external surface of the aggregate (Megharaj et al. 2011). A reduced bioavailability of contaminants in soil is caused by the slow mass transfer to the degrading microorganisms. Contaminants become unavailable when the rate of mass transfer is zero. The decrease of the bioavailability in the course of time is often referred to as aging or weathering. It may result from (1) chemical oxidation reactions incorporating contaminants into natural organic matter, (2) slow diffusion into very small pores and absorption into organic matter, and (3) the formation of semirigid films around nonaqueous-phase liquids (NAPL) with a high resistance toward NAPL-water mass transfer.

These bioavailability problems can be overcome by the use of food-grade surfactants (Boopathy and Manning 1999), which increase the availability of contaminants for microbial degradation.

Sorption which influences the bioavailability of a contaminant is a critical factor, yet a poorly understood process in bioremediation. There are two schools of thought concerning bioavailability and the consequent biodegradation of organic contaminants (Singh et al. 2008): (1) the prerequisite release of contaminant from sorbed phase to aqueous phase for its degradation by microorganisms (Harms and Zehnder 1995; Shelton and Doherty 1997), and (2) biodegradation of the contaminant in the sorbed phase, without being desorbed, by the enzymes (Singh et al. 2003). The degradation of sorbed contaminants can presumably occur via microbially mediated desorption of contaminants through production of biosurfactants and the development of a steep gradient between solid phase and interfacial contaminant (Tang et al. 1998). Thus, these reports suggest that bioavailability is even species specific (i.e., the ability of certain species to desorb the contaminant and then degrade). The organic contaminants can also be degraded without prior desorption. Singh et al. (2003) demonstrated that a soil bacterium, *Brevibacterium* sp., degraded the pesticide fenamiphos which was intercalated into the cationic-surfactant-modified montmorillonite clay (CTMA-Mt-fenamiphos complex). The interlayer space is otherwise inaccessible to the bacterium due to its size of several orders lower than that of the bacteria. The scanning electron microscope analysis showed the surface attachment of bacteria to the surface of the CTMA-Mt-fenamiphos complex, suggesting the involvement of extracellular enzyme in the degradation of fenamiphos, without its prior desorption. The degradation of sorbed contaminants depends on the enrichment and isolation procedures used for obtaining the culturable bacteria. As against the conventional approach of providing the contaminant as a sole carbon source in aqueous medium, the provision of phenanthrene sorbed on a polyacrylic porous resin to the bacterial cultures led to faster degradation of phenanthrene than that isolated by the conventional technique (Tang et al. 1998; Grosser et al. 2000).

5.2.4 *Surfactants: Bioavailability Enhancers*

The process of bioremediation depends on the metabolic potential of microorganisms to detoxify or transform the pollutant molecule, which is dependent on both accessibility and bioavailability (Antizar-Ladislao 2010). Following entry into the soil environment, pollutants rapidly bind to the mineral and organic matter (solid phases) via a combination of physical and chemical processes. Sorption, complexation, and precipitation constitute the pollutant-soil interaction. The ability of soils to release (desorb) pollutants determines its susceptibility to microbial degradation, thereby influencing effectiveness of the bioremediation process (Megharaj et al. 2011). Since water solubility of many organic contaminants is the controlling removing mechanism, additives are used to enhance it. Surfactants (surface-active agents) are chemical compounds that generally are included in the formulation of detergents (Kreinfeld and Stoll 1997; Smulders et al. 2001). Chemical and biosurfactants are amphiphilic compounds which can reduce surface and interfacial tensions by accumulating at the interface of immiscible fluids and increase the solubility and mobility of hydrophobic or insoluble organic compounds (Prince 1997; Mulligan 2005). The surfactants contain hydrophilic and hydrophobic moieties; hydrophilic groups can be anionic, cationic, zwitterionic, and nonionic. The synthetic surfactants contain sulfate, sulfonate, or carboxylate group (anionic); quaternary ammonium group (cationic); polyoxyethylene, sucrose, or polypeptide (nonionic); and the hydrophobic parts of paraffins, olefins, alkylbenzenes, alkylphenols, or alcohols (Mulligan et al. 2001; Mishra et al. 2009). The presence of both hydrophobic and hydrophilic groups in each molecule is a fundamental physical property of surfactants, which allows these compounds to form micelles in solution. It is the formation of micelles in solution that gives surfactants their detergency and solubilization properties. The concentration of surfactants in water at which surfactant molecules aggregate into clusters (micelles) is known as the critical micelle concentration (CMC) (Rosen and Kunjappu 2012).

Solubilization of hydrophobic contaminants is attributed to the incorporation of the molecule into the hydrophobic core of micelles in solution (Guha and Jaffé 1996). The salient mechanisms which are involved in the surfactant-amended remediation are (1) lowering of interfacial tension, (2) surfactant solubilization of hydrophobic organic compounds, and (3) the phase transfer of organic compounds from soil-sorbed to pseudo-aqueous phase (Laha et al. 2009). Surfactants enhance mobilization and biodegradation of polycyclic aromatic hydrocarbons (PAHs) in soils (Tiehm et al. 1997). Enhanced rates of degradation of naphthalene and phenanthrene in the presence of some nonionic surfactants at applications below their CMC were observed by Aronstein et al. (1991). Similarly, significant solubility enhancements of dicloro difenil tricloroetano (DDT) in Triton and Brij 35 surfactants were noticed by Kile and Chiou (1989) below their CMC. Factors such as cost, effectiveness at concentrations lower than 3 %, low toxicity to humans, animals, and plants, low adsorption to soil, low soil dispersion, and low surface tension determine the selection of surfactants for field application (Mulligan et al. 2001). Microorganisms also produce surfactants (surface-active amphiphilic metabolites such as glycolipids, phospholipids, lipopeptides, lipoproteins, and lipopolysaccharides). These low- and high-molecular-weight biosurfactants find their uses in food processing, cosmetic, and pharmaceutical industries, in addition to bioremediation efforts (Christofi and Ivshina 2002). The biosurfactant kind and microbial species which can produce them are numerous, leading to continuous search for the novel biosurfactants (Satpute et al. 2010).

For the incensement of the bioavailability of pesticides in soil bioremediation approaches, several studies have used biosurfactants and its combinations with chemical surfactants. The most studied biosurfactants for bioremediation are rhamnolipids, synthetized by *Pseudomonas aeruginosa*, surfactine (Olivera et al. 2000), lipopeptides (Tecon and Van der Meer 2010), and sophorolipids (Schippers et al. 2000). In lab-scale studies, Yin et al. (2008) proved that the use of rhamnolipids is more efficient for the oil-polluted effluent remediation than Triton or Tween commercial surfactants. On the other hand, Guo and Mulligan (2006) reported that rhamnolipids are able to remove more than 90 % of the adsorbed styrene in a composed mixture of soil and sand.

There are few reports regarding the application of biosurfactants in pesticide-polluted sites. However, in a lab study, the BSP3 *Burkholderia cenocepacia* strain produces a biosurfactant identified as a glucolipid. This product showed a remarkable improvement in the solubilization of the pesticides methyl parathion, ethyl parathion, and trifluraline. Due to this biosurfactant surface, active properties, and good performance in the improvement of pesticide solubilization, Wattanaphon et al. (2008) suggest using it as a solubilizing agent in the environmental remediation of pesticide-polluted soils. In another study made by Zhang et al. (2011), the *Pseudomonas aeruginosa* CH7 strain, isolated from activated sludge, proved to be capable of not only degrading ß-cipermetrine, but also using this pesticide, which was identified as a rhamnolipid, as a unique source of carbon and energy for the production of biosurfactants. For this reason, the authors suggest to use this strain and the obtained biosurfactant in the bioremediation of ß-cipermetrine-polluted soils and waters. On the other hand, Sekhon et al. (2011) report the isolation of a bacterium from endosulfan-polluted soils, which was identified as *Bacillus subtilis* SK320. This bacterium produces a biosurfactant belonging to the lipopeptides, whose production is mediated by esterase enzymes. It also proved to have bioemulsifier activity on different substrates, olive oil being the best inducer on the production of this biosurfactant. In other studies where chemical surfactants have been used, their efficiency on removing pesticides from soils is reported. Iglesias-Jiménez et al. (1996) reported data concerning pesticide adsorption in soil-water systems in the presence of the surfactants Tween 80 y Tetradecil trimethyl ammonium bromide on the pesticides diazinon, acephate, atrazine y etofumesate. Beigel et al. (1998) published another similar report about desorption presented by a soil submitted to low concentrations of anionic and nonionic surfactants in the presence of triticonazole fungicide. Furthermore, Jayashree et al. (2006) developed research works about the improvement of endosulfan-polluted soils, with the help of surfactants Tween 80, Triton X-100, and surfactin. In Table 5.1 some examples of biosurfactants are listed.

However, the in situ application of surfactants to enhance bioavailability of persistent organic pollutants requires careful planning and selection based on the prior information about the fate and behavior of the surfactant and the target pollutant. Caution is required to prevent groundwater contamination via leaching and consequent toxicity to microorganisms. Hence, a good strategy will be to select bacteria that are capable of not only catabolizing the target contaminant but also producing surfactant. More knowledge on the mechanisms of pollutant-surfactant interactions with regard to diffusion, in and out of the micelles, and modeling of pollutant's transport at the field site can help to design efficient remediation strategy (Megharaj et al. 2011).

A pesticide will have limited solubility when a physic-chemical barrier affects its degradation rate by microorganisms. When a pesticide is in the soil, different reactions with the microorganisms take place, which has an influence in bioavailability. Figure 5.3 shows the main interactions conditioning bioavailability. The conversion of chemical compounds by microbial cells during bioremediation is ruled by absorption speed, metabolism (the cell intrinsic activity), and the transference rate to the cell (mass transference). These factors regulate the chemical compound's bioavailability. The knowledge of these interactions may increase the possibilities of success in the process of bioremediation.

5.3 Bioremediation Technologies

Bioremediation technologies can be generally classified as in situ or ex situ. In situ bioremediation involves treating the contaminated material at the site, while ex situ involves the removal of the contaminated material to be treated elsewhere (Chowdhury et al. 2012). Examples of bioremediation technology developed until now are shown in Table 5.2. Bioremediation processes may be either aerobic (Bedard and May 1995; Wiegel and Wu 2000) or anaerobic (Komancová et al. 2003). However, due to the problem associated with either of this method to treat highly complex compounds, sometime sequential anaerobic-aerobic bioremediation processes are also adopted to remediate contaminated sites (Master et al. 2002).

Bioremediation of pesticides utilizing biodegradation abilities of microorganisms includes the natural attenuation, although it may be enhanced by engineered techniques, either by addition of selected microorganisms (bioaugmentation) or by biostimulation, where nutrients are added. Genetic engineering is also used to improve the biodegradation capabilities of microorganisms by genetically modified organisms (GMO). Nevertheless, there are many factors affecting the efficiency of this process and risks associated to the use of GMO in the field (Joutey et al. 2013).

Table 5.1 Biosurfactants of microbial origin (adapted from Mulligan 2005)

Type of surfactant	Microorganism
Trehalose lipids	*Arthrobacter paraffineus*
	Corynebacterium spp.
	Mycobacterium spp.
	Rhodococcus erythropolis
	Nocardia sp.
Rhamnolipids	*Pseudomonas aeruginosa*
	Pseudomonas sp.
	Serratia rubidaea
Sophorose lipids	*Candida apicola*
	Candida bombicola
	Candida lipolytica
	Candida bogoriensis
Glycolipids	*Alcanivorax borkumensis*
	Arthrobacter sp.
	Corynebacterium sp.
	Rhodococcus erythropolis
	Serratia marcescens
	Tsukamurella sp.
Cellobiose lipids	*Ustilago maydis*
Polyol lipids	*Rhodotorula glutinis*
	Rhodotorula graminis
Diglycosyl diglycerides	*Lactobacillus fermentii*
Lipopolysaccharides	*Acinetobacter calcoaceticus* (RAG1)
	Pseudomonas sp.
	Candida lipolytica
Arthrofactin	*Arthrobacter* sp.
Lichenysin A, lichenysin B	*Bacillus licheniformis*
Surfactin	*Bacillus subtilis*
	Bacillus pumilus
Viscosin	*Pseudomonas fluorescens*
Ornithine, lysine peptides	*Thiobacillus thiooxidans*
	Streptomyces sioyaensis
	Gluconobacter cerinus
Phospholipids	*Acinetobacter* sp.
Sulfonylipids	*Thiobacillus thiooxidans*
	Corynebacterium alkanolyticum
Fatty acids (corynomycolic acids, spiculisporic acids, etc.)	*Capnocytophaga* sp.
	Penicillium spiculisporum
	Corynebacterium lepus
	Arthrobacter paraffineus
	Talaromyces trachyspermum
	Nocardia erythropolis

(continued)

Table 5.1 (continued)

Type of surfactant	Microorganism
Alasan	*Acinetobacter radioresistens*
Streptofactin	*Streptomyces tendae*
Particulate surfactant (PM)	*Pseudomonas marginalis*
Biosur PM	*Pseudomonas maltophilia*

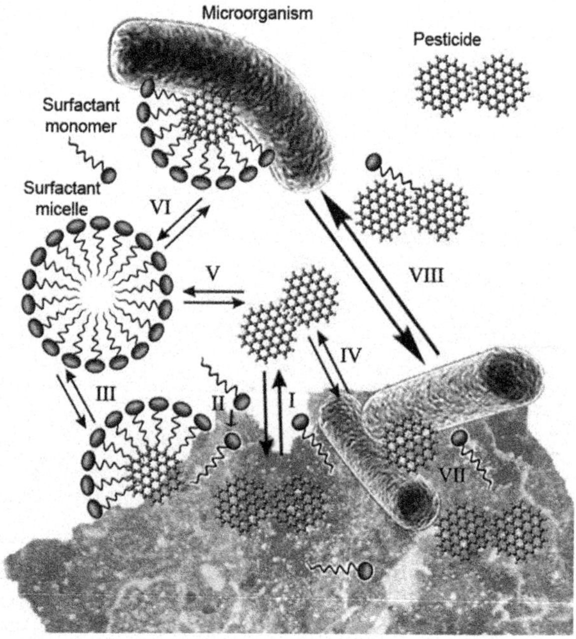

Fig. 5.3 Interactions among microorganisms, soil, pesticides, and surfactants. *I*, Adsorption of a pesticide into the soil; *II*, adsorption of the surfactant molecule into the soil; *III* pesticide solubilization; *IV*, pesticide in aqueous phase and available to microorganisms; *V*, pesticide partition between the aqueous phase and the micelle; *VI*, micelle sorption towards the microorganism; *VII*, pesticide in solid phase interacting with the microorganism; *VIII*, microorganism sorption to the soil (adapted from Volkering et al. 1998 and Riojas-González et al. 2010)

5.4 Experiences of Bioremediation in México

5.4.1 Pesticide Situation

Mexico has an extension of 200 million hectares, which between 18 and 22 are used for agriculture and around 112 for livestock, which has remained largely stable over at least two decades. In 2010 the Mexican agriculture was 3.7 % of gross domestic product (INEGI 2011), and occupied 16 % of the economically active population and income produced for about 20 million people (FAOSTAT 2012) with corn planting, beans, sorghum, wheat, barley, potatoes, and vegetables being the most important in the country (Albert 2004).

However, to obtain these results in Mexican agriculture, it has become necessary to use pesticides. The Food and Agriculture Organization (FAO 2002) defines pesticides as any substance or mixture of substances destined for preventing, destroying, or controlling any pest. The intensive use of pesticides was in the beginning as a

Table 5.2 Methods applied in bioremediation (Boopathy 2000; Juwarkar et al. 2010; Shukla et al. 2010; Banasiak et al. 2011)

Site	Technology	Principle	Applications
In situ	Air-sparging	It involves the injection of air below the water to increase oxygen concentrations and enhance biological degradation of contaminants	Biodegradative abilities of indigenous microorganisms Presence of metals and other inorganic Environmental parameters Biodegradability of pollutants Chemical solubility Geological factors Distribution of pollutants
	Phytoremediation	Using plants in soil and groundwater remediation	
	Bioventing	Method of treating contaminated soils by drawing oxygen through the soil to stimulate microbial activity	
	Bioattenuation	Natural attenuation is a proactive approach that focuses on the verification and monitoring of natural remediation processes	
	Bioaugmentation	Addition of bacterial cultures to a contaminated medium; frequently used in both in situ and ex situ systems	
	Biostimulation	Stimulation of indigenous microbial populations in soils and/or ground water through nutrient addition mainly	
	Microbe-assisted phytoremediation	Convergence of phytoremediation and microbial biodegradation strategies led to a more successful approach to remediation of contaminants	
Ex situ	Land farming	Solid-phase treatment system for contaminated soils: may be done in situ or ex situ	Surface application, aerobic process, application of organic materials to natural soils followed by irrigation and tilling To make plants healthier good alternative to land filling Surface application, agricultural to municipal waste
	Biopiling	Includes a treatment bed, i.e., mound of contaminated soil, an aeration system, an irrigation/nutrient system, and a leachate collection system. Factors such as moisture, heat, nutrients, oxygen, and pH are controlled to enhance biodegradation	
	Composting	Aerobic, thermophilic treatment process in which contaminated material is mixed with a bulking agent; can be done using static piles or aerated piles	
	Bioreactors	Biodegradation in a container or reactor; may be used to treat liquids or slurries	Bioaugmentation Toxicity of amendments Toxic concentrations of contaminants
	Biofiltration	Use of microbial stripping columns to treat air emissions	Air filtration in order to avoid air pollution

quick and safe way to modernize the Mexican agriculture, thus achieving higher agricultural productivity, although this varies depending on the crop.

In Mexico, until the middle of the last century about 40 botanical compounds were used such as tobacco, guava, and some flower extracts (Villavicencio and Pérez 2010) and other inorganic compounds as lead arsenate, copper arsenate (Paris green), and a mixture of copper sulfate and lime called **Bordeaux mixture**. The use of synthetic pesticides began in the mid-1940s, with the introduction of DDT and other organochlorine pesticides (Ortiz-Hernández et al. 2011). Pesticides most widely used in Mexico are herbicides followed by insecticides and fungicides.

Due to the problems that this situation presents, physicochemical methods for the treatment of these wastes have been implemented. Incineration at high temperatures undertaken in special ovens and alkaline hydrolysis are the most used; however these procedures generate more pollution because they only move the problem from one medium (soil) to another (air) (Ortiz-Hernández 2002). Another alternative treatment that has been analyzed for degradation of these pollutants is the application of biotechnological methods. The application of pure strains and microbial consortia to degradation of pesticides has been carried out for a wide variety of substrates (Malato et al. 2012). There are several reports which have been isolated microorganisms capable of degrading this waste, besides having the potential to prevent new accumulations.

According to available data, currently the regions with greatest use of pesticides are Sinaloa, Chiapas, Veracruz, Jalisco, Nayarit, Colima, Sonora, Baja California, Tamaulipas, Michoacán, Tabasco, State of Mexico, Puebla, and Oaxaca, applying the 80 % of total pesticides used in the country. Therefore, in these regions and in their existing crops, higher concentrations of pesticide are presented (Albert 2004).

According to the Ministry of Environment and Natural Resources [SEMARNAT (2012), SEMARNAT for its acronym in Spanish], in Mexico 86,000 tons of pesticides and insecticides were produced only in 2009. However, the FAO reports that in the same year 126,000 tons of pesticides (fungicides, bactericides, herbicides, and insecticides) were consumed (FAOSTAT 2012). As shown in Fig. 5.4, the trend of consumption has remained virtually unchanged since 2005.

Derived from years of applying, there are obsolete pesticides, understanding these as those pesticides that are not used today because they have been banned, are

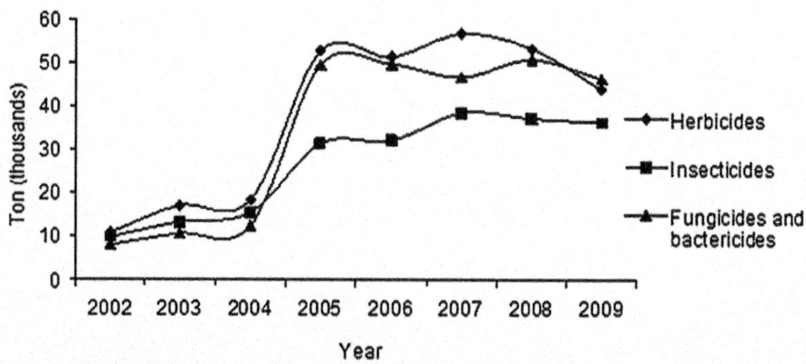

Fig. 5.4 Use of pesticides in Mexico (FAOSTAT 2012)

impaired or damaged, have expired expiration date, and cannot be used for any other reason or just are no longer relevant to the current owners. This leads to poor storage conditions, such as corroded drums, causing leakage, and seepage to the ground and bodies of water and spills.

In Mexico obsolete pesticides have been reported in both liquid and solid form. According to data reported by the SEMARNAT, in 2011 a total of 44,584,165 L and 262,474,444 kg of obsolete or expired pesticides throughout the country were recorded. These obsolete pesticides cover a wide group of compounds ranging from extremely toxic to slightly toxic. Besides, 500 m^3 of highly polluted soils, 28 reports of pesticide-contaminated sites in 15 states of the Mexican Republic, and empty pesticide containers can be about 7,000 tons annually (SEMARNAT 2012).

5.4.2 Soil Remediation

The improper management of hazardous waste and leakage caused by accidents and environmental crime can affect the health of the population, pollution of soil, water, and air, and damage to ecosystems. In Mexico, 582 contaminated sites are identified, which were caused by waste disposal 55 %, mining 13 %, industrial 11 %, and for oil and its derivatives 3.4 %. So far, 1.5 % of the sites have been remediated, 3.4 % are in the process of remediation, and 95 % have not been addressed (SEMARNAT 2012).

In Mexico, the remediation of contaminated soils is regulated by the Law for the Prevention and Management of Waste (LGPGIR for its acronym in Spanish), whose provisions stipulate that when a soil is contaminated by hazardous chemicals, immediate remedial action to prevent or reduce imminent health risks and the environment should be performed. Any person or corporation, who directly or indirectly contaminates a site or causes environment affectation through the generation, handling, release, discharge, seepage or incorporation of environmentally hazardous materials or wastes, shall be responsible for site remediation. Otherwise, they may be subjected to sanctions according to the previously established laws in national regulations. In Section 5 Fraction XXVIII of the LGPGIR defines remediation as a set of measures that must be applied to contain, eliminate, or reduce contaminants until a safe level for health and environment or prevent their dispersion, allowing its use according to the land use regulation and programs and ecological planning and development applicable. The last paragraph of the LGPGIR, section 149, recognizes sludge, sewage tanks, sludge and sediment sumps, storage tanks, among others, as materials whose chemical, mechanical, and physical properties have similarities with contaminated soils, and therefore should be remedied too. Therefore carrying out engineering works whose purpose is to decrease the risk in some contaminated sites is also considered as site remediation.

To carry out remediation of contaminated soils it is necessary that the enterprises are registered at the SEMARNAT, who must give the necessary authorization in order to offer their services for the remediation of contaminated soils. This is established in the Regulation of the LGPGIR. Until 2010, there were 56 authorized enterprises to remediate soils and materials contaminated mainly with hydrocarbons, but in 2014, 227 are registered (http://tramites.semarnat.gob.mx/images/stories/menu/empresas/

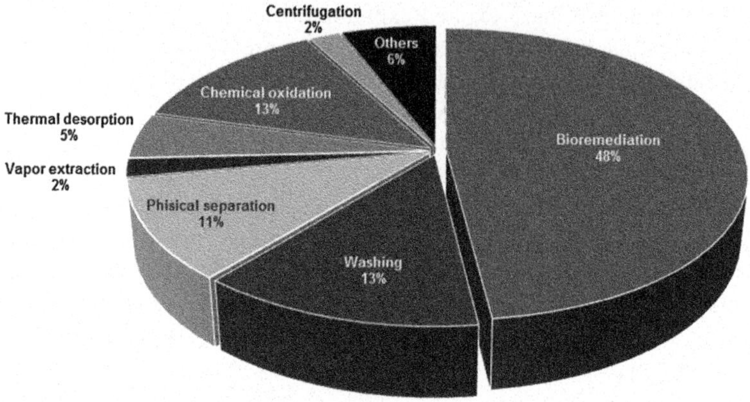

Fig. 5.5 Main remediation technologies utilized in Mexico by authorized enterprises (adapted from Volke-Sepúlveda and Velasco-Trejo 2002)

rubro15.pdf). Authorization for remediation of contaminated soil is given for a period of 10 years (SEMARNAT 2012).

Remediation technologies applied in Mexico for these companies are shown in Fig. 5.5.

5.4.3 *Some Studies on Bioremediation and/or Biodegradation of Soil Contaminated with Pesticides*

Mexico's efforts in bioremediation of contaminated soil have been directed mainly for removing petroleum hydrocarbons from soil, sediment and sludge, as it is the most demanded area for being an oil producer country. However, in the case of bioremediation of soils contaminated with pesticides, there are no reports of works carried out in the field, besides companies that provide services and have authorization from SEMARNAT and do not have the expertise to remediate soils contaminated with pesticides.

However, various efforts for pesticide biodegradation with native microorganisms isolated from agricultural soils and sewage sludge have been carried out. In addition, there are reports about bioremediation of contaminated soils at laboratory scale, using microcosms. Studies have been conducted on different pesticides that are of interest to the country, which are those with greater use or that have been included in Annex A of the Stockholm Agreement which means that Mexico has to eliminate stockpiles of these pesticides and remediate the environment where they are polluting. Examples of these pesticides are organochlorine pesticides such as DDT, aldrin, and endosulfan, among others.

Examples of studies conducted in Mexico are presented in Tables 5.3 and 5.4. Table 5.3 shows the experiences of bioremediation and Table 5.4 shows the experiences about the isolation of microorganisms in order to carry out the pesticide biodegradation. In the same tables we describe the main results of those studies.

Table 5.3 Some examples of bioremediation studies in contaminated soils with pesticides in Mexico

Pesticide	Microorganism	Results	Reference
DDT and its metabolites	Native microorganisms	Stimulation of native microbial populations in soil by the addition of small amounts of secondary carbon sources (cosubstrates) and its effect on the degradation and theoretical mineralization of DDT [l,l,l-trichloro-2,2-bis(p-chlorophenyl)ethane] and its main metabolites, DDD [2,2-dichloroethane-1,1-diyl-bis(pchlorobenzene)] and DDE [2,2-dichloroethene-1,1-bis(pchlorophenyl)], were evaluated. Microbial activity in soil polluted with DDT, DDE, and DDD was increased by the presence of phenol, hexane, and toluene as cosubstrates. The consumption of DDT was increased from 23 % in a control (without cosubstrate) to 67 %, 59 %, and 56 % in the presence of phenol, hexane, and toluene, respectively. DDE was completely removed in all cases, and DDD removal was enhanced from 67 % in the control to 86 % with all substrates tested, except for acetic acid and glucose substrates. In the latter cases, DDD removal was either inhibited or unchanged from the control. The CO_2 produced was higher than the theoretical amount for complete cosubstrate mineralization indicating possible mineralization of DDT and its metabolites. Bacterial communities were evaluated by denaturing gradient gel electrophoresis, which indicated that native soil and the untreated control presented a low bacterial diversity. In the presence of toluene a bacterium related to *Azoarcus*, a genus that includes species capable of growing at the expense of aromatic compounds such as toluene and halobenzoates under denitrifying conditions, was detected	Ortız et al. (2013)
Atrazine	*Trichoderma* sp.	The use of local strains of native fungi isolated from the *Phaseolus vulgaris* L. rhizosphere present in cultivable soils for atrazine degradation in contaminated soils was evaluated. *Trichoderma* sp. was isolated, identified, and characterized as a native strain of the rhizosphere, from agro ecosystem of the valley of Tulancingo, Hidalgo, Mexico, and showed resistance to atrazine herbicide up to 10,000 mg L^{-1}. It is able to degrade up to 89 % of the atrazine in 500 mg kg^{-1} of soil under laboratory conditions	Islas-Pelcastre et al. (2013)
Parathion and cadusafos	–	The sorption behavior of two organophosphorous insecticides was studied in three agricultural soil samples from central Mexico, Vertisols, and Andosols. Using ^{14}C-labeled substances, we assessed sorption and desorption properties in classical batch equilibrium and static soil incubation experiments. Our results show that cadusafos was less sorbed by the various soils (K_d values 7.6–12.7 L kg^{-1}) compared with parathion (K_d values 38.6–74.9 L kg^{-1}), despite similar log K_{ow} values. Cadusafos exhibited a greater reversibility of sorption than parathion in both soil types. Time-dependent sorption was quantitatively significant, leading to a rapid decrease in the concentration of available insecticide. This finding is partly due to the formation of non-extractable, bound residues. The decrease in the available concentration of both insecticides was greater in the Andosol compared with the Vertisols. Soil organic matter clearly influenced the sorption behavior and availability of parathion. On the other hand, the sorption of cadusafos was more influenced by other soil properties such as clay content and cation exchange capacity. Calculation of residual insecticide levels in the soil solution suggests that both insecticides may have persistent toxic effects in the studied soils	Olvera-Velona et al. (2008)

(continued)

Table 5.3 (continued)

Pesticide	Microorganism	Results	Reference
Methyl parathion	Native microorganisms	The effect of adding methyl parathion on soil characteristics, CO_2 emissions, mineral N and 4-nitrophenol, a degradation product of methyl parathion, and the degradation of methyl parathion was investigated in an aerobic incubation experiment when using chinampa soils of Mexico City. Treatments were sterilized and non-sterilized soil amended with or without methyl parathion at 280 mg kg^{-1}, a concentration sometimes found in the top soil. Methyl parathion removal followed a first-order kinetic with a half-life ranging from 16 to 34 days, while small amounts of 4-nitrophenol were detected after 3 days. No abiotic processes affected the methyl parathion concentration and no 4-nitrophenol was formed in the sterilized soil. It was found that methyl parathion was removed rapidly from soil and less than 35 mg kg^{-1} of 4-nitrophenol was found in soil	Chávez-López et al. (2011)
2,4-D	Aerobic and sulfate-reducing 2,4-D clastic native microorganisms	The removal of 2,4-dichlorophenoxyacetic acid (2,4-D) from an agricultural mineral soil with high contents of clay and organic matter was evaluated in lab-scale slurry bioreactors under both aerobic and anaerobic (sulfate-reducing) conditions. Also, the effect of an additional carbon source (sucrose) on 2,4-D removal was assessed. A soil of 48 % clay and 4 % organic matter was sterilized and contaminated with 300 mg 2,4-D kg^{-1} dry matrix and subsequently treated in both aerobic (A-SB) and sulfate-reducing slurry bioreactors (SR-SB), with and without 1 g L^{-1} sucrose. Both SBs received a seed (20 %, v/v) acclimated to 2,4-D from aerobic and sulfate-reducing continuous, complete mix reactors. Aerobic conditions had a removal efficiency of 93 %, compared to 25 % under sulfate-reducing conditions at 14-day treatment. Aerobic removal was not affected by the sucrose supplementation; however, the SR-SB without sucrose still showed an important 2,4-D removal (up to 18 %) suggesting that 2,4-D could be used as substrate. The 2,4-D-clastic bacteria were present in all reactors during the incubation: aerobic 2,4-D-clastic bacteria ranged between 6 and 8 logs, whereas the sulfate-reducing 2,4-D clastic population ranged between 5 and 7 logs	Robles-González et al. (2006)
Lindane	Aerobic, methanogenic, and sulfate-reducing lindane-clastic native microorganisms	In this study the effect of dominant electron acceptor [either aerobic, methanogenic, or sulfate-reducing slurry bioreactor (SB)] and biostimulation with sucrose on lindane removal from heavy soil was evaluated. Besides the effect of the type of combined environments [partially aerated methanogenic (PAM) and simultaneous methanogenic-sulfate reducing (M-SR)] and addition of silicone oil as solvent on lindane removal from a clayish agricultural soil with high levels of organic matter was assessed. In the first experiment, the main effect of electron acceptor was significant; lindane removals followed the order SR>A-M SBs. On the other hand, cosubstrate sucrose was not significant. Yet, the interaction was moderately significant; cosubstrate influence was distinct depending on the type of electron acceptor. Cosubstrate slightly improved lindane removal in both anoxic SBs (SR and M units), whereas lindane removal in A-SB with sucrose was lower than A-SB without sucrose. In the second experiment, both factors [simultaneous electron acceptor (SEA) combination and solvent addition] were significant. Removal of lindane in SEA-SBs, PAM, and M-SR without silicone oil was low (~16 %). On the other hand, the order of lindane removals in SBs with oil silicone M-SR SB was significantly superior (65 %) to that of PAM SB (39 %). Finally, SBs with SEA where one of the anaerobic metabolites is methanogenic were not as successful as SBs with single-electron acceptors for removal of lindane from heavy soil	

Lindane	*Ocimum basilicum L. Ocimum minimum L*	To study a potential way to bioremediate soils contaminated with this pesticide, two plant species of the genus *Ocimum* were studied: *Ocimum basilicum* L. and *Ocimum minimum* L., since they are economically feasible and well adapted to the climatic conditions of the Nayarit zone (Mexican pacific coast). Young plants were transplanted into soil experimentally polluted with endosulfan. Growth of both species was not affected by endosulfan, the plants grew, flourished, and produced seeds; 30 days later, endosulfan concentration was lower in the soil with *O. basilicum* than in the soil without plants. On day 90, no differences in endosulfan concentrations were found between soil with or without *O. minimum*. At day 1, plants in the polluted soil showed lipoperoxidation, as measured by thiobarbituric acid-reactive species (TBARS). Interestingly, a higher TBARS value was observed at day 3 in transplanted plants as compared to non-transplanted plants. In conclusion, both species can endure endosulfan pollution (as high as 1 g kg^{-1}) in soils. *O. basilicum* seems to be an adequate candidate for bioremediation of soils polluted with endosulfan	Ramírez-Sandoval et al. (2011)
Endosulfan	*Ocimum basilicum*	For phytoremediation purposes, we have previously observed that the presence of *Ocimum basilicum* decreased the concentration of endosulfan in experimentally polluted soil by 37 % after 30 days. To study the possible mechanism, we evaluated whether endosulfan could affect (1) the activity of glutathione S transferase (GST) of *O. basilicum* and (2) microorganisms from rhizosphere. Young plants were added to experimentally polluted soil with endosulfan. Rhizosphere microorganisms were exposed to several concentrations of endosulfan and cultured in Luria Bertani (LB) broth or agar; their growth was determined by triplicate either spectrophotometrically or by plate counts. After exposure to the pesticide endosulfan in *O. basilicum* and its rhizosphere, three effects were observed: (1) In LB broth, optimal growth of microorganisms was observed at 72 and 48 h after exposure to endosulfan 2.4 and 3.4 mg/10 mL. (2) Optimal growth of microorganisms in LB agar was observed at 0.3 and 2.4 mg/10 mL. (3) GST was increased after exposure to these pesticides over its control. These observations suggest that phytostimulaton and phytotransformation could be involved as possible mechanisms of the phytoremediatory effect of *O. basilicum*	Ramírez-Sandoval et al. (2013)

Table 5.4 Some examples of pesticide biodegradation studies in Mexico

Pesticide	Isolation media	Microorganism	Results	Reference
Lindane	*Agave tequilana* leaves	*Fusarium verticillioides*	The fungal strain is able to use lindane as a carbon and energy source under aerobic conditions. Lindane biodegradation was higher in the presence of limited amounts of nitrogen and phosphorus, with addition of agave leaves to the culture medium and the use of higher concentrations of lindane, copper, and yeast extract. Under these conditions the strain was able to degrade the 83 % of lindane. The analysis of the metabolites identified g-pentachlorocyclohexene and benzoic acid derivatives. This finding suggests that there is an aerobic carboxylation step, reported for the first time, in the lindane biodegradation pathway	Guillén-Jiménez et al. (2012)
Endosulfan and DDT	Green coffee beans	*Flavimonas oryzihabitans Pseudomonas aeruginosa*	Bacterial strains were examined for their capacity to grow and to remove 1,1,1-trichloro-2,2-bis (4-chlorophenyl) ethane (DDT) and 1,2,5,6,7,7-hexachloro-5-norbornene-2,3-dimethanol cyclic sulfite (endosulfan) in a liquid medium using defective green coffee beans, glucose, or peptone as cosubstrates. Coffee bean was an adequate nutrient source for bacterial growth and a 63.2 % DDT removal was observed at 7 days of incubation with *Flavimonas oryzihabitans*, while in the same system *Pseudomonas aeruginosa* was able to remove both endosulfan and DDT (51.2 and 67.4 %, respectively)	Barragán-Huerta and Rodríguez-Vázquez (2010)
Tetrachorvinphos	Soil	Bacterial consortium	A bacterial consortium which degrades tetrachlorvinphos was isolated from agricultural soil in central Mexico. This consortium was composed of six pure strains which were characterized based on their morphological and biochemical characteristics. The strains were presumptively identified as *Stenotrophomonas malthophilia*, *Proteus vulgaris*, *Vibrio metschinkouii*, *Serratia ficaria*, *Serratia* spp. and *Yersinia enterocolitica.* The consortium and the six bacteria were assessed in order to discover their ability to degrade tetrachlorvinphos (TCV) in mineral medium and in rich medium. Growth curve experiments showed that the bacterial consortium was able to grow in mineral medium containing TCV as the only carbon source. However, only one pure strain was able to remove TCV in mineral medium, while all of them removed it in rich medium. Hydrolysis products were detected and identified by gas chromatography-mass spectrometry. These data indicate that the isolated strains can be used for waste biodegradation or bioremediation of TCV-contaminated soil or water	Ortiz-Hernández and Sánchez-Salinas (2010)

Methyl parathion and tetrachlorvinphos		Bacterial consortium	A bacterial consortium was immobilized with two supports consisting of alginate beads or stones of tezontle colonized by biofilm. Removal kinetics were recorded for suspended and immobilized consortium using a mineral salt medium supplemented with both pesticides at 25 mg L^{-1}, with 0.1 % (w/v) glucose as a cosubstrate. The viability of the consortium cultivated in suspension was maintained for 6 days, whereas the viability of the consortium immobilized in alginate and tezontle supports was maintained for up to 11 and 13 days, respectively. The percentage of removal in the present study was 41 %, 72 %, and 66 % for methyl parathion and 53 %, 65 %, and 47 % for tetrachlorvinphos with the suspended, alginate immobilized, and tezontle-immobilized consortium, respectively	Yáñez-Ocampo et al. (2009)
Methyl parathion and tetrachlorvinphos	Soil	Bacterial consortium	With the aim of developing a tool for pesticide biodegradation, a tezontle-packed up-flow reactor (TPUFR) with an immobilized bacterial consortium for biological treatment of methyl-parathion and tetrachlorvinphos was evaluated. Different four flow rates and hydraulic residence times were evaluated. In the bioreactor was obtained a 75 % efficiency in the removal of methyl-parathion and tetrachlorvinphos. Their adsorptions in the volcanic rock were 9 % and 6 %, respectively. It was demonstrated that the removal of pesticides was due to the biological activity of the immobilized bacterial consortium. Immobilization of a bacterial consortium using tezontle as a support is innovative and an economical tool for the treatment of mixtures of organophosphorus pesticide residues	Yáñez-Ocampo et al. (2011)
Phorate, tetrachlorvinphos, methyl-parathion, terbufos, trichloronate, ethoprophos, phosphamidon, fenitrothion, dimethoate and DEF	–	*Flavobacterium* sp. *ATCC 27551*	The biotransformation by *Flavobacterium* sp. of ten organophosphate pesticides was experimentally and theoretically studied. The *Flavobacterium* sp. ATCC 27551 strain bearing the organophosphate-degradation gene was used. Bacteria were incubated in the presence of each pesticide for a duration of 7 days. The percent of degradation for each pesticide was as follows: Ethoprophos 83.0, dimethoate 72.0, fenitrothion 67.6, def 64.0, phosphamidon 58.0, trichloronate 48.0 methyl parathion 45.7, terbufos 42.0, phorate 37.3, and tetrachlorvinphos 7.8	Ortiz-Hernández et al. (2003)

(continued)

Table 5.4 (continued)

Pesticide	Isolation media	Microorganism	Results	Reference
Methyl-parathion	Soil	*Burkholderia glatei*, *Burkholderia tuberum*	These bacteria were isolated from agricultural soils in Mexico and both showed high efficiency in the hydrolysis of methyl parathion and its metabolite p-nitrophenol. In contrast to other reported bacteria, these bacteria perform the hydrolysis of both methyl parathion and its major metabolite in a period of 15 h	Popoca-Ursino (2012)
Methyl parathion	–	*E. coli cloned with opd* gene of *Flavobacterium* sp. ATCC 27551	The goal of this study was to optimize methyl parathion (*O,O-dimethyl-O-4-p-nitrophenyl phosphorothioate*) degradation using a strain of *Escherichia coli DH5α* expressing the *opd* gene. The results indicate that this strain had lower enzymatic activity compared to the *Flavobacterium* sp. ATCC 27551 strain from which the *opd* gene was derived. Both strains were assessed for their ability to degrade methyl parathion (MP) in a mineral salt medium with or without the addition of glucose either as suspended cells or immobilized on tezontle, a volcanic rock. MP was degraded by both strains with similar efficiencies, but immobilized cells degraded MP more efficiently than cells in suspension. However, the viability of *E. coli* cells was much higher than that of the *Flavobacterium* sp. The decrease in toxicity from the treated effluents through acetylcholinesterase activity tests was confirmed, indicating the potential of this method for the treatment of solutions containing MP	Abdel-Razek et al. (2013)
Methyl parathion	Soil	Bacterial consortium	A methyl parathion degrading consortium of bacteria was isolated from agricultural soils in central Mexico (Morelos State), using methyl parathion as the only carbon source. The consortium ability to degrade the pesticide was assessed with a mineral medium containing 15 mg L^{-1} of pesticide. From this process, 11 genera were founded in the consortium, which were evaluated for their degradation efficiency. Only five of them show enzymatic activity on methyl parathion, which could be considered as a potential enzyme source to reduce environmental pollution with this pesticide and its residues	Ortiz-Hernández et al. (2001)

Methyl parathion and coumaphos	Soil	Bacterial consortium	In this study, a bacterial consortium was used that was isolated from agricultural soils to degrade a mixture of the organophosphate pesticides methyl parathion and coumaphos. The efficiency of removal was evaluated using mineral salt medium supplemented with glucose, and the bacterial consortium was cultivated as free cells and immobilized on *Luffa cylindrica* fibers. To improve the structure of the fibrous network and to achieve greater retention of microorganisms, removal was also tested prior to fibers treatment with sodium hydroxide (NaOH). The results indicate that the microorganisms used had better growth as free cells. A removal of 54.88 % and 62 % for MP and COU, respectively, was observed using the free cells; but when the cells were immobilized on loofa sponge fibers, the removal was increased to 98 and 100 % of those pesticides. This pesticide removal was the result of a combined effect among the activity of the microorganisms, the adhesion to the bacterial cells, and the adsorption on the support material. We observed that a strong and fast adsorption on the loofa sponge fiber, since removal obtained only with loofa fiber, did not present significant difference with immobilized microorganisms	Moreno-Medina et al. (2014)

5.5 Concluding Remarks

In Mexico, as in many countries, the pesticide use has brought benefits such as the increment of agricultural production, soil productivity, and product quality, which are reflected in economic benefits, vector disease control, and, in general, in public health. However, due to their high toxicity and persistence, these chemical compounds constitute a serious menace for the human health and for the environment, when they are found in soil as pollutants. The elimination of pesticides from crop fields is an important not totally solved task in Mexico, despite the existence of certified companies specialized in the remediation of highly polluted soils. As was reviewed here, the bioremediation strategies, specially the use of specific microorganisms or microbial consortia for the biodegradation of pesticides in soil, are an environmental friendly and economically feasible technology that is gaining acceptance for the successful conditioning and detoxification of contaminated sites. In Mexico many studies report the isolation of microorganisms capable of degrading pesticides, besides studies of soil bioremediation performed in microcosms. The results of these works have the potential to be applied in the pesticide degradation of residues that are stored, in addition to the bioremediation of contaminated soils that exist in Mexico.

References

Abdel-Razek SA, Folch-Mallol JL, Perezgasga-Ciscomani L et al (2013) Optimization of methyl parathion biodegradation and detoxification by cells in suspension or immobilized on tezontle expressing the opd gene. J Environ Sci Heal B Pestic Food Contam Agric Wastes 48:449–461

Abraham WR, Nogales B, Golyshin PN et al (2002) Polychlorinated biphenyl-degrading microbial communities and sediments. Curr Opin Microbiol 5:246–253

Aislabie J, Lloyd-Jones G (1995) A review of bacterial-degradation of pesticides. Aust J Soil Res 33:925–942

Albert L (2004) Panorama de los Plaguicidas en México. 7° Congreso de Actualización en Toxicología Clínica, Nayarit, México, pp 2–8

Alexander M (1994) Biodegradation and bioremediation. Academic Press, San Diego, CA

Alexander M (2000) Aging, bioavailability, and overestimation of risk from environmental pollutants. Environ Sci Tech 34:4259–4265

Anjum R, Rahman M, Masood F, Malik A et al (2012) Bioremediation of pesticides from soil and wastewater. In: Malik A (ed) Environmental protection strategies for sustainable development. Springer, Amsterdam, pp 295–328

Antizar-Ladislao B (2010) Bioremediation: working with bacteria. Elements 6:389–394

Aronstein BN, Calvillo YM, Alexander M (1991) Effect of surfactants at low concentrations on the desorption and biodegradation of sorbed aromatic compounds in soil. Environ Sci Technol 25:1728–1731

Bailey GW, White JL (1970) Factors influencing the adsorption, desorption and movement of pesticides in soil. Residue Rev 32:29–92

Banasiak L, Schafer A, Van der Bruggen B (2011) Sorption of pesticide endosulfan by electrodialysis membranes. Chem Eng J 166:233–239

Banat IM, Makkar RS, Cameotra SS (2000) Potential commercial application of microbial surfactants. Appl Microbiol Biotechnol 53:495–508

Barragán-Huerta BE, Rodríguez-Vázquez R (2010) Green bean coffee as nutrient source for pesticide degrading-bacteria. In: Mendez-Vilas A (ed) Current research, technology and education topics in applied microbiology and microbial biotechnology. Microbiology boock series, No 2, vol 2. FORMATEX Reserch Center, Extremadura, pp 1322–1327

Bedard DL, May RJ (1995) Characterization of the polychlorinated biphenyls in the sediments of Woods Pond: evidence for microbial dechlorination of Aroclor 1260 *in situ*. Environ Sci Technol 30:237–245

Beigel C, Barriuso E, Calvet R (1998) Sorption of low levels of nonionic and anionic surfactants on soil: effects on sorption of triticonazole fungicide. Pest Sci 54:52–60

Bollag JM, Bollag WB (1995) Soil contamination and the feasibility of biological remediation. In: Skipper HD, Turco RF (eds) Bioremediation: science and applications. Soil Science Society of America, Madison, WI, pp 1–10, Special publication

Boopathy R (2000) Factors limiting bioremediation technologies. Biores Technol 74:63–67

Boopathy R, Manning J (1999) Surfactant-enhanced bioremediation of soil contaminated with 2, 4, 6-trinitrotoluene in soil slurry reactors. Water Environ Res 74:119–124

Bustamante M, Durán N, Diez MC (2012) Biosurfactants are useful tools for the bioremediation of contaminated soil: a review. J Soil Sci Plant Nutr 12:667–687

Cameotra SS, Makkar RS (2010) Biosurfactant-enhanced bioremediation of hydrophobic pollutants. Pure Appl Chem 82:97–116

Chakraborty R, Wu CH, Hazen TC (2012) Systems biology approach to bioremediation. Curr Opin Biotechnol 23:483–490

Chávez-López C, Blanco-Jarvio A, Luna-Guido M et al (2011) Removal of methyl parathion from a chinampa agricultural soil of Xochimilco Mexico: a laboratory study. Eur J Soil Biol 47:264–269

Chino-Flores C, Dantán-González E, Vázquez-Ramos A et al (2011) Isolation of the *opdE* gene that encodes for a new hydrolase of *Enterobacter* sp. capable of degrading organophosphorus pesticides. Biodegradation 23:387–397

Chowdhury S, Bala NN, Dhauria P (2012) Bioremediation – a natural way for cleaner environment. Int J Pharmaceut Chem Biol Sci 2:600–611

Christofi N, Ivshina IB (2002) Microbial surfactants and their use in field studies of soil remediation. J Appl Microbiol 93:915–929

Commandeur LCM, Parsons JR (1994) Biodegradation of halegonated aromatic compounds. In: Ratledge C (ed) Biochemistry of microbial degradation. Kluwer Academic Publishers, Dordrecht, pp 423–458

Crawford RL, Rosenberg E (2013) Bioremediation. In: Rosenberg E, DeLong EF, Lory S et al (eds) The prokaryotes. Springer-Verlag, Heidelberg, pp 295–307

Csizer Z (2002) Persistent organic pollutants (POPs). UNIDO programmes on persistent organic pollutants (POPs). International forum on strategies and priorities for environmental industries. UNIDO, Bratislava, p 17

Dua M, Singh A, Sethunathan N et al (2002) Biotechnology and bioremediation: successes and limitations. Appl Microbiol Biotechnol 59:143–152

Edwards CA, Bohlen PJ (1992) The effects of toxic chemicals on earthworms. In: Ware GW (ed) Reviews of environmental contamination and toxicology. Spriger-Verlag, New York, pp 23–99

FAO (2002) Código internacional de conducta para la distribución y utilización de plaguicidas. Food and Agriculture Organization, Roma, p 7

FAOSTAT (2012) División estadística. http://faostat3.fao.org/home/index_es.html?locale=es#DOWNLOAD. Accessed Aug 2013

Gevao B, Semple KT, Jones KC (2000) Bound pesticide residues in soils: a review. Environ Pollut 108:3–14

Grosser RJ, Friedrich M, Ward DM et al (2000) Effect of model sorptive phases on phenanthrene biodegradation: different enrichment conditions influence bioavailability and selection of phenanthrene-degrading isolates. Appl Environ Microbiol 66:2695–2702

Guha S, Jaffé PR (1996) Bioavailability of hydrophobic compounds partitioned into the micellar phase of nonionic surfactants. Environ Sci Technol 30:1382–1391

Guillén-Jiménez FM, Cristiani-Urbina E, Cancino-Díaz JC et al (2012) Lindane biodegradation by the *Fusarium verticillioides* AT-100 strain, isolated from *Agave tequilana* leaves: kinetic study and identification of metabolites. Int Biodet Biodeg 74:36–47

Guo Y, Mulligan CN (2006) Combined treatment of styrene-contaminated soil by rhamnolipid washing followed by anaerobic treatment. In: Hudson RC (ed) Hazardous materials in soil and atmosphere, treatment, removal and analysis. Nova Science Publishers, New York, pp 1–38

Haritash AK, Kaushik CP (2009) Biodegradation aspects of polycyclic aromatic hydrocarbons (PAHs): a review. J Hazard Mater 169:1–15

Harms H, Bosma TNP (1997) Mass transfer limitation of microbial growth and pollutant degradation. J Ind Microbiol Biotechnol 18:97–105

Harms H, Zehnder A (1995) Bioavailability of sorbed 3-chlorodibenzofuran. Appl Environ Microbiol 61(1):27–33

Ibekwe AM, Papiernik SK, Gan J et al (2001) Impacts of fumigants on soil microbial communities. Appl Environ Microbiol 67:3245–3257

Iglesias-Jiménez E, Sanchez-Martin MJ, Sánchez-Camazano M (1996) Pesticide adsorption in a soil-water system in the presence of surfactants. Chemosphere 32:1771–1782

INEGI (2011) Instituto Nacional de Estadística y Geografía, México. Producto Interno Bruto. http://www.inegi.org.mx/sistemas/bie/cuadrosestadisticos/GeneraCuadro.aspx?s=est&nc=785&c=24393. Accessed June 2013

Islas-Pelcastre M, Villagómez-Ibarra JR, Madariaga-Navarrete JA et al (2013) Bioremediation perspectives using autochthonous species of *Trichoderma* sp. for degradation of atrazine in agricultural soil from the Tulancingo Valley, Hidalgo, Mexico. Trop Subtrop Agroecosyst 16: 265–276

Jain RK, Kapur M, Labana S et al (2005) Microbial diversity: application of microorganisms for the biodegradation of xenobiotics. Curr Sci 89:101–112

Jayashree R, Vasudevan N, Chandrasekaran S (2006) Surfactants enhanced recovery of endosulfan from contaminated soils. Int J Environ Sci Technol 3:251–259

Joutey NT, Bahafid W, Sayel H, El Ghachtouli N (2013) Biodegradation: involved microorganisms and genetically engineered microorganisms. In: Chamy R (ed) Biodegradation-life of science. InTech, Rijeka, pp 289–320

Juwarkar AA, Dubey KV, Nair A et al (2008) Bioremediation of multi-metal contaminated soil using biosurfactant-a novel approach. Ind J Microbiol 48:142–146

Juwarkar AA, Singh SK, Mudhoo A (2010) Comprehensive overview of elements in bioremediation. Rev Environ Sci Biotechnol 9:215–288

Kaufman DD (1983) Fate of toxic organic compounds in land-applied wastes. In: Parr JF, Marsh PB, Kla JM (eds) Land treatment of hazardous wastes. Corp, Park Ridge, New Jersey, pp 77–151

Kile DE, Chiou CT (1989) Water solubility enhancements of DDT and trichlorobenzene by some surfactants below and above the critical micelle concentration. Environ Sci Technol 23: 832–838

Klose S, Acosta-Martinez V, Ajwa HA (2006) Microbial community composition and enzyme activities in a sandy loam soil after fumigation with metal bromide or alternative biocides. Soil Biol Biochem 38:1243–1254

Komancová M, Jurčová I, Kochánková L et al (2003) Metabolic pathways of polychlorinated biphenyls degradation by *Pseudomonas* sp. 2. Chemosphere 50:537–543

Kreinfeld F, Stoll G (1997) Surfactants in consumer products and raw materials situation – a brief survey. In: Hill K, von Rybinski W, Stoll F (eds) Alkyl polyglycosides technology, properties, and applications. VCH Verlagsgesellschaft, Weinheim, pp 225–233

Laha S, Tansel B, Ussawarujikulchai A (2009) Surfactant-soil interactions during surfactant-amended remediation of contaminated soils by hydrophobic organic compounds: a review. J Environ Manage 90:95–100

Leahy JG, Colwell RR (1990) Microbial degradation of hydrocarbons in the environent. Microbiol Rev 54:305–315

Li X, Schuler MA, Berenbaum MR (2007) Molecular mechanisms of metabolic resistance to synthetic and natural xenobiotics. Annu Rev Entomol 52:231–253

Madrigal-Monárrez I, Benoit P, Barriuso E et al (2008) Caracterización de la sorción y desorción de herbicidas en suelos con diferentes tipos de ocupación. Rev Ing Inv 28:96–104

Malato S, Blanco J, Estrada C et al (2012) Tratamiento y disposición final. Degradación de Plaguicidas. http://www.estrucplan.com.ar/Producciones/entrega.asp?IdEntrega=2960. Accessed Dec 2013

Martikainen E, Haimi J, Ahtiainen J (1998) Effects of dimethoate and benomyl on soil organisms and soil processes – a microcosm study. Appl Soil Ecol 9:381–387

Master ER, Lai VW, Kuipers B et al (2002) Sequential anaerobic-aerobic treatment of soil contaminated with weathered Aroclor 1260. Environ Sci Technol 36:100–103

McCarthy JF, Zachara JM (1989) Subsurface transport of contaminants. Environ Sci Technol 23:496–502

Megharaj M, Ramakrishnan B, Venkateswarlu K (2011) Bioremediation approaches for organic pollutants: a critical perspective. Environ Int 37:1362–1375

Mishra M, Muthuprasanna P, Prabha KS et al (2009) Basics and potential applications of surfactants. A review. Int J Pharm Tech Res 1:1354–1365

Moreno-Medina D, Sanchez-Salinas E, Ortiz-Hernández ML (2014) Removal of methyl parathion and coumaphos pesticides by a bacterial consortium immobilized in *Luffa cylindrical*. Rev Int Contam Ambient 30:51–63

Mörner J, Bos R, Fredrix M (2002) Reducing and eliminating the use of persistent organic pesticides. Inter-organization programme for the sound management of chemicals (IOMC). http://www.chem.unep.ch/Publications/pdf/POPred_E.pdf. Accessed Nov 2013

Mosleh YY, Paris-Palacios S, Couderchet M et al (2003) Acute and sublethal effects of two insecticides on earthworms (*Lumbricus terrestris* L.) under laboratory conditions. Environ Toxicol 18:1–8

Mulchandani A, Kaneva I, Chen W (1999) Detoxification of organophosphate nerve agents by immobilized *Escherichia coli* with surface-expressed organophosphorus hydrolase. Biotechnol Bioeng 63:216–223

Mulligan CN (2005) Environmental applications for biosurfactants. Environ Pollut 133:183–198

Mulligan CN, Yong RN, Gibbs BF (2001) Surfactant-enhanced remediation of contaminated soil: a review. Eng Geol 60:371–380

Nawaz K, Hussain K, Choudary N et al (2011) Eco-friendly role of biodegradation against agricultural pesticides hazards. Afr J Microbiol Res 5:177–183

Niti C, Sunita S, Kamlesh K et al (2013) Bioremediation: an emerging technology for remediation of pesticides. Res J Chem Environ 17:88–105

Odukkathil G, Vasudevan M (2013) Toxicity and bioremediation of pesticides in agricultural soil. Rev Environ Sci Biotechnol 12:421–444

Olivera NL, Commendatore MG, Moran AC et al (2000) Biosurfactant-enhanced degradation of residual hydrocarbons from ship bilge wastes. J Ind Microbiol Biotechnol 25:70–73

Olvera-Velona A, Benoit P, Barriuso E et al (2008) Sorption and desorption of organophosphate pesticides, parathion and cadusafos, on tropical agricultural soils. Agron Sustain Dev 28: 231–238

Ortız I, Velasco A, Le Borgne S et al (2013) Biodegradation of DDT by stimulation of indigenous microbial populations in soil with cosubstrates. Biodegradation 4:215–225

Ortiz-Hernández ML (2002) Biodegradación de plaguicidas organofosforados por nuevas bacterias aisladas del suelo. Universidad Autónoma del Estado de Morelos, Dissertation

Ortiz-Hernández ML, Sánchez-Salinas E (2010) Biodegradation of the organophosphate pesticide tetrachlorvinphos by bacteria isolated from agricultural soils in Mexico. Rev Int Contam Ambient 26:27–38

Ortiz-Hernández ML, Monterrosas-Brisson M, Yáñez-Ocampo G et al (2001) Biodegradation of methyl-parathion by bacteria isolated of agricultural soil. Rev Int Contam Ambient 17: 147–155

Ortiz-Hernández ML, Quintero-Ramirez R, Nava-Ocampo A et al (2003) Study of the mechanism of *Flavobacterium* sp. for hydrolyzing organophosphate pesticides. Fund Clin Pharmacol 17:717–723

Ortiz-Hernández ML, Sánchez-Salinas E, Olvera-Velona A et al (2011) Pesticides in the environment: impacts and its biodegradation as a strategy for residues treatment, pesticides. In: Stoytcheva M (ed) Formulations, effects, fate. InTech, Rijeka, pp 551–574

Pandey S, Singh DK (2004) Total bacterial and fungal population after chlorpyrifos and quinalphos treatments ingroundnut (*Arachis hypogaea* L.) soil. Chemosphere 55:197–205

Pongrac P, Vogel-Mikuš K, Kump P et al (2007) Changes in elemental uptake and arbuscular mycorrhizal colonisation during the life cycle of *Thlaspi praecox* Wulfen. Chemosphere 69:1602–1609

Popoca-Ursino C (2012) Aislamiento de bacterias con capacidad de degradación de plaguicidas organofosforados y caracterización genómica y molecular de sus actividades. Dissertation, Universidad Autónoma del Estado de Morelos

Prince RC (1997) Bioremediation of marine oil spills. Trends Biotechnol 15:158–160

Ramírez-Sandoval M, Melchor-Partida GN, Muñiz-Hernández S et al (2011) Phytoremediatory effect and growth of two species of Ocimum in endosulfan polluted soil. J Hazard Mater 192:388–392

Ramírez-Sandoval M, Muñiz-Hernández S, Velázquez-Fernández JB (2013) Mechanisms of phytoremediatory effect of *Ocimum basilicum* L. and its rhizosphere exposed to different concentrations of the organochlorine pesticide endosulfan. Chem Eng Trans 34:73–78

Rayu S, Karpouzas DG, Singh BK (2012) Emerging technologies in bioremediation: constraints and opportunities. Biodegradation 23:917–926

Richins RD, Kaneva I, Mulchandani A et al (1997) Biodegradation of organophosphorus pesticides by surface-expressed organophosphorus hydrolase. Nat Biotechnol 15:984–987

Riojas-González HH, Torres BLG, Mondaca FI et al (2010) Efectos de los surfactantes en la biorremediación de suelos contaminados con hidrocarburos. Rev Quim Viva 3:120–145

Robles-González I, Ríos-Leal E, Ferrera-Cerrato R et al (2006) Bioremediation of a mineral soil with high contents of clay and organic matter contaminated with herbicide 2,4-dichlorophenoxyacetic acid using slurry bioreactors: Effect of electron acceptor and supplementation with an organic carbon source. Proc Biochem 41:1951–1960

Rosen MJ, Kunjappu JT (2012) Surfactants and interfacial phenomena. Wiley, Hoboken, NJ, 600

Rubilar O, Tortella G, Cea M et al (2011) Bioremediation of a Chilean Andisol contaminated with pentachlorophenol (PCP) by solid substrate cultures of white rot fungi. Biodegradation 22:31–41

Satpute SK, Banat IM, Dhakephalkar PK et al (2010) Biosurfactants, bioemulsifiers and exopolysaccharides from marine microorganisms. Biotechnol Adv 28:436–450

Schippers C, Gessner K, Müller T et al (2000) Microbial degradation of phenanthrene by addition of a sophorolipid mixture. J Biotechnol 83:189–198

Sekhon KK, Khanna S, Cameotra SS (2011) Enhanced biosurfactant production through cloning of three genes and role of esterase in biosurfactant release. Microb Cell Fact 10:49

SEMARNAT (2012) Secretaría de Medio Ambiente y Recursos Naturales. Base de datos estadísticos. Producción de insecticidas y plaguicidas. http://dgeiawf.semarnat.gob.mx:8080/ibi_apps/WFServlet?IBIF_ex=D2_AGRIGAN05_04&IBIC_user=dgeia_mce&IBIC_pass=dgeia_mce. Accessed Sept 2013

Semple KT, Doick KJ, Jones KC, Burauel P et al (2004) Defining bioavailability and bioaccessibility of contaminated soil and sediment is somplicated (view point). Environ Sci Tech 15: 229–231

Shelton DR, Doherty MA (1997) A model describing pesticide bioavailability and biodegradation in soil. Soil Sci Soc Am J 61:1078–1084

Shukla KP, Singh NK, Sharma S (2010) Bioremediation: developments, current practices and perspectives. Gen Eng Biotechnol J 3:1–20

Singh BK, Walker A (2006) Microbial degradation of organophosphorus compounds. FEMS Microbiol Rev 30:428–471

Singh N, Megharaj M, Gates WPJ (2003) Bioavailability of an organophosphorus pesticide, fenamiphos, sorbed on an organo clay. J Agric Food Chem 51:2653–2658

Singh N, Sethunathan N, Megharaj M (2008) Bioavailability of sorbed pesticides to bacteria: an overview. In: Naidu R et al (eds) Chemical bioavailability in terrestrial environment. Developments in soil science series, vol 32. Elsevier, Amsterdam, pp 73–82

Smulders E, von Rybinski W, Sung E et al (2001) Laundry detergents. Ullmann's encyclopedia of industrial chemistry. Wiley, Weinheim, pp 38–98

Suyala A, Chauhana SS, Srivastava A et al (2013) Adsorption-desorption behaviour of chlorimuron-ethyl herbicide on homoionic clays. Eur J Soil Sci 2:28–34

Tang WC, White JC, Alexander M (1998) Utilization of sorbed compounds by microorganisms specifically isolated for that purpose. Appl Microbiol Biotechnol 49:117–121

Tecon R, Van der Meer JR (2010) Effect of two types of biosurfactants on phenanthrene availability to the bacterial bioreporter *Burkholderia sartisoli* strain RP037. Appl Microbiol Biotechnol 85:1131–1139

Tejeda M, Gómez I, Del Toro M (2011) Use of organic amendments as a bioremediation strategy to reduce the bioavailability of chlorpyrifos insecticide in soils. Effects on soil biology. Ecotoxicol Environ Saf 74:2075–2081

Tiehm A, Stieber M, Werner P et al (1997) Surfactant-enhanced mobilization and biodegradation of polycyclic aromatic hydrocarbons in manufactured gas plant soil. Environ Sci Technol 31:2570–2576

Van Hamme JD, Singh A, Ward OP (2003) Recent advances in petroleum microbiology. Microbiol Mol Biol Rev 67:503–549

Vasseur P, Bonnard M, Palais F et al (2008) Bioavailability of chemical pollutants in contaminated soils and pitfalls of chemical analyses in hazard assessment. Environ Toxicol 23:652–656

Vidali M (2001) Bioremediation. An overview. Pure Appl Chem 73:1163–1172

Villavicencio M, Pérez B (2010) Plantas tradicionalmente usadas como plaguicidas en el Estado de Hidalgo, México. Polibotánica 30:193–238

Volkering F, Breure AM, Rulkens WH (1998) Microbiological aspects of surfactant use for biological soil remediation. Biodegradation 8:401–417

Volke-Sepúlveda T, Velasco-Trejo JA (2002) Tecnologías de remediación para suelos contaminados. INE-SEMARNAT, México

Wattanaphon HT, Kerdsin A, Thammacharoen C et al (2008) A biosurfactant from Burkholderia cenocepacia BSP3 and its enhancement of pesticide solubilization. J Appl Microbiol 105:416–423

Wiegel J, Wu Q (2000) Microbial reductive dehalogenation of polychlorinated biphenyls. FEMS Microbiol Ecol 32:1–15

Worrall F, Fernandez-Perez M, Johnson AC et al (2001) Limitations on the role of incorporated organic matter in reducing pesticide leaching. J Contam Hydrol 49:241–262

Yáñez-Ocampo G, Sanchez-Salinas E, Jimenez-Tobon GA et al (2009) Removal of two organophosphate pesticides by a bacterial consortium immobilized in alginate or tezontle. J Hazard Mater 168:1554–1561

Yáñez-Ocampo G, Sánchez-Salinas E, Ortiz-Hernández ML (2011) Removal of methyl parathion and tetrachlorvinphos by a bacterial consortium immobilized on tezontle-packed up-flow reactor. Biodegradation 22:1203–1213

Yeomans J, Carrillo E, Ruiz A (2004) Biodegradación de Plaguicidas. BIOtecnia (Sonora, México) 6:3–62

Yin J, Qiang Y, Jia Y et al (2008) Characteristics of biosurfactant produced by *Pseudomona aeruginosa* S6 from oil containing wastewater. Proc Biochem 44:302–308

Zhang L, Hong S, Zhou L et al (2003) Fate and assessment of persistent organic pollutants in water and sediment from Minjian River Estuary, Southeast China. Chemosphere 52:1423–1430

Zhang C, Wang S, Yan Y (2011) Isomerization and biodegradation of beta-cypermethrin by Pseudomonas aeruginosa CH7 with biosurfactant production. Biores Technol 102:7139–7146

Zheng G, Wong JW (2010) Application of microemulsion to remediate organochlorine pesticides contaminated soils. Proc Annu Int Conf Soils Sediment Water Energ 15:21–35

Chapter 6
Bioremediation and Biotransformation of Carbon Nanostructures Through Enzymatic and Microbial Systems

Nelson Durán, Amauri J. Paula, Diego Stéfani T. Martinez, and Amedea B. Seabra

Abstract Nanomedicine, environmental sciences, waste water and soil technologies intensively use many carbon nanostructures, such as carbon nanotubes, graphenes, and fullerenes. Possibly, due to the several perspectives of use of these nanomaterials, approaches on toxicology and safety management have become the focus of intense interest, as the industrial production of these materials has grown enormously in the last few years; besides that short- and long-term behaviors are not yet fully understood. Our concerns involving these carbon-based nanomaterials are their stability and potential effects of their life cycles on environment. Following this focus, this review discusses the literature related to the biodegradability of these nanomaterials, mainly through enzymes, microorganisms and cells, in order to understand the actual status, and contributes to the uses of biocompatible and biodegradable functionalized carbon nanostructures. Moreover, this review address crucial aspects towards the use of these nanomaterials with reduced impact for animals and environment.

N. Durán (✉)
Institute of Chemistry, Biological Chemistry Laboratory – NanoBioss, Universidade Estadual de Campinas, CP 6154, Campinas, CEP 13083-970, São Paulo, SP, Brazil

Center of Natural and Human Sciences, Universidade Federal do ABC, São Paulo, SP, Brazil
e-mail: duran@iqm.unicamp.br

A.J. Paula
Department of Physics, Universidade Federal do Ceará, Fortaleza, CE, Brazil

D.S.T. Martinez
Brazilian National Nanotecnology Laboratory (LNNano), National Center for Research on Energy and Materials (CNPEM), Campus CNPEM, Campinas, São Paulo, SP, Brazil

A.B. Seabra
Exact and Earth Sciences Department, Universidade Federal de São Paulo, Diadema, SP, Brazil

A. Alvarez, M.A. Polti (eds.), *Bioremediation in Latin America: Current Research and Perspectives*, DOI 10.1007/978-3-319-05738-5_6

6.1 Introduction

In the last years, peroxidases, mainly from plants, appeared as important enzymes to degrade carbon nanotubes and graphenes (Seabra et al. 2013). Indeed, peroxidases have been extensively studied and used in many applications (Ryan et al. 2006; Battistuzzi et al. 2010; Zakharova et al. 2011), such as on bioremediation (O'Brien et al. 1978; Durán et al. 1981, 1988, 2002; Husain and Ulber 2011), biocatalysis (Bromberg and Durán 2001; Uyama et al. 2002; Xu et al. 2005; Kadokawa and Kobayashi 2010) and biosensors (Azevedo et al. 2003; Fornera et al. 2011; Kosman and Juskowiak 2011). Oxidases, such as laccase, were also investigated in different types of applications (Durán et al. 2002; Minussi et al. 2007; Witayakran and Ragauskas 2009; Gitsov et al. 2008a, b; Mita et al. 2003; Jeon et al. 2012; Margot et al. 2013).

This chapter discusses the interactions of peroxidases and oxidases, free or within cells, with nanostructures at the nanotechnological level. Interactions of nanomaterials with proteins and biomolecules, in particular the manifestation of the nanoparticle-protein corona effect, are an important issue in the field of biotransformations (Asuri et al. 2007; Saptarshi et al. 2013). In this context, the actual status of relevant aspects of peroxidases and laccases on the biodegradation of carbon-related nanomaterials will be summarized and discussed. In particular, the importance of these enzymes on the degradation of carbon nanotubes (CNTs), graphene (Fisher et al. 2012a, b; Shen et al. 2012; Bao et al. 2012; Sanchez et al. 2012; Liu et al. 2012; Seabra et al. 2013) and fullerenes (Chawla et al. 2010; Johnston et al. 2010; Seabra et al. 2013; Yue et al. 2013) will be presented. In addition, data related to microorganisms and cell interactions with carbon nanostructures at the nanotechnological level will be discussed.

6.2 Peroxidase Biodegradation of Carbon Nanotubes

Single carbon nanotubes (SWCNTs) have been elected as a revolutionary nanomaterial for biological systems due to important applications involving biosensing and drug delivery (Iijima 1991; Rajavel et al. 2012; Lee et al. 2012; Rolfe 2012; Yamashita et al. 2012; Li et al. 2012). Not only SWCNTs but also the multi wall carbon nanotubes (MWCNTs) motivated the researchers to act proactively towards the evaluation of the toxicity of these materials. The concerns consider the diversity of carbon nanotube morphologies as well as different kinds of functionalization which may exert several effects on eco-environment (Petersen and Henry 2012; Yang et al. 2012).

It is known that in biological medium, CNTs may have harmful effects, such as inflammation responses, oxidative stress, accumulation as aggregates, etc. However, we have to be aware that the reported toxic effects of carbon nanotubes are mainly related to the presence of impurities rather than to the carbon nanotubes themselves (Cellot et al. 2010). This is an indication of the necessity of using

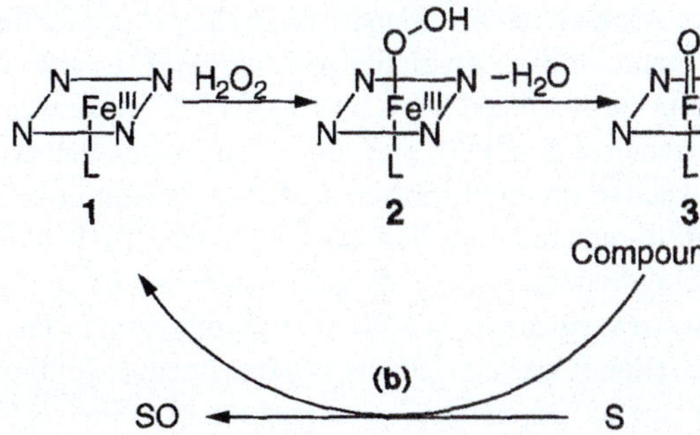

Fig. 6.1 Mechanistic representation of CNTs degradation by HRP at low H_2O_2 concentration. (Reproduced by permission of Elsevier, van Rantwijk and Sheldon 2000) *S = SWCNTs and SO = HO-SWCNTs*

efficient purification methods for these materials before their applications in biological systems (Umbuzeiro et al. 2011).

SWCNTs underwent degradation by horseradish peroxidase (HRP) at a low hydrogen peroxide concentration at pH 7 during 8 weeks, and the average length of carbon nanotubes decreased enormously after 16 weeks of incubation. This was the first indication that CNTs can be degraded in the environment (Allen et al. 2008). Several methods were used to corroborate these results and the mechanisms of the biodegradation process are schematically represented in Eq. 6.1 (van Rantwijk and Sheldon 2000).

Allen et al. (2009) followed several strategies to study the differences between functionalyzed and pristine SWCNTs degraded by HRP, and compared with the chemical degradation, as a model peroxidase (hemin and iron chloride). Pristine SWCNTs were not affected by HRP incubation, though showing significant degradation with either hemin or iron chloride. The authors explained these results by considering that upon treatment of SWCNTs with HRP/H_2O_2, a heterolytic cleavage of hydrogen peroxide with HRP occurred and probably pristine SWCNTs did not undergo any transformation. However, Fenton catalysis resulted in a homolytic cleavage of hydrogen peroxide that oxidizes pristine SWCNTs. Then, hydrogen peroxide oxidized Fe^{2+} to Fe^{3+}, producing hydroxyl radical, and Fe^{3+} is reduced back to Fe^{2+} by additional hydrogen peroxide, producing a peroxy radical, according to the equations below (Seabra et al. 2013):

$$Fe^{2+} + H_2O_2 \rightarrow Fe^{3+} + \cdot OH^{+} + {}^{-}OH \quad (6.1)$$

$$Fe^{3+} + H_2O_2 \rightarrow Fe^{2+} + \cdot OOH + H^{+} \quad (6.2)$$

The authors supported the hypothesis that, even considering that HRP-Compound I (Fig. 6.1) and hydroxyl/hydroperoxyl radicals are important as oxidizing species, the capacity of HRP-Compound I to degrade SWCNTs probably depends on the proximity to the active heme site. Radicals reactivity were limited only by the diffusion rate and their half-lives (Fig. 6.1).

A study based on the comparison of the potential effects of different peroxidases (HRP, hemoglobin, myeloperoxidase (MPO), and lactoperoxidase (LPO)) in degrading carboxylated SWCNTs (c-SWCNTs) showed a minor c-SWCNTs degradation (Allen et al. 2008). The authors suggested that possibly the destruction of the enzyme active site in the presence of H_2O_2, a Fenton-like reaction, could explain the peroxidase-induced degradation of the SWCNTs. In comparison the ability of MPO (producing hypochlorite) to degrade CNTs and LPO (producing hypobromite) were found to be very active on SWCNTs degradation (Vlasova et al. 2011a).

In parallel, it was carried out an experiment to mimic the phagosome actions and inflammatory sites, in which a suspension of c-SWCNTs in blood plasma was incubated with high concentrations of MPO/H_2O_2 or hypochlorite. The results showed an efficient degradation of CNTs. Vlasova et al. (2011b) hypothesized that hypochlorite is a principal candidate for the oxidative degradation of CNTs in vivo.

The investigation of the ability of HRP and hydrogen peroxide (H_2O_2) to degrade chemically functionalized MWCNTs showed that HRP/H_2O_2 incubated with o-MWCNTs (oxidation for 5 h) and o-MWCNTs (oxidation for 8 h) shortened the sizes of CNTs in both suspensions, as indicated by the light scattering and absorbance. A layer of debris (carbonaceous sheets) was observed after 60 days and the less degraded material was found to be p-MWCNTs (pristine). In the case of nitrogen-doped MWCNTs, the authors observed a total degradation by the HRP/H_2O_2 system. On bases of these data, the authors suggested a layer-by-layer degradation mechanism, inducing a complete degradation of the CNTs (Zhao et al. 2011a, b).

Recently, a study comparing the biodegradation of both SWCNTs and MWCNTs using two different oxidative conditions has been published (Russier et al. 2011). Firstly, the oxidized SWCNTs and MWCNTs were submitted to the degradation action of an enzyme-free reaction buffer, corresponding to the phagolysosomal simulant fluid (PSF), and to the HRP/H_2O_2 system. SWCNTs were efficiently degraded by PSF (30 days of incubation) as well as by HRP/H_2O_2, as previously showed by reports (Allen et al. 2008, 2009; Kagan et al. 2010). In the case of oxidized MWCNTs, transmission electron microscopy (TEM) analysis revealed changes in the morphological structure of MWCNTs, mainly due to the incubation with the HRP system. This result can be explained by considering that the material surface exhibited defects. This is an indication that the carbon nanotube degradation mechanism is different between SWCNTs and MWCNTs. MWCNTs appeared extensively shortened rather than unzipped, besides that some CNTs were peeled off showing small pieces and debris fragments (Russier et al. 2011; Bianco et al. 2011).

Taken together, it can be summarized that both SWCNTs and MWCNTs are degraded in the presence of the HRP/H_2O_2 system, and other peroxidases such as myelo- and lacto- or a model peroxidase system (hemin); Fenton reagent ($FeCl_3$) or even in simulated biological fluids. From these data it is possible to visualize two mechanistic schemes. One that involved the redox cycle of peroxidase and the second based on morphological changes during peroxidation of CNTs. It is important that these facts demonstrate CNTs degradation by enzymes, and this probably could minimize the harmful effects of these nanomaterials on the environment (Seabra et al. 2013).

6.3 Biodegradation of Carbon Nanotubes: In Vivo and In Vitro

By using SWCNTs capped with phosphatidylserine, it was possible to measure the signal of the initiation of CNTs digestion by macrophages, primary monocytes, dendritic cells, and microglia (Konduru et al. 2009). Alveolar macrophages in vivo showed a similar behavior. This strategy probably improves the targeting and uptake of CNTs by phagocytes and can be beneficial to induce intracellular degradation of functionalized CNTs (Konduru et al. 2009; Stern et al. 2012). Lysosome is an important organelle, since it is considered the intracellular site of nanomaterials interaction and degradation (Stern et al. 2012). As discussed before, important advances towards the understanding of CNTs biotransformation was achieved by Kagan et al. (2010) and Fadeel et al. (2011) by studying the possibility of peroxidase intermediates inside human cells or fluids to be involved in the degradation of SWCNTs. Besides the reactive radical intermediates from an oxidant (hypochlorous acid) and also from the human neutrophil myeloperoxidase enzyme catalyzed the in vitro biodegradation of SWCNTs.

Kagan et al. (2010) showed that two important intermediates such as hypochlorite and reactive radical intermediates of the human neutrophil enzyme myeloperoxidase (hMPO) catalyze the biodegradation of single-walled carbon nanotubes in vitro, in neutrophils and, in a minor degree, in macrophages. Molecular calculations suggested that occur strong interactions betweem amino acids (basics) of the enzyme and the carboxyls on the CNTs sites near the catalytic group. One important site was located at the proximal end of the haem group, related to the tyrosine residues 293 and 313 (known to be catalytically active in hMPO, Fig. 6.2a, left). Microscopic measurements demonstrated the attachment of hMPO to the surface of the nanotubes (Fig. 6.2d). The other binding site close to the distal end of the haem group, far away from catalytically competent tyrosine residues 293 and 313 (Fig. 6.2a, right), may be involved in the binding of pristine CNTs. It is important to point out that pristine nanotubes underwent less effective biodegradation by hMPO. Indeed, pristine nanotubes have shown to be susceptible to HOCl oxidation. However, the short-CNTs were similarly susceptible to oxidation, indicating a nonspecific reaction of hypochlorite with both types of CNTs. Kagan et al. (2010) speculate that HOCl might be essential for triggering and accumulating carboxylated sites on pristine nanotubes, which permits a more effective positioning of hMPO and posterior biodegradation of the tubes by enzymes reactive intermediates (Fig. 6.2).

Recently, it has been attempted to explain enzymatic oxidation in terms of oxi-reduction potentials using chloride, which is a substrate for MPO to produce hypochlorite and antioxidants, such as L-glutathione and ascorbic acid, that have lower redox potentials than CNTs. These two antioxidants, with or without chloride, significantly inhibited MPO-catalyzed biodegradation of carboxylated-SWCNTs (Kotchey et al. 2013).

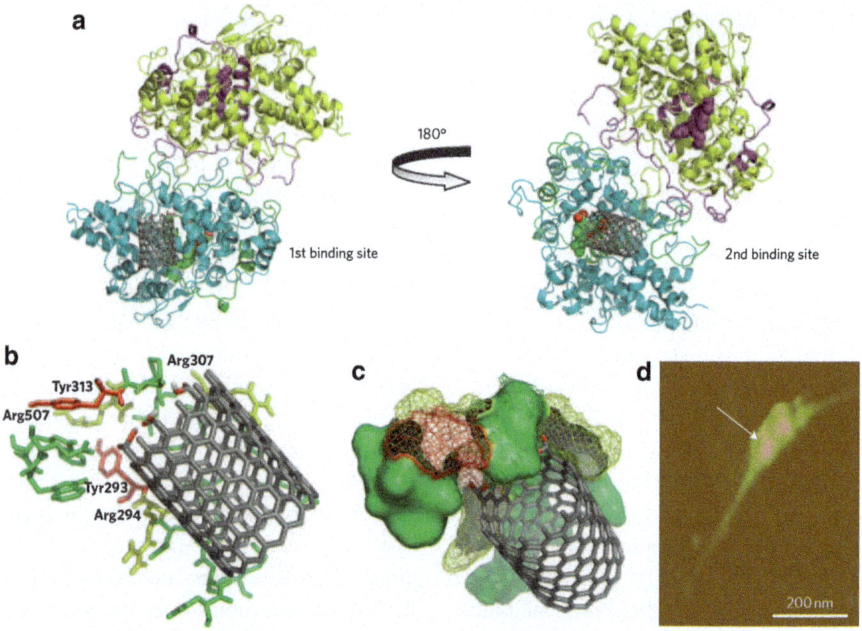

Fig. 6.2 Modeling calculations demonstrating possible nanotube interaction sites on hMPO. (**a**) Two putative binding sites for nanotubes on the hMPO monomer surface are located on each side of the protein. The best model site with lowest binding energy is shown for carboxylated short-cut nanotubes (*left*) and pristine nanotubes (*right*). (**b**) Stabilization of the carboxylated ends of the nanotubes by arginine (*yellow*) and tyrosine residues (*red*) in hMPO in the first binding pocket (*left*). All residues shown are within 5 Å of nanotube and are predicted to participate in the degradative catalysis of nanotubes. This includes Tyr293, Arg294, Arg307, Tyr313, and Arg507. (**b**) and (**c**), using space-fill representation. (**d**) AFM image confirming the binding of a dimer of hMPO with a single nanotube. (Reproduced by permission of Nature Publishing Group, from Kagan et al. 2010)

As it was not observed toxicity for biodegraded SWCNTs, it has been suggested that the degree of degradation is important to define the severity of the associated inflammatory responses observed after exposition to these nanomaterials.

Liu et al. (2010) described the bio-persistence of functionalized SWCNTs evaluated in vitro with simulated phagolysosomal conditions. These conditions are important, since macrophages generate superoxide anion, hydrogen peroxide, and hydroxyl radical, which contribute to microbial killing in the phagolysosome. They observed that carboxylic functionalities and defects on the surface of SWCNTs increase the biodegradation of the CNTs. In contrast, unfunctionalized and other surface functionalized CNTs did not suffer morphological variations.

Cellular degradation of double-walled CNTs (DWCNTs) showed evidences of intracellular modifications of the CNTs. Defects were firstly accumulated on the outer surface of the CNTs after their internalization by human prostate adenocarcinoma (PC-3) or by human cervical carcinomas (HeLa) cells, while DWCNTs inner surface chemical structure was preserved up to several hours of incubation (Neves et al. 2010).

The biomedical uses of CNTs, such as in neurological applications on components of implants, electrodes or as delivery vehicles are important issues (Seabra et al. 2013). Amino-functionalized CNTs (MWCNTs-NH_3^+) were administrated to the mouse brain cortex and monitored. Between 2 and 14 days after the cortical administration, the authors reported severe CNTs structural deformation leading to partial degradation of MWCNTs-NH_3^+ in vivo. Internalization within microglia was observed at the initial periods after MWCNTs-NH_3^+ application (Nunes et al. 2012). The cellular internalization of MWCNTs-NH_3^+ by phagocytic RAW 264.7 murine macrophages and nonphagocytic A459 human lung carcinoma cells were influenced by the ability of CNTs to cross cellular membranes. In a short time, free CNTs were detected in the cytoplasm after their incubation with cells, meaning that MWCNTs-NH_3^+ were able to escape from phagosomes, avoiding their elimination by endo-lysosomal or/and phago-lysosomal vesicles. Wrapped CNTs formation into endosome-like structures after 12 h was observed, and this is probably due to the formation of endolysosomal vesicles, which leads to the interaction of internalized CNTs with lysosomal enzymes, allowing an enzymatic degradation (Lacerda et al. 2012).

In vivo studies using myeloperoxidase knockout B6.129X1-MPO (MPO k/o) and wild-type C57Bl/6 (w/t) mice exposed to SWCNTs were reported by Kagan et al. 2010; Shvedova et al. 2012). After 28 days of exposure, it was observed significant differences between SWCNTs degradation in wild type mice compared to MPO knockout animals (Shvedova et al. 2012).

Besides this, it was shown that the oxidation and clearance of SWCNTs from the lungs of these animals (MPO k/o) after pharyngeal aspiration were significantly less effective, whereas the inflammatory response was more consistent in comparison to wild-type C57Bl/6 mice. The authors suggested new approaches to control the bio-persistence of nanomaterials through pharmacological manipulations, and maybe also by genetic management (Shvedova et al. 2012).

By comparing two types of CNTs, such as SWCNTs and MWCNTs, it was demonstrated that MWCNTs were found to aggregate within living tissues, in contrast, smaller particles consisting of SWCNTs were easily phagocytized by macrophages and transported to local lymph nodes. The authors showed that metabolic processes in the fibers regeneration, by histoenzymatic staining and the detection of oxidative enzymes confirmed their histological observations (Fraczek et al. 2008). These results indicate that in the case of SWCNTs in the tissues, CNTs underwent oxidative degradation with almost no aggregation, leading to cell penetrations.

Acid-functionalized SWCNTs (AF-SWCNTs) on the immune system, on primarily cultured murine peritoneal macrophages and purified splenic T cells did not show any significant cytotoxixity acting on macrophages. However, analyses of the mitochondrial membrane potential and proteasome subunit gene expression showed that AF-SWCNTs damaged mitochondrial function and proteasome formation, in a concentration-dependent manner. However, AF-SWCNTs at low concentrations inhibited the phagocytic efficiency of latex beads in macrophages and biased naıve T-cell differentiation to Th1 types (IFN-c and TNF induction), implying the risk of Th1-associated diseases. The lengths of AF-SWCNTs found in lysosomes/exhibited a sixfold shorterting of the size. This is in accordance with the biodegradation of carbon nanotubes by myeloperoxidase (MPO) in neutrophils and macrophages, as reported by Kagan et al. (2010). Dong et al. (2013) suggested that the shorterting of carbon nanotube in macrophages may be a result of a selective uptake of AF-SWCNTs, since size-dependent cellular uptake and expulsion of SWCNTs occurred, as previously demonstrated in NIH-3T3 cells (Jin et al. 2009). Biodegradation of CNTs is, in part, a defense mechanism which could mitigate potential toxic effects of these nanomaterials.

All the facts from in vitro studies indicate that CNTs are susceptible to enzymatic biodegradation. The study of interactions of carbon nanotubes with immune-component cells that focused on cellular recognition of nanotubes and their enzymatic degradation is one of the first steps to evaluate the safe applications of these nanostructures in nanomedicine.

It is known that cell internalization processes are possible through endocytosis, which is the most common way of transporting nano-objects into cells by enclosing them in vesicles or vacuoles formed by the cytoplasmic membrane, and comprises three mechanisms which are: phagocytosis, pinocytosis, and caveolae-dependent or clathrin-mediated endocytosis (Nel et al. 2009). There are several correlations with morphological and chemical characteristics of CNTs that described the occurrence of specific mechanisms of endocytosis. However, up to now, much data reported from different groups are sometimes inconsistent (Zhao et al. 2011a, b) (Fig. 6.3). Although knowing that the subcellular localization of CNTs also depends on the

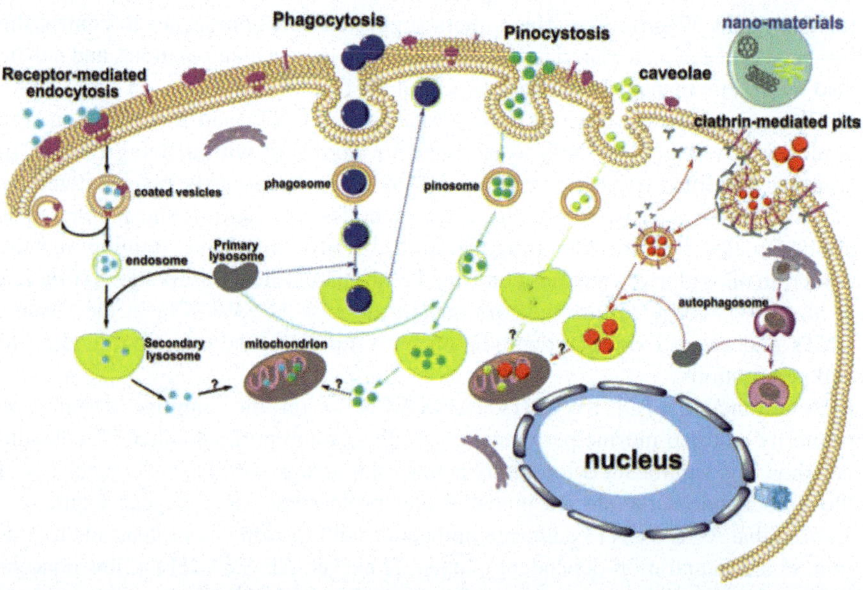

Fig. 6.3 Schematic of the known pathways for intracellular uptake of nanoparticles. (Reproduced by permission of John Wiley and Sons WILEY-VCH Verlag GmbH & Co. KGaA, Weinheim, from Zhao et al. 2011b)

internalization pathway, morphology and physicochemical characteristics (Fisher et al. 2012a, b), the diffusion of these nanomaterials in biological systems is now an object of debate by the scientific community.

Andon et al. (2013) described the interaction of one oxidized SWCNTs with two molecules of eosinophil peroxidase (EPO), one of the major human oxidant generating enzymes and showed that human EPO (in vitro) and murine EPO of ex vivo activated eosinophils catalyze the oxidative biodegradation of SWCNTs. Through a molecular modeling it was found two binding sites for SWCNTs on EPO, possessing the same side as the catalytic site and located on the distal side of EPO. This report is important in terms of understanding the type of interactions, since eosinophils are key players of the innate immune system and these results are relevant to potential respiratory exposures to carbon nanotubes.

Bussy et al. (2013) provided the first evidence of SWCNTs biological transformation inside macrophages. This process was governed by pH acidification, inducing the detachment of catalyst nanoparticles (iron nanoparticles) from CNTs backbones. These results demonstrated that although iron catalyst residues were initially isolated from biological media due to their surrounding carbon shell, they are more available to the biological system after the uptake and residence of CNTs bundles in acidic cell compartment.

6.4 Enzymatic Degradation, Microbial Biodegradation and Biotransformation of Graphenes

Among the carbon nanostructures, graphene (single atomic layer of sp^2-hybridized carbon) has been rapidly inserted in the context of nanobiotechnology in different fields of application, such as a novel biosensing platform for the detection of biological molecules (Chan et al. 2009). However, unfortunately, graphene has been relatively less explored in comparison with CNTs (Novoselov et al. 2004). In the toxicological aspects of graphene, many experimental protocols which have been used for CNTs are now being applied to graphene. Following this aspects, an article evaluated the capacity of HRP to degrade graphene-based nanomaterials. Kotchey et al. (2011) using graphene oxide, which was obtained from graphite oxide, observed that at low hydrogen peroxide concentration, HRP catalyzed the oxidation of graphene oxide, resulting in the formation of holes in its basal plane. An interesting observation was that the reduced graphene was not affected by HRP. Theoretical studies showed that HRP was preferentially bound to the basal plane rather than the edge of both graphene oxide and reduced graphene, being HRP localized closer to graphene oxide compared to its reduced form. This fact explains the high susceptibility of graphene oxide to suffer carbon–carbon bond cleavage, facilitating the oxidation at its basal planes (Kotchey et al. 2011). A green approach for the reduction of graphene oxide was also studied on bilayer graphene oxide which was synthesized by a modified Hummers method (Hummers and Offeman 1958) from natural graphite. This approach used wild carrot root with endophytic microorganisms present, and it was able to reduce exfoliated graphene oxide to graphene (Kuila et al. 2012).

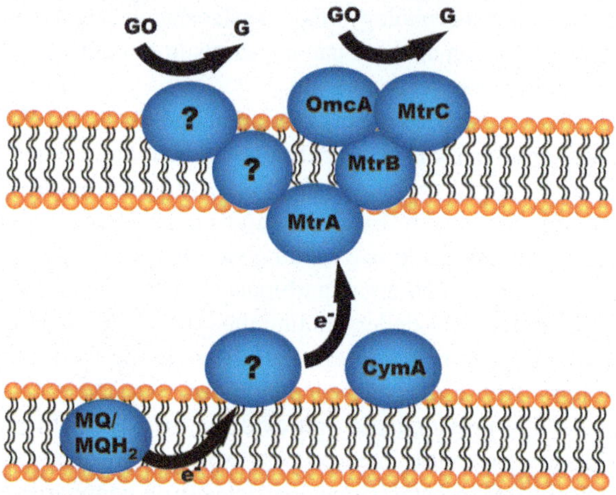

Fig. 6.4 MtrA mediates electron transfer from the inner membrane to the outer membrane in the case of GO reduction. MtrA, MtrB, MtrC/OmcA, and CymA are known to be involved in the reduction of solid materials, such as iron and manganese oxides. Data presented indicates that, in the case of GO reduction, *Shewanella* does not use CymA to mediate electron flow from the quinone pool to the periplasmic cytochrome, MtrA. Additionally, other outer membrane cytochromes may be involved in GO reduction. (Reproduced by permission of American Chemical Society, from Salas et al. 2010)

Besides this, graphene oxide can act as a terminal electron acceptor for heterotrophic, metal-reducing, environmental bacteria (*Shewanella*) (Salas et al. 2010). The electron transfer in this case was mediated by the cytochromes MtrA, MtrB, and MtrC/OmcA. However, mutants lacking CymA, another cytochrome associated with the extracellular electron transfer, conserved the ability to reduce graphene oxide (Fig. 6.4).

All of these facts demonstrated that the biotransformation of graphene oxide can occur under ambient conditions, in a relatively rapid kinetics. The capacity of microorganisms to convert graphene oxide to graphene exhibits a positive prospect for the bioprocessing of graphene (Salas et al. 2010).

A study dealing with an important issue such as biodegradability of graphene under in vivo conditions (lateral size ~200 nm) has recently been discussed (Girish et al. 2013). Raman signatures of graphene to three dimensional (3D) image was used for the localization in lung, liver, kidney and spleen of mouse and identified gradual development of structural disorder. After 24 h postinjection, larger aggregates of sizes up to 10 μm was detected in various organs. Graphene was biodegraded and phagocytized by tissue-bound macrophages, and the gene expression studies of pro-inflammatory cytokines suggested the possibility of phagocytic immune responses. The in vitro studies conducted on macrophage cell lines demonstrated the structural disorder in the entrapped graphene, corroborating the role of macrophages in biodegradation. Girish et al. (2013) stated that this is the first report that provided evidences of in vivo biodegradation of graphene and probably these results will promote a great change in the perspective on potential biological applications of graphenes.

It is clear from all of these results that peroxidases are able to transform only graphene oxide but not graphene. However, this fact does not exclude the possibility of new research indicating the graphene biodegradation in vivo. Theoretical calculations showed that the HRP binding to the basal plane of the graphene structure is closer for graphene oxide rather than for its reduced form, thus facilitating the oxidation of graphene oxide. At this moment, it is important to point out that reduction of graphene oxide by many biological organisms and environmental moiety is an important advance to understand the possible behavior of this material in the environment (Seabra et al. 2013).

6.5 Biodegradation and Biotransformation of Fullerenes: In Vitro and In Vivo

Fullerenes (termed C60, buckminsterfullerene or buckyballs) have a spherical cage-like structure and exhibits small diameter (~1 nm). Naturally produced fullerene can be released during combustion processes such as forest fires. C60-fullerenes production was expanded from 1990 (Kratschmer et al. 1990), after 5 years of their discovery (Kroto et al. 1985) and is increasing with the development of nanotechnology.

Fullerenes have potential for cancer treatment due to their ability to trap singlet oxygen species resulting in an efficient antioxidant agent (Gitsov et al. 2012). In parallel, several toxicological concerns regarding this nanostructure were raised once it was reported that pristine C60 could have a long biological half-life (Nielsen et al. 2008). Then, the biotransformation of fullerene and its derivatives by enzymes was studied. Laccase was used in the biotransformation of fullerene, as shown in Fig. 6.5.

Laccase complex systems (such as dendritic copolymers/laccase/mediator N-hydroxy-5-norbornene-2,3-dicarboxylic acid imide/1-hydroxybenzotriazole or 2,20-azinobis (3-ethylbenzothiazoline-6-sulfonic acid)/1-hydroxybenzotriazole) biotransformed fullerene into epoxide- and hydroxyl-derivatives under mild and environmental friendly reaction conditions (Gitsov et al. 2012).

Macrophage-like cells (TIP-1 cells) treated with phorbol 12-myristate 13-acetate were exposed to fullerene nanowhiskers (C60 NWs), being that over 70 % of these nanoparticles (C60 NWs) were internalized after 48 h of exposure. After this long term coculture, a degradation of C60 NWs was observed in the cells and the number of short NWs increased after the exposition. Nudejima et al. (2009, 2010) showed that macrophages may be able to decompose C60 NWs into C60 molecules, as the primary immune response. Due to dissolution recrystallization processes, during the coculture or probably the enzymatic treatment, it was observed the presence of granular substances after the interaction with fullerene whiskers.

In another study conducted by Okuda-Shimazaki et al. (2010) it was shown that C60 NWs exhibited very weak cytotoxic activity and induction of gene expression on human acute monocytic leukemia cell line THP-1, as compared to MWCNTs.

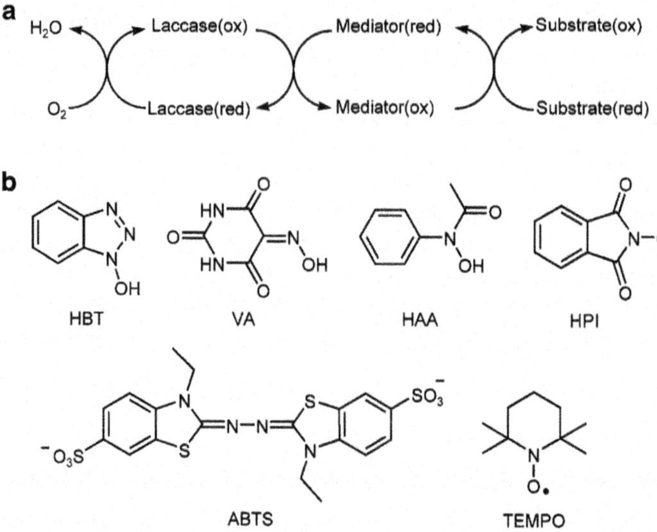

Fig. 6.5 (**a**) Schematic representation of laccase-mediator redox cycle; (**b**) chemical structures of laccase mediators. (Reproduced from Witayakran and Ragauskas 2009, by permission of WILEY-VCH Verlag GmbH & Co. KGaA, Weinheim)

In vitro cytotoxicity of fullerenes towards human derived macrophages (monocytes) was evaluated and no cytotoxic effect was observed when using a C60 dispersion in THF. The dispersion presented single and poly crystals of varied shapes and sizes, and these crystals aggregate within lysosomes due to the internalization by phagocytosis or endocytosis, in which C60 was degraded into smaller structures (Porter et al. 2006). These data showed that aggregation and disaggregation phenomena that occur within the cells may be ruled by some enzymatic processes.

Schreiner et al. (2009) observed that two white rot basidiomycete fungi (*Phlebia tremellosa* and *Trametes versicolor*) metabolized and degraded C60-fullerol (hydroxyl-functionalized). These fungi produce oxidative enzymes, such as lignin peroxidase, manganese peroxidase and laccases (Durán et al. 2000, 2002) that normally act against lignin as well as polycyclic aromatic compounds (Marques-Rocha et al. 2000; Gibson et al. 2005; Chen et al. 2012). Both species after 32 weeks could bleach fullerol as well as oxidize a small portion to CO_2, and incorporate it into fungal biomass (Schreiner et al. 2009).

Mice (p53-heterozygos (p53(+/−)) were submitted to intraperitoneal injection of fullerenes and it was observed small brownish black plaques on the serosal surface. In this study, histological analyses demonstrated that plaques contained polygonal clefts and lacunae surrounded by a thin layer of foamy cells and separated by thin fibrous septa. Clefts/lacunae morphology corresponded to the injected fullerene aggregates in both size and shape. It was also noted that the edge of the clefts was brown pigmented, indicating a possible biodegradation of C60 particles by phagocytic cells, blending proteins and/or other organic components (Takagi et al. 2008).

Gharbi et al. (2005) showed that intraperitoneal administration of fullerenes in rats led to a maximum accumulation, which reached about 24 % of the injected amount during the first week following the injection. By measuring the concentration during 21 days, the concentrations decreased around 1 % of the values measured at day 7, indicating that C60 could be eliminated and probably also transformed by the rat livers.

In conclusion, fullerenes whiskers acting on human acute monocytic leukemia cell line (THP-1) could decompose into short fullerenes nanoparticles via dissolution-recrystallization during a long-term coculture or enzymatic treatment, decreasing their cytotoxic effects. Regarding the degradation of fullerenes by monocytes, similar observations were reported.

In the case of in vivo experiments with C60, a possible biodegradation by phagocytic cells, enzymes or other organic components, as well as transformations in the liver are reported. Laccase action, as reported to be the main fullerene degrading enzyme, indicates the occurrence of multi-oxidation processes in the presence of mediators. Another aspect to be pointed is the efficient bleaching and oxidation of fullerol to CO_2 by fungi. This is an extremely important result from the environmental point of view.

6.6 Morphological and Surface Chemical Aspects on the Biotransformation

Fullerene has a characteristic molecular behavior resulted from its defined morphology and chemical structure. On the other hand as previously discussed, CNTs and graphene possess important size and morphological differences, as well a surface microchemical environment extremely complex, a result from the synthetic process used in their production and/or from the chemical reactions performed during the functionalization. Therefore, nanotoxicology must consider these features in its protocols and models which will evaluate the biological effect of CNTs- and graphene-based nanomaterials. Then, it is imperative to identify length and diameter differences (morphology), and quantity and types of functionalization groups (chemical) before any biological studies, as small differences may induce distinct biological effects (Seabra et al. 2013).

The literature is extensive regarding the morphological impacts on several bioprocessing and biodegradation of carbonaceous nanomaterials, but mainly for carbon nanotubes (Russier et al. 2011). Furthermore, evidences suggested that the surface chemistry of CNTs plays an important role on the biological interaction. For instance, by considering only the surface charge of SWCNTs, it was observed that those with a negatively charged surface have a faster uptake by macrophages (Dumortier et al. 2006) compared to pristine SWCNTs, which did not induce any activation responses, such as superoxide and NO production (Kagan et al. 2006). Furthermore, there are evidences that graphitic structure oxygenation has a great influence on the biodegradation of carbonaceus materials by enzymes, microorganisms, or even in cells (Allen et al. 2008, 2009; Fadeel et al. 2011; Kagan et al. 2010;

Liu et al. 2010; Neves et al. 2010; Russier et al. 2011; Shvedova et al. 2012; Vlasova et al. 2011a, b, 2012; Zhao et al. 2011a, b). It is known from the literature that the surface microchemical environment of oxygenated carbon nanotubes and graphenes contains different chemical structures, such as alcohols, ketones, ethers, esters, carboxylic acids, and anhydrides (Johari and Shenoy 2011; Paula et al. 2011), and the debris, that are small oxidized fragments which are adsorbed on the graphitic surface (Rourke et al. 2011; Stefani et al. 2011; Vlasova et al. 2012). At this point, is necessary to be aware that is a hard task to probe the influence of each oxygenated group on the bioprocessing. However, besides oxidation to induce surface defects, the nitrogen doping was also efficient to enhance biodegradability features for CNTs (Zhao et al. 2011a, b). Probably in this case, the degradation is related to the transformation of the polyaromatic structure of the graphitic moiety.

There are evidences that agglomeration process of pristine CNTs (Fraczek et al. 2008), graphene (Kuila et al. 2012), and fullerenes (Porter et al. 2006; Okuda-Shimazaki et al. 2010) have an important impact on their ability to interact with enzymes and also on cell interactions. Another important aspect is the large variation in the results from the use of different characterization protocols used for nanomaterials; besides morphological and surface variations from the processing and functionalization methods for carbonaceus nanostructures (Roebben et al. 2011). In light of this discussion, is unquestionable the importance of sample standardization and the use of robust protocols for experimental analyses, since, it was clear from the significant variations among the results in the literature, mainly related to the size, surface charge and chemical stability of nanoparticles obtained in different studies around the world. Then, it is of paramount importance the optimization of both synthesis and characterization methods for carbonaceous nanomaterials in order to understand better the biotransformation and biodegradation processes resulted from biointeractions.

6.7 Influence of Protein Corona on the Biotransformation of Nanostructures

Protein corona is an emergent concept in nanobiotechnology and nanotoxicology (Lundqvist et al. 2008; Walczyk et al. 2010; Paula et al. 2013). Nanomaterials in contact with complex biological fluids (plasma, culture cell medium, dissolved organic matter, etc.) will be coated by the biomolecules (e.g., proteins, lipids, and carbohydrates) presented in these biofluids. This biomolecular coating, the so-called protein corona, has important implications during the assessment of signaling, biodistribution and toxicity of nanomateriais to biological systems. For example, there is a positive correlation between the plasma protein binding capacity of a nanomaterial and the rate at which it is taken up by cells in vitro and in vivo (Walkey and Chan 2012). Therefore, protein corona must be considered as a new entity interacting with cells (Lynch et al. 2009; Monopoli et al. 2011; Shannahan et al. 2013).

The protein corona is composed of a "hard" and "soft" region with strong and weak binding to the nanomaterial surface, respectively. Moreover, the formation and composition of protein corona critically depend on the physicochemical properties of nanomaterials (size, surface chemistry, morphology, composition, etc). The bio-related factors (e.g., protein source, cell lines and their uptake mechanisms) as well as the colloidal aspects (exposure time, gradient of proteins, adsorption affinity, etc.) have also critical influence on the protein corona, finally reflecting on the biological behavior of nanoparticles, such as the nanoparticle cell-penetrating capacity and cytotoxic effects. In fact, the protein corona provides the biological identity of the nanosized material in biological milieu (Monopoli et al. 2012).

Among the nanoparticle parameters which affect the protein corona, the surface properties such as hydrophobicity and surface charge have more significant role than others parameters. Therefore, a better understanding of the role of each parameter on the protein corona formation is promising for the design of targeting nanomaterials, long-circulating drug carriers, for decreasing the toxicity of nanostructures or for remediation technologies (Rahman et al. 2013). Recently, it has been reported the effect of Dulbecco's Modified Eagle's Medium (DMEM) containing 10 % heat inactivated fetal bovine serum (FBS) on CNT-corona proteins (Shannahan et al. 2013). In this study, it was evaluated several CNTs, such as unmodified, carboxylated-SWCNTs, MWCNTs and polyvinylpyrrolidone (PVP)-coated MWCNTs (MWCNTs-PVP). These carbonaceous nanomaterials were incubated in simulated cell culture media, and then analyzed for their associated protein corona. All nanotubes were associated to albumin, titin, and apolipoproteins. Either carboxylated-SWCNTs and MWCNTs were found to bind to a large number of proteins (181 and 133 respectively) compared to pristine nanotubes (<100). Due to these results, Shannahan et al. (2013) concluded that hydrogen bonding and electrostatic interactions were important in protein corona processing. In this sense, the unique constituents of protein corona in cell culture medium (DMEM plus FBS) cause specific cellular effects on the in vitro toxicity of CNTs (Shannahan et al. 2013). Similarly, it has been reported that protein corona also has critical consequences during the assessment of cytotoxicity of graphene oxide (Hu et al. 2011; Mao et al. 2013).

Therefore, it is now evident that protein corona plays a key role on the biotransformation of nanostructures. In this way, to understand the fate of nanostructures in the biological context it is imperative to analyze the intricate factors involved in the nano–bio interface (Nel et al. 2009; Cai et al. 2013; Orts-Gil et al. 2013; Saptarshi et al. 2013).

6.8 Concluding Remarks

It is clear from this chapter book that important enzymes, cells, and microorganisms play a key role in mediating biodegradation of carbon nanostructures. In particular, peroxidases and oxidases, such as laccases, are reported to be the main enzymes for the biodegradation of carbon-related nanomaterials. Moreover, the chemical nature of functionalized carbon-based materials dictates the extension of biodegradation. In this scenario, there is an increasing interest in the development of new approaches of surface modifications of carbon nanostructures, aiming to greatly reduce their toxicity. Biodegradation of CNTs, graphenes and fullerenes could greatly minimize the harmful effects imposed on the environment by these nanomaterials. From the actual information, it should be noted that studies focused on investigating the subcellular localization of carbon-related nanomaterials are highly demanded, due to the debate by the scientific community regarding internalization of these materials. Due to the biomedical and technological importance of carbon-related nanomaterials, the detailed investigation of their cytotoxicity and biodegradation represents a crucial issue to be further and deeply investigated. In this context, this chapter aims to be a source of inspiration for new and exciting works in this field.

Acknowledgment Support from FAPESP, CNPq, INOMAT (MCT/CNPq), NanoBioss (MCTI), and the Brazilian Network in Nanotoxicology (MCT/CNPq) is acknowledged.

References

Allen BL, Kichambare PD, Gou P et al (2008) Biodegradation of single-walled carbon nanotubes through enzymatic catalysis. Nano Lett 8:3899–3903

Allen BL, Kotchey GP, Chen Y et al (2009) Mechanistic investigations of horseradish peroxidase-catalyzed degradation of single-walled carbon nanotubes. J Am Chem Soc 131:17194–17205

Andon FT, Kapralov AA, Yanamala N et al (2013) Carbon nanotubes: iodegradation of single-walled carbon nanotubes by eosinophil peroxidase. Small 9:2721–2729

Asuri P, Bale SS, Pangule RC et al (2007) Structure, function, and stability of enzymes covalently attached to single-walled carbon nanotubes. Langmuir 23:12318–12321

Azevedo AM, Martins VC, Prazeres DM et al (2003) Horseradish peroxidase: a valuable tool in biotechnology. Biotechnol Annu Rev 9:199–247

Bao H, Pan Y, Li L (2012) Recent advances in graphene-based nanomaterials for biomedical applications. Nano Life 2:1230001

Battistuzzi G, Bellei M, Bortolotti CA et al (2010) Redox properties of heme peroxidases. Arch Biochem Biophys 500:21–36

Bianco A, Kostarelos K, Prato M (2011) Making carbon nanotubes biocompatible and biodegradable. Chem Commun 47:10182–10188

Bromberg N, Durán N (2001) Violacein transformation by peroxidase and oxidases: implications on its biological properties. J Mol Catal B Enzymatic 11:463–467

Bussy C, Paineau E, Cambedouzou J et al (2013) Intracellular fate of carbon nanotubes inside murine macrophages: pH-dependent detachment of iron catalyst nanoparticles. Particle Fibre Toxicol 10:1–12

Cai X, Ramalingam R, Wong HS et al (2013) Characterization of carbon nanotube protein corona by using quantitative proteomics. Nanomedicine 9:583–593

Cellot G, Ballerini L, Prato M et al (2010) Neurons are able to internalize soluble carbon nanotubes: new opportunities or old risks? Small 6:2630–2633

Chawla P, Chawla V, Maheshwari R et al (2010) Fullerenes: from carbon to medicine. Mini Rev Med Chem 10:662–677

Chen YR, Sarkanen S, Wang YY (2012) Lignin-degrading enzyme activities. Methods Mol Biol 908:251–268

Dong PX, Wan B, Wang ZX et al (2013) Exposure of single-walled carbon nanotubes impairs the functions of primarily cultured murine peritoneal macrophages. Nanotoxicology 7: 1028–1042

Dumortier H, Lacotte S, Pastorin G et al (2006) Functionalized carbon nanotubes are non-cytotoxic and preserve the functionality of primary immune cells. Nano Lett 6:1522–1528

Durán N (2003) Application of oxidative enzymes in wastewater treatment. In: Sakurai A (ed) Wastewater treatment using enzymes, vol 2. Research Signpost, Trivandrum, pp 41–51

Durán N, Baeza J, Freer J et al (1981) Dimethylsulfoxide as biological probe. Conformational effect on peroxidase system. Biochem Biophy Res Commun 103:131–138

Durán N, Mansilla H, Leite LCC et al (1988) Peroxidase-hydrogen peroxide acting on lignin. J Inorg Biochem 34:105–115

Durán N, Rosa MA, D'Annibale A et al (2002) Applications of laccases and tyrosinases (phenoloxidases) immobilized on different supports: a review. Enzyme Microb Technol 31: 907–931

Fadeel B, Shvedova AA, Kagan VE (2011) Interactions of carbono nanotubes with the immune system: focus on mechanisms of internalization and biodegradation. In: Alexiou C (ed) Nanomedicine—basic and clinical applications in diagnostics and therapy, vol 2. Erlangen, Else Kroner-Fresenius Symposia, pp 80–87

Fisher C, Rider AE, Han ZJ et al (2012) Applications and nanotoxicity of carbon nanotubes and graphene in biomedicine. J Nanomat Article Number: 315185

Fisher C, Rider AE, Han ZJ et al (2012b) Applications and nanotoxicity of carbon nanotubes and graphene in biomedicine. J Nanomater 315185:1–19

Fornera S, Yazawa K, Walde P (2011) Spectrophotometric quantification of lactose in solution with a peroxidase-based enzymatic cascade reaction system. Anal Bioanal Chem 401:2307–2310

Fraczek A, Menaszek E, Paluszkiewicz C et al (2008) Comparative in vivo biocompatibility study of single and multi-wall carbon nanotubes. Acta Biomater 4:1593–1602

Gharbi N, Pressac M, Hadchouel M et al (2005) Fullerene is a powerful antioxidant in vivo with no acute or subacute toxicity. Nano Lett 5:2578

Girish CM, Sasidharan A, Gowd GS et al (2013) Confocal Raman imaging study showing macrophage mediated biodegradation of graphene in vivo. Adv Healthc Mater Doi. doi:10.1002/adhm.201200489

Gitsov I, Lambrych K, Lu P et al (2005) Nondestructive regioselective modification of laccase by linear-dendritic copolymers: enhanced oxidation of benzo-apyrene in water. In: Cheng HN, Gross RA (eds) Polymer biocatalysis and biomaterials. American Chemical Society, Washington, DC, pp 80–94

Gitsov I, Simonyan A, Krastanov A et al (2008a) Green oxidation of steroids in nano-reactors assembled from laccase and linear-dendritic copolymers. In: Cheng HN, Gross RA (eds) Polymer biocatalysis and biomaterials II. American Chemical Society, Washington, DC, pp 110–128

Gitsov I, Hamzik J, Ryan J et al (2008b) Enzymatic nanoreactors for environmentally benign biotransformations. I. Formation and catalytic activity of supramolecular complexes of laccase and linear-dendritic block copolymers. Biomacromolecules 9:804–811

Gitsov I, Simonyan A, Wang L et al (2012) Polymer-assisted biocatalysis: unprecedented enzymatic oxidation of fullerene in aqueous medium. J Polym Sci A Polym Chem 50:119–126

Hu W, Peng C, Lv M et al (2011) Protein corona-mediated mitigation of cytotoxicity of graphene oxide. ACS Nano 5:3693–3700

Hummers WS, Offeman RE (1958) Preparation of graphitic oxide. J Am Chem Soc 80:1339

Husain Q, Ulber R (2011) Immobilized peroxidase as a valuable tool in the remediation of aromatic pollutants and xenobiotic compounds: a review. Crit Rev Environ Sci Technol 41: 770–804

Iijima S (1991) Helical microtubules of graphitic carbon. Nature 354:56–58

Jeon JR, Baldrian P, Murugesan K et al (2012) Laccase-catalyzed oxidations of naturally occurring phenols: from in vivo biosynthetic pathways to green synthetic applications. Microb Biotechnol 5:318–332

Jin H, Heller DA, Sharma R et al (2009) Size-dependent cellular uptake and expulsion of single-walled carbon nanotubes: single particle tracking and a generic uptake model for nanoparticles. ACS Nano 3:149–158
Johari P, Shenoy VB (2011) Modulating optical properties of graphene oxide: role of prominent functional groups. ACS Nano 5:7640–7647
Johnston HJ, Hutchison GR, Christensen FM et al (2010) The biological mechanisms and physicochemical characteristics responsible for driving fullerene toxicity. Toxicol Sci 114: 162–182
Kadokawa JI, Kobayashi S (2010) Polymer synthesis by enzymatic catalysis. Curr Opin Chem Biol 14:145–153
Kagan VE, Tyurina YY, Tyurin VA et al (2006) Direct and indirect effects of single walled carbon nanotubes on RAW 264.7 macrophages: Role of iron. Toxicol Lett 165:88–100
Kagan VE, Konduru NV, Feng W et al (2010) Carbon nanotubes degraded by neutrophil myeloperoxidase induce less pulmonary inflammation. Nat Nanotechnol 5:354–359
Konduru NV, Tyurina YY, Feng W et al (2009) Phosphatidylserine targets single-walled carbon nanotubes to professional phagocytes in vitro and in vivo. PLoS One 4:e4398
Kosman J, Juskowiak B (2011) Peroxidase-mimicking DNAzymes for biosensing applications: a review. Anal Chim Acta 707:7–17
Kotchey GP, Allen BL, Vedala H et al (2011) The enzymatic oxidation of graphene oxide. ACS Nano 5:2098–2108
Kotchey GP, Gaugler JA, Kapralov AA (2013) Effect of antioxidants on enzyme-catalysed biodegradation of carbono nanotubes. J Mater Chem B Mater Biol Med 1:302–309
Kratschmer W, Lamb LD, Fostiropoulos K et al (1990) Solid C60: a new form of carbon. Nature 347:354–358
Kroto HW, Heath JR, O'Brien SC et al (1985) C60-Buckminsterfullerene. Nature 318:162–163
Kuila T, Bose S, Khanra P et al (2012) A green approach for the reduction of graphene oxide by wild carrot root. Carbon 50:914–921
Lacerda L, Russier J, Pastorin G et al (2012) Translocation mechanisms of chemically functionalized carbono nanotubes across plasma membranes. Biomaterials 33:3334–3343
Lee SH, Sung JH, Park TH (2012) Nanomaterial-based biosensor as an emerging tool for biomedical applications. Ann Biomed Eng 40:1384–1397
Li X, Wang L, Fan Y et al (2012) Biocompatibility and toxicity of nanoparticles and nanotubes. J Nanomater 548389:1–19
Liu X, Hurt RH, Kane AB (2010) Biodurability of single-walled carbono nanotubes depends on surface functionalization. Carbon 48:1961–1969
Liu J, Tang J, Gooding JJ (2012) Strategies for chemical modification of graphene and applications of chemically modified grapheme. J Mater Chem 22:12435–12452
Lundqvist M, Stigler J, Elia G (2008) Nanoparticle size and surface properties determine the protein corona with possible implications for biological impacts. Proc Natl Acad Sci U S A 105: 14265–14270
Lynch I, Salvati A, Dawson KA (2009) Protein-nanoparticle interactions: what does cell see? Nat Nanotechnol 4:546–547
Mao H, Chen W, Laurent S et al (2013) Hard corona composition and cellular toxicities of the graphene sheets. Coll Surf B: Biointerf 109:212–218
Margot J, Bennati-Granier C, Julien Maillard J et al (2013) Bacterial versus fungal laccase: potential for micropollutant degradation. AMB Express 3:63
Marques-Rocha FJ, Hernandez-Rodriguez VZ, Vazquez-Duhalt R (2000) Biodegradation of soil-adsorbed polycyclic aromatic hydrocarbons by the white rot fungus *Pleurotus ostreatus*. Biotechnol Lett 22:469–472
Minussi RC, Miranda MA, Silva JA et al (2007) Purification, characterization and application of laccase from Trametes versicolor for color and phenolic removal of olive mill wastewater in the presence of 1-hydroxybenzotriazole. Afric J Biotechnol 6:1248–1254

Mita N, Tawaki SI, Uyama H et al (2003) Laccase-catalyzed oxidative polymerization of phenols. Macromol Biosci 3:253–257

Monopoli MP, Walczyk D, Campbell A et al (2011) Physical-chemical aspects of protein corona: relevance to in vitro and in vivo biological impacts of nanoparticles. J Am Chem Soc 133: 2525–2534

Monopoli MP, Aberg C, Salvati A et al (2012) Biomolecular coronas provide the biological identity of nanosized materials. Nat Nanotechnol 7:779–786

Nel AE, Madler L, Velegol D et al (2009) Understanding biophysicochemical interactions at the nano-bio interface. Nat Mater 8:543–557

Neves V, Heister E, Costa S et al (2010) Uptake and release of double-walled carbon nanotubes by mammalian cells. Adv Funct Mater 20:3272–3279

Nielsen GD, Roursgaard M, Jensen KA et al (2008) In vivo biology and toxicology of fullerenes and their derivatives. Basic Clin Pharmacol Toxicol 103:197–208

Novoselov KS, Geim AK, Morozov SV et al (2004) Electric field effect in atomically thin carbon films. Science 306:666–669

Nudejima S, Miyazawa K, Okuda-Shimazaki J et al (2009) Observation of phagocytosis of fullerene nanowhiskers by PMA-treated THP-1 cells. J Phys Conf Ser 159:012008

Nudejima SI, Miyazawa K, Okuda-Shimazaki J et al (2010) Biodegradation of C(60) fullerene nanowhiskers by macrophage-like cells. In: Rossi M, Pham TD, Falugi C, Bussing A, Koukkou M (eds) Advances in biomedical research, proceedings, recent advances in biology and biomedicine. WSEAS Press, Cambridge, MA, pp 89–94

Nunes A, Bussy C, Gherardini L et al (2012) *In vivo* degradation of functionalized carbon nanotubes after stereotactic administration in the brain cortex. Nanomedicine (Lond) 7: 1485–1494

O'Brien PJ, Bechara EJH, O'Brien CR et al (1978) Generation of bio-electronic energy by electron transfer: reduction of peroxidase compounds I and compound II by eosin. Biochem Biophys Res Commun 81:75–81

Okuda-Shimazaki J, Nudejima S, Takaku S et al (2010) Effects of fullerene nanowhiskers on cytotoxicity and gene expression. Health 2:1456–1459

Orts-Gil G, Natte K, Thiermann R et al (2013) On the role of surface composition and curvature on biointerface formation and colloidal stability of nanoparticles in a protein-rich model system. Colloids Surf B 108:110–119

Paula AJ, Stefani D, Souza AG et al (2011) Surface chemistry in the process of coating mesoporous SiO(2) onto carbon nanotubes driven by the formation of Si-O-C bonds. Chem Eur J 17:3228–3237

Paula AJ, Araújo-Júnior RT, Martinez DST et al (2013) Influence of protein corona on the transport of molecules into cells by mesoporous silica nanoparticles. ACS Appl Mater Interfaces 5:8387–8393

Petersen EJ, Henry TB (2012) Methodological considerations for testing the ecotoxicity of carbon nanotubes and fullerenes: review. Environ Toxicol Chem 31:60–72

Porter AE, Muller K, Skepper J et al (2006) Uptake of C60 by human monocyte macrophages, its localization and implications for toxicity: studied by high resolution electron microscopy and electron tomography. Acta Biomater 2:409–419

Rahman M, Laurent S, Tawil N et al (2013) Protein-nanoparticle interactions. In: Martinac B (ed) Springer series in biophysics, vol 15. Springer, Berlin, p 84

Rajavel K, Minitha CR, Ranjith KS et al (2012) Recent progress on the synthesis and applications of carbon based nanostructures. Recent Patents Nanotechnol 6:99–104

Roebben G, Ramirez-Garcia S, Hackley VA et al (2011) Interlaboratory comparison of size and surface charge measurements on nanoparticles prior to biological impact assessment. J Nanopart Res 13:2675–2687

Rolfe P (2012) Micro- and nanosensors for medical and biological measurement. Sens Mater 24:275–302

Rourke JP, Pandey PA, Moore JJ et al (2011) The real graphene oxide revealed: stripping the oxidative debris from the graphene-like sheets. Angew Chem Int Ed 50:3173–3177

Russier J, Menard-Moyon C, Venturelli E et al (2011) Oxidative biodegradation of single- and multi-walled carbon nanotubes. Nanoscale 3:893–896

Ryan BJ, Carolan N, O´Fagain C (2006) Horseradish and soybean peroxidases: comparable tools for alternative niches? Trends Biotechnol 24:355–363

Salas EC, Sun Z, Luttge A et al (2010) Reduction of graphene oxide via bacterial respiration. ACS Nano 4:4852–4856

Sanchez VC, Jachak A, Hurt RH et al (2012) Biological interactions of graphene-family nanomaterials: an interdisciplinary review. Chem Res Toxicol 25:15–34

Saptarshi SR, Duschl A, Lopata AL (2013) Interaction of nanoparticles with proteins: relation to bio-reactivity of the nanoparticle. J Nanobiotechnol 11:26

Schreiner KM, Filley TR, Blanchette RA et al (2009) White-rot basidiomycete-mediated decomposition of C60 fullerol. Environ Sci Technol 43:3162–3168

Seabra AB, Paula AJ, Durán N (2013) Redox-enzymes, cells and micro-organisms acting on carbon nanostructures transformation: a mini-review. Biotechnol Prog 29:1–10

Shan C, Yang H, Song J et al (2009) Direct electrochemistry of glucose oxidase and biosensing for glucose based on graphene. Anal Chem 81:2378–2382

Shannahan JH, Brown JM, Chen R et al (2013) Comparison of nanotube-protein corona composition in cell culture media. Small 12:2171–2181

Shen H, Zhang L, Liu M et al (2012) Biomedical applications of graphene. Theranostics 2: 283–294

Shvedova AA, Kapralov AA, Feng WH et al (2012) Impaired clearance and enhanced pulmonary inflammatory/fibrotic response to carbon nanotubes in myeloperoxidase-deficient mice. PLOS One 7:e30923

Stefani D, Paula AJ, Vaz BG et al (2011) Structural and proactive safety aspects of oxidation debris from multiwalled carbon nanotubes. J Hazard Mater 189:391–396

Stern ST, Adiseshaiah PP, Crist RM (2012) Autophagy and lysosomal dysfunction as emerging mechanisms of nanomaterial toxicity. Particle Fiber Toxicol 9:20

Takagi A, Hirose A, Nishimura T et al (2008) Induction of mesothelioma in p53(+/−) mouse by intraperitoneal application of multi-wall carbono nanotube. J Toxicol Sci 33:105–116

Umbuzeiro GA, Coluci VR, Honorio JG et al (2011) Understanding the interaction of multi-walled carbon nanotubes with mutagenic organic pollutants using computational modeling and biological experiments. Trends Anal Chem 30:437–446

Uyama H, Mariuchi N, Tonami H et al (2002) Peroxidase catalyzed oxidative polymerization of bisphenols. Biomacromolecules 3:187–193

van Rantwijk F, Sheldon RA (2000) Selective oxygen transfer catalyzed by heme peroxidases: synthetic and mechanistic aspects. Curr Opin Biotechnol 11:554–564

Vlasova II, Vakhrusheva TV, Sokolov AV et al (2011a) Peroxidase-induced degradation of single-walled carbon nanotubes: hypochlorite is a major oxidant capable of in vivo degradation of carbon nanotubes. J Phys Conf Ser 291:012056

Vlasova II, Sokolov AV, Chekanov AV et al (2011b) Myeloperoxidase-induced biodegradation of singlewalled carbon nanotubes is mediated by hypochlorite. Russian J Bioorg Chem 37:453–463

Vlasova II, Vakhrusheva TV, Sokolov AV et al (2012) PEGylated single-walled carbon nanotubes activate neutrophils to increase production of hypochlorous acid, the oxidant capable of degrading nanotubes. Toxicol Appl Pharmacol 264:131–142

Walczyk D, Bombelli FB, Monopoli MP et al (2008) What the cell "Sees" in bionanoscience. J Am Chem Soc 132:5761–5768

Walkey CD, Chan WCW (2012) Understanding and controlling the interaction of nanomaterials with proteins in a physiological environment. Chem Soc Rev 41:2780–2799

Witayakran S, Ragauskas AJ (2009) Synthetic applications of laccase in green chemistry. Adv Synth Catal 351:1187–1209

Xu P, Uyama H, Whitten JE, Kobayashi S, Kaplan DL et al (2005) Peroxidase-catalyzed in situ polymerization of surface orientated caffeic acid. J Am Chem Soc 127:11745–11753

Yamashita T, Yamashita K, Nabeshi H et al (2012) Carbon nanomaterials: efficacy and safety for nanomedicine. Materials 5:350–363

Yang ST, Luo J, Zhou Q et al (2012) Pharmacokinetics, metabolism and toxicity of carbon nanotubes for bio-medical purposes. Theranostics 2:271–282

Yue FN, Luo SM, Zhang CD (2013) Degradation and transformation of engineering carbon nanomaterials in the environment: a review. Chinese J Appl Ecol 24:589–596

Zakharova GS, Uporov IV, Tishkov VI (2011) Horseradish peroxidase: modulation of properties by chemical modification of protein and heme. Biochemistry (Moscow) 76:1391–1401

Zhao Y, Allen BL, Star A (2011a) Enzymatic degradation of multiwalled carbon nanotubes. J Phys Chem A 115:9536–9544

Zhao F, Zhao Y, Liu Y et al (2011b) Cellular uptake, intracellular trafficking, and cytotoxicity of nanomaterials. Small 7:1322–1337

Chapter 7
Phytoremediation: Strategies of Argentinean Plants Against Stress by Heavy Metals

Mariana Rosa, Fernando Prado, Mirna Hilal, Eduardo Pagano, and Carolina Prado

Abstract Environmental pollution by organic and inorganic contaminants is an increasing global problem, but mostly affects both poor and developing countries. The effects of environmental pollution on biota and human health are reaching intolerable limits, so urgent actions are required to address their remediation before it is too late to generate a less polluted world and better life quality to next generations. In Argentina the environmental pollution has already been installed in regions once considered non-contaminated sites such as the Andean mountain range and fertile plains (Pampas). For a long time several Argentinean laboratories are working on phytoremediation technologies to clean polluted soils and water bodies. However, most of these studies are carried out at small-scale under laboratory conditions and/or in small-sized confined environments. In fact to date no more than one full-scale experiment has been carried out to decontaminate sewage and industrial effluents from a metallurgical factory. Strikingly, most studies have focused on a few toxic heavy metals, i.e., chromium (Cr), cadmium (Cd), lead (Pb), zinc (Zn), copper (Cu), manganese (Mn), and nickel (Ni), but ongoing investigations have not considered the most widespread of all: arsenic (As). It affects more than half of the Argentine territory, including large areas of the Chaco-Pampean plains and also numerous rivers that provide water for both human and animal consumption, as well as for crop irrigation. Studies on phytoremediation of organic pollutants are less developed.

M. Rosa • F. Prado • M. Hilal • C. Prado (✉)
Laboratorio de Fisiología Vegetal, Facultad de Ciencias Naturales e IML, Universidad Nacional de Tucumán, Miguel Lillo 205, 4000 Tucumán, Argentina
e-mail: caroprado02@hotmail.com

E. Pagano
Departamento de Biología Aplicada y Alimentos, Facultad de Agronomía, Universidad de Buenos Aires, 1417 Buenos Aires, Argentina

A. Alvarez, M.A. Polti (eds.), *Bioremediation in Latin America: Current Research and Perspectives*, DOI 10.1007/978-3-319-05738-5_7

7.1 Introduction

Traditionally, elements that are toxic to animals and plants have been known as "heavy metals." These include transition metals mainly corresponding to the fourth period and some from both the fifth and sixth periods of the Periodic Table. Some non-transition metals such as aluminum (Al) and lead (Pb), metalloids such as arsenic (As), selenium (Se), and boron (B), and some lanthanides and actinides with metallic properties are also included, but alkaline earth metals [beryllium (Be), magnesium (Mg), calcium (Ca), strontium (Sr), barium (Ba), and radium (Ra)], are not included (Babula et al. 2009). Both, the high density and high atomic weight have been used to define heavy metals, but there is considerable disagreement about whether both or only one of the parameters must be high and about where to set the cutoff values for the parameter(s). Values of 5 and 6 g/cm^3 have been used to define the high density, whereas 23, 40, and 63 have been used as cutoff values for the atomic weight. However, the gold with atomic number of 79 and density of 18.88 g/cm^3 is not considered as a heavy metal, while the cadmium (Cd) with atomic number of 48 and density of 8.65 g/cm^3 is considered a typical heavy metal. Furthermore, since heavy metals comprise metals that are not particularly heavy, and sometimes to elements that are only somewhat metallic, it is considered a meaningless and misleading term (Duffus 2002). Even, the term "metal" is commonly misused in both toxicological literature and in legislation to mean the pure metal and all the chemical species in which it may exist. This usage implies that the pure metal and all its compounds have the same physicochemical, biological, and toxicological properties with highly toxic or ecotoxic properties, which is untrue (Madrid 2010). In order to avoid the use of the term "heavy metal," the term "trace elements" it has been suggested to identify all metal(loid)s that can affect the metabolism of living organisms including animals, plants, fungi, and bacteria (Kopittke et al. 2010). However, trace element is also an imprecise term and does not reflect the chemical characteristics of elements. In order to solve this imprecise definition, Appenroth (2010) based on the Periodic Table of Chemical Elements, clustered elements that affect the living organisms into three subgroups: (1) include all transition elements excepting lanthanum (La) and actinium (Ac), (2) include elements from lanthanide and actinide series, including La and Ac, and (3) comprises a heterogeneous group of elements commonly named lead-group that includes metals, i.e., bismuth (Bi), metalloids such as As, Se, B, germanium (Ge), and tellurium (Te), and amphoterous oxide forming elements such as Al, Pb, gallium (Ga), indium (In), thallium (Tl), tin (Sn), and polonium (Po). Despite this more specific classification that better reflects the chemical basis of elements, allowing a more precise prediction of their toxic effects, there is a generalized and accepted tendency to use the term heavy metal.

In the environment heavy metals are present in trace (10 mg/kg, or mg/L) or in ultra-trace (1 μg/kg, or μg/L) quantities, but only when they are present in bioavailable forms and at excessive levels have the potential to become toxic to animals and plants (Nagajyoti et al. 2010). Adverse effects of heavy metals have been known for a long time, but the increasing industrialization and urbanization has given that exposure to heavy metals continues and even increases in many regions of the world (Rai 2009). Even in developed countries such as Japan, Germany, Sweden, and Italy the heavy metal contamination became in a serious environmental problem (Kudo and Miyahara 1991; Lindqvist et al. 1991; Zietz et al. 2003; Tamasi and Cini 2004). Furthermore, the increasing problem of heavy metals has led that these are listed as priority pollutants by the US Environmental Protection Agency (EPA). Lead, Hg, As, and Cd are ranked first, second, third, and sixth, respectively, on the list of the US Agency for Toxic Substances and Disease Registry (ATSDR), which lists all hazards present in toxic waste sites according to their prevalence and toxicity (Rai 2009). Heavy metals are really difficult to remove from the environment and the low levels observed pose a high risk of accumulation in the food chain, and thus many of them constitute a global environmental hazard. Therefore the problem of heavy metal pollution is emerging as a matter of concern at local, regional, and global scales. Though it is difficult to generalize what is the more toxic heavy metal, cleaning of the contaminated sites requires a fast and efficient solution. In this chapter we examine benefits and disadvantages of the use of plants to remove heavy metals from both aquatic and terrestrial environments.

7.2 Heavy Metals and Plant Tolerance

The uptake of mineral elements by plants occurs either by: (1) non-energy-consuming process (non-metabolism-dependent), or (2) energy-consuming process (metabolism-dependent) (Siedlecka 1995). The first one is called passive transport and takes place when heavy metals are present at high concentrations in the root environment. In these conditions, metals can diffuse down the concentration gradient with the aid of membrane carriers or through ionic channels and even through water channels (aqueous pores). Whilst the last is an active mechanism involved in the uptake of ions against their concentration gradients. In this mechanism, ATP-driven proton (H^+) pumps generate a pH-dependent H^+ motive force, which promotes the ion passing to the cell through selective ion channels or carriers (Leontiadou et al. 2007). The lack of selectivity during the absorption of ions by plants leads to uptake of practically all known heavy metals. The excess of heavy metals affect the plant physiology by inducing oxidative stress, but many plants have adapted to avoid the damaging effects of heavy metal toxicity, using different strategies. Major strategies used by plants against high external heavy metal concentrations are: (1) avoidance and (2) tolerance (Dalvi and Bhalerao 2013). The avoidance (restriction to metal uptake) reduces the concentration of heavy metal that enters into plant tissues through the roots and/or leaves by: extracellular complexation with root exudates, sorption onto cell walls, immobilization by mycorrhizal association, extracellular precipitation by modification of rhizosphere pH, reduced uptake, or increased efflux (Krzesłowska 2011; Javed et al. 2013). The tolerance allows plants to survive in the presence of high internal metal concentration. Heavy metals are intracellularly detoxified by chelation with amino acids, organic acids, and metal-thiol binding ligands (glutathione, metallothioneins, phytochelatins), or sequestered by compartmentation within the vacuole (Hall 2002). In addition, tolerant plants are able to upregulate antioxidant and glyoxalase defense systems to counter deleterious effects caused by metal-induced reactive oxygen species and methylglyoxal (Hossain et al. 2012). According to tolerance mechanisms, plants can be grouped in three types: (1) accumulators when heavy metals are concentrated in above-ground plant parts from low or high soil levels, (2) indicators when heavy metal concentrations in shoots are approximately equal to soil concentrations, and (3) excluders when metal concentrations in shoots are maintained at constant and low levels over a wide range of soil concentrations (Baker 1981). It is important recall that heavy metal-tolerant plants are mostly excluders which grow and reproduce on natural metalliferous soils such as calamine soils (enriched in Zn and Pb) and serpentine soils (derived from Fe and Mg-rich ultramafic rocks, also enriched in Ni, Cr, and Co) (Vassilev et al. 2004). Plants naturally occurring in metalliferous soils are also known as metallophytes (Rascio and Navari-Izzo 2011). Species that colonize a metalliferous environment develop genetically divergent populations when the gene flow between populations is limited (genetic island). Genetic selection is very intense and leads to the evolution of locally adapted metal-tolerant populations within a very short evolution-time. For instance in California some species have evolved to Cu tolerant races in times as short as 70 years (Bondada and Ma 2003). These metal-tolerant populations are named metallicolous ecotypes to differentiate the non-metallicolous populations that not tolerate higher levels of heavy metals. Although a metallicolous population that evolved in continuous contact with a determined metalliferous soil develops an acquired resistance to the predominant soil metal, the same one is not restrictive to this sole metal. A well documented example of this trait are populations of the metallicolous *Agrostis tenuis* and *A. capillaris* species growing in a Ni-contaminated mining area can simultaneously tolerate Ni, Pb, and Zn (Bradshaw et al. 1965). It is noteworthy that metallicolous plants are absent in non-metalliferous soils, indicating that they are less competitive than non-metallicolous ones from non-metalliferous soils. To explain this less fitness to a successful development it has been assumed that the tolerance to heavy metals evolved with a physiological cost, namely cost of tolerance (Dechamps et al. 2008). Two hypotheses have been formulated to explain the cost of tolerance: (1) trade-off hypothesis, and (2) metal requirement hypothesis. The first one proposes that resources needed for the tolerance are diverted away from other traits such as growth rate, seed production or competitive ability, while the second one postulates that tolerance mechanisms, i.e., heavy metal sequestration lead to essential metal deficiency when metal-tolerant plants grow on non-metalliferous soils (Harper et al. 1997). In agreement with the hypothesis of tolerance cost, metallicolous plants naturally occurring are mainly small and have slow growth, with low biomass yield (Lassat 2002). However, metallicolous plants are adapted not only to a high metal contamination, but they also have to cope with other abiotic stresses normally occurring in metalliferous environments such as drought, mineral deficiency, and intense solar irradiation. Thus metallicolous plants seem to have a lower sensitivity to noxious pollutants compared with plants inhabiting in non-metalliferous sites (Wright et al. 2006). Agreeing with this assumption, the above-ground biomass of metallicolous individuals is significantly bigger than that biomass of non-metallicolous individuals (Deram et al. 2006). So this trait would be useful to implementing phytoremediation technologies.

7.3 Phytoremediation of Heavy Metals

Cleanup of heavy metal-contaminated sites is utmost necessary in order to minimize negative impact on ecosystems and also to minimize the entry of potentially toxic metals into the food chain. For many years different physical, chemical, and biological methods, have been used to remove heavy metals. These methods include, among others, in situ vitrification of metals, soil incineration, excavation and landfill, soil flushing, solidification, and stabilization of electrokinetic systems (Wuana and Okieimen 2011). These technologies often are appropriate only for small areas, require special equipment and are labor-intensive, being also costly. Furthermore, they cause irreversible soil changes, and produce secondary pollution problems (Pulford and Watson 2003). Moreover, chemical and physical methods are not readily accepted by people. From an environmental point of view, chemical technologies are unaffordable for many poor countries, even though most of them have higher levels of heavy metal contamination. Therefore, recent researches have driven to develop alternative remediation methods efficient, cost-effective, and environment friendly for decontamination of heavy metal-polluted sites (Wu et al. 2010). One of the most promissory methods is the phytoremediation, also called green technology that uses plants to remove heavy metals. The phytoremediation requires relatively low technology, can be accomplished in situ and offers the site decontamination. In addition, it maintains the biological activity and physical structure of soils, is potentially cheap and visually unobtrusive, and also provides the possibility of metal recovery (Ali et al. 2013). The phytoremediation basically refers to the use of plants to reduce the concentrations or toxic effects of metal contaminants in the environment, but it can also be used to remove organic pollutants such as polynuclear aromatic hydrocarbons, polychlorinated biphenyls, and pesticides (Pilon-Smits and Freeman 2006).

Although the phytoremediation commonly remediates both soil and water in situ, avoiding either the dramatic disruption of landscapes or the perturbation of water bodies to preserve the ecosystems, it has also several disadvantages and constraints that restrict its applicability. These disadvantages include long time required (usually more than one year), depth limitations (3 ft for soil and 10 ft for groundwater), climatic or geologic constraints (e.g., flooding, drought), possibility of produce secondary impacts on water quality, and plant–herbivore deleterious interactions, among others (Garbisu and Alkorta 2001). In addition, plant species or even populations react differently when exposed to elevated levels of heavy metals in nonnatural habitats. Then, a wide range of responses according to contaminant type, concentration, uptake or metabolism, and ability of plants to grow under specific soil and climatic conditions, may be occurring (Chaney et al. 2007).

In spite of the phytoremediation appears to be a promissory inexpensive technology to clean up polluted environments, there are very few full-scale demonstrations of the phytoextraction of metals and metalloids from soils and water bodies beyond demonstrative pilot-scale. Most of these trials were performed in the USA, India, China, and Europe (Green and Hoffnagle 2004), being very scarce in South America. Argentina, Chile, and Brazil are only countries with pilot-scale trials for heavy metal phytoremediation (Hadad et al. 2006; Philippi et al. 2006; Guerra et al. 2011). Since, there are very few phytoremediation trials conducted during long term (Hammer et al. 2003; Lewandowski et al. 2006), so particular uncertainties over the longer-term effectiveness of phytoremediation and associated environmental risks must be also (Angle and Linacre 2005; Scholz and Hansmann 2007). According to Dickinson et al. (2009) an additional focus on biomass energy, improved biodiversity, watershed management, soil protection, carbon sequestration, and improved soil health is also required to justify advancement of the phytoremediation technology.

7.4 Phytoremediation in Argentina: An Overview

In Argentina studies on the phytoremediation of heavy metals were carried out using well known aquatic and terrestrial phytoremediator species such as *Salvinia herzogii*, *Pistia stratiotes*, *Eichhornia crassipes*, *Salvinia minima*, *Typha dominguensis*, *Potamogeton pusillus*, *Myriophyllum aquaticum*, *Schoenoplectus americanus* (syn. *Scirpus americanus*), *Medicago sativa*, *Phragmites australis*, *Spirodela intermedia*, *Lemna minor* (Miretzky et al. 2004; Arreghini et al. 2006; Hadad et al. 2010; Prado et al. 2010a; Mufarrege et al. 2010; Monferrán et al. 2012; Harguinteguy et al. 2013), but few of them are also performed with shrubs (aquatic and terrestrial) and trees that have not been described as efficient phytoremediator species. Many of these plants are native species *Egeria densa*, *Ricciocarpus natans*, *Schoenoplectus californicus*, *Hydrocotyle ranunculoides*, *Sesbania virgata*, *Jacaranda mimosifolia*, *Tecoma stans*, *Nicotiana glauca*, *Acacia caven*, *Canna indica*, *Erythrina crista-galli*, *Aspidosperma quebracho blanco* (Albarracín Franco et al. 2008; de Viana and Albarracín Franco 2008; Albarracín Franco and de Viana 2009; Chiodi Boudet et al. 2011; Branzini et al. 2012; Rizzo et al. 2012), while others correspond to introduced ones [e.g., *Pinus ponderosa*, *P. murrayana*, *Populus nigra*, *Festuca arundinacea*, *Pelargonium hortorum*, *Lolium multiflorum*, *Beta vulgaris*, *Spinacia oleracea*, *Citrumelo* (*Citrus paradisi* × *Poncirus trifoliata*)] (López et al. 2005; Orroño et al. 2012; Podazza et al. 2012; Cordero et al. 2013). However, it is noteworthy that most phytoremediation studies were conducted on a small scale (mainly in laboratory and/or greenhouse). Until today, both pilot- and full-scale trials were only performed by the Maine's group in Santa Fe province. This group used an artificial wetland to remediate heavy metals, phosphate, nitrite, nitrate, and ammonium from wastewater. Prior to full-scale wetland operation, the Maine's group studied this process for 1 year in a pilot-scale wetland (6×3×0.5 m) using different aquatic macrophytes. In this way they achieved a deep knowledge on wetland functionality to efficient removal of Cr, Ni, Fe, and Zn (Hadad et al. 2006).

The full-scale heavy metal phytoremediation trial was carried out to clean up both sewage and industrial effluents from a metallurgical factory. The process was conducted for 6 years under natural conditions in an artificial wetland (50×40×0.8 m) built in the field of the factory Bahco Argentina S.A located in the city of Santo Tomé (Santa Fe, province). Two well-known phytoremediator aquatic macrophytes, i.e., *E. crassipes* and *T. dominguensis* in consortium with another one, i.e., *Panicum elephantipes* were used for metal removal (Maine et al. 2006; 2009). According to obtained data, during wetland operation *E. crassipes* showed a fast growth and it soon became dominant, attaining 80 % cover of the wetland surface, while *T. domingensis* and *P. elephantipes* were developed as accompanying species, attaining a cover of 14 and 4 %, respectively. Wetland macrophytes were able to remove up to 86 % of Cr and 67 % of Ni, as well as 70 and 60 % of the incoming nitrate and nitrite. In addition phosphate and ammonium were not retained within the wetland (Maine et al. 2006). To our knowledge no other pilot- or full-scale heavy metal phytoremediation trials were carried out in Argentina. In a similar way, Miretzky et al. (2004) in a laboratory trial studied the simultaneous removal of several heavy metals (Fe, Cu, Zn, Mn, Cr, and Pb), added to waters from the Chascomús Lake, Vitel Lake and Vitel Stream at different concentrations in order to simulate the anthropic pollution occurring in these water bodies. Macrophytes *P. stratiotes*, *S. intermedia* and *L. minor* collected from the same sites were used to metal removal. High percentages of metal removal were obtained for the three species and metals, but *L. minor* did not survive probably due to the low dissolved oxygen concentration in the water of pots at final of the experiment. Interestingly, the less toxic metals, i.e., Fe and Mn were less removed than the more toxic ones, probably due to macrophytes had already removed these metals from the water bodies. The rate of metal uptake was dependent on the metal concentration for the three macrophytes during the first hours of the treatment, but after it decreased gradually until became negligible. This suggests that during treatment process the resultant plant biomass might be harvested regularly, so that the water purification becomes a continuous operation. Furthermore these authors also analyzed the heavy metal removal by using dead macrophytes (Miretzky et al. 2006). Similarly, in two studies the Prado's group working in pot experiments, analyzed the accumulation of Cr by the aquatic fern *S. minima* in contrasting seasons (summer and winter) under field conditions. These authors demonstrated that the uptake of metal was reduced by five times during the winter season whereas the relative growth rate (RGR) decreased by three times indicating that metabolic activities related to Cr accumulation are more affected than the plant growth by climatic conditions (temperature and solar day length) (Prado et al. 2010b, 2012). Thus, in treatment wetlands both the macrophytes combinations to be used and the period of macrophytes replacement, as well as the climatic conditions of pond-located sites should be seriously taken account to get a more efficient heavy metal removal. On the other hand, an interesting study on heavy metal remediation from the argentinean Pampas soils was performed using the native tree species *S. virgata* (Branzini et al. 2012). In a pot experiment, Cu, Zn, and Cr were added either individually or in binary mixture to a representative Pampas soil from Buenos Aires province at both low and high doses. In binary mixture the accumulation of Cu and Zn was higher than that each individual accumulation, suggesting a synergistic or additive metal effect on the extraction capacity of *S. virgata*. Contrarily, Cr was more accumulated in the individual treatment. Furthermore, the co-presence of metals mainly Cu and Zn at high concentrations resulted in higher reduction of *S. virgata* biomass when comparing with single metal treatments. Strikingly, most studies have been focused on a few toxic heavy metals, i.e., Cr, Cd, Pb, Zn, Cu, Mn, and Ni, but ongoing investigations have not considered the most widespread of all: As. It affects more than half of the argentine territory, including large areas of the Chaco-Pampean plains and also numerous rivers that provide water for both human and animal consumption, as well as for crop irrigation (Bundschuh et al. 2010). Although most of the phytoremediation studies in Argentina are conducted to heavy metal remediation of polluted sites, a few of them are also channelized to the removal of organic contaminants (e.g., phenol, phenolics, herbicides, pesticides) too (Merini et al. 2009; Ibáñez et al. 2012; Sosa Alderete et al. 2012; Talano et al. 2012; Basílico et al. 2013). Like to in other countries, in Argentina many studies on phytoremediation were performed to understand underlying molecular and cellular mechanisms in plants' tolerance to inorganic and organic pollutants (Wevar Oller et al. 2005; Podazza et al. 2006; Merini et al. 2009; Ibáñez et al. 2011; Prado et al. 2013).

7.5 Concluding Remarks

In Argentina, the phytoremediation research to date has largely focused on heavy metal accumulation by plants. However to make real progress, a wide approach by using complete metabolic pathways, targeted to know tolerant mechanisms and coordinated gene expression in specific organs (e.g., root, stem, leaf, seed), needs to be developed in both controlled and field-grown plants. Moreover, special studies on metal-enriched biomass should also be performed to get its safe disposal without risks to the environment.

Finally, according to Gasic and Korban (2006) "The success of phytoremediation will require an improved understanding of the biological processes involved in metal acquisition, movement across plant cells within a plant, shoot accumulation, and metal detoxification in both metal hyperaccumulator and non accumulator plants. Once our knowledge of these intricate processes is further enhanced, we will be able to fully take advantage of plants for phytoremediation of numerous and vast areas of agricultural land and/or natural habitats."

References

Albarracín Franco S, de Viana M (2009) *Aspidosperma quebracho blanco* and *Lolium multiflorum*: two species able to remediate boron from contaminated soils. Avd Energy Ren Med Amb Biomed 13:107–112

Albarracín Franco S, Larenas Parada G, Quiroga M et al (2008) Boron reduction in substrates by native species action. Avd Energy Ren Med Amb Biomed 12:127–132

Ali H, Khan E, Sajad MA (2013) Phytoremediation of heavy metals: concepts and applications. Chemosphere 91:869–881

Angle JS, Linacre NA (2005) Metal phytoextraction: a survey of potential risks. Int J Phytorem 7:241–254

Appenroth KJ (2010) What are "heavy metals" in plant sciences? Acta Physiol Plant 32:615–619

Arreghini S, De Cabo L, de Iorio AF (2006) Phytoremediation of two types of sediment contaminated with Zn by *Schoenoplectus americanus*. Int J Phytorem 8:223–232

Babula P, Adam V, Opatrilova R et al (2009) Uncommon heavy metals, metalloids and their plant toxicity: a review. In: Lichtfouse E (ed) Organic farming, pest control and remediation of soil pollutants, sustainable agriculture reviews. Springer, Heidelberg, pp 275–317

Baker AJM (1981) Accumulators and excluders: strategies in the response of plants to heavy metals. J Plant Nutr 3:643–654

Basílico G, De Cabo L, Faggi A (2013) Impacts of composite wastewater on a Pampean stream (Argentina) and phytoremediation alternative with *Spirodela intermedia* Koch (Lemnaceae) growing in batch reactors. J Environ Manage 115:53–59

Bondada BR, Ma LQ (2003) Tolerance of heavy metals in vascular plants: arsenic hyperaccumulation by chinese brake fern (*Pteris vittata* L.). In: Chandra S, Srivastava M (eds) Pteridology in the new millenium. Kluwer, Amsterdam, pp 397–420

Bradshaw AD, McNeilly T, Gregory RPG (1965) Industrialization, evolution and development of heavy metal tolerance in plants. In: Goodman GT, Edwards RW, Lambert JM (eds) Ecology and the industrial society. The British ecological society symposium. Blackwell, Oxford, pp 327–344

Branzini A, González RS, Zubillaga M (2012) Absorption and translocation of copper, zinc and chromium by *Sesbania virgata*. J Environ Manage 102:50–54

Bundschuh J, Litter M, Ciminelli VST et al (2010) Emerging mitigation needs and sustainable options for solving the arsenic problems of rural and isolated urban areas in Latin America: a critical analysis. Water Res 44:5828–5845

Chaney RL, Angle JS, Broadhurst CL et al (2007) Improved understanding of hyperaccumulation yields commercial phytoextraction and phytomining technologies. J Environ Qual 36:1429–1443

Chiodi Boudet L, Escalante A, von Haeften G et al (2011) Assessment of heavy metal accumulation in two aquatic macrophytes: a field study. J Braz Soc Ecotoxicol 6:57–64

Cordero C, Incignieri K, Chiacchiarini PA et al (2013) Evaluation of plant species for phytoremediation of a mining waste from Andacollo, Neuquén-Argentina. Adv Mater Res 825:512–515

Dalvi AA, Bhalerao SA (2013) Response of plants towards heavy metal toxicity: an overview of avoidance, tolerance and uptake mechanism. Ann Plant Sci 2:362–368

de Viana ML, Albarracín Franco S (2008) Experimental tolerance to boron of the plant species *Nicotiana glauca*, *Jacaranda mimosifolia*, *Tecoma stans*, *Medicago sativa* and *Spinacea oleracea* in Argentina. Rev Biol Trop 56:1053–1061

Dechamps C, Noret N, Mozek R et al (2008) Cost of adaptation to a metalliferous environment for *Thlaspi caerulescens*: a field reciprocal transplantation approach. New Phytol 177:167–177

Deram A, Denayer FO, Petit D et al (2006) Seasonal variations of cadmium and zinc in *Arrhenatherum elatius*, a perennial grass species from highly contaminated soils. Environ Pollut 140:62–70

Dickinson NM, Baker AJM, Doronila A et al (2009) Phytoremediation of inorganics: realism and synergies. Int J Phytorem 11:97–114

Duffus JH (2002) Heavy metals-a meaningless term? (IUPAC technical report). Pure Appl Chem 74:793–807

Garbisu C, Alkorta I (2001) Phytoextraction: a cost-effective plant-based technology for the removal of metals from the environment. Bioresour Technol 77:229–236

Gasic K, Korban SS (2006) Heavy metal stress. In: Madhava Rao KV, Raghavendra AS, Janardhan Reddy K (eds) Physiology and molecular biology of stress tolerance in plants. Springer, Heidelberg, pp 219–254

Green C, Hoffnagle A (2004) Phytoremediation field studies database for chlorinated solvents, pesticides, explosives, and metals. U.S. Environmental Protection Agency, Office of Superfund Remediation and Technology Innovation, Washington, DC, p 168

Guerra F, Gainza F, Pérez R et al (2011) Phytoremediation of heavy metals using poplars (*Populus spp*): a glimpse of the plant responses to copper, cadmium and zinc stress. In: Golubev IA (ed) Handbook of phytoremediation. Nova Publishers, New York, pp 387–414
Hadad HR, Maine MA, Bonetto CA (2006) Macrophyte growth in a pilot-scale constructed wetland for industrial wastewater treatment. Chemosphere 63:1744–1753
Hadad HR, Mufarrege MM, Pinciroli M et al (2010) Morphological response of *Typha domingensis* to an industrial effluent containing heavy metals in a constructed wetland. Arch Environ Contam Toxicol 58:666–675
Hall JL (2002) Cellular mechanisms for heavy metal detoxification and tolerance. J Exp Bot 53:1–11
Hammer D, Kayser A, Keller C (2003) Phytoextraction of Cd and Zn with *Salix viminalis* in field trials. Soil Use Manage 19:187–192
Harguinteguy CA, Schreiber R, Pignata ML (2013) *Myriophyllum aquaticum* as a biomonitor of water heavy metal input related to agricultural activities in the Xanaes River (Córdoba, Argentina). Ecol Indic 27:8–16
Harper FA, Smith SE, Macnair MR (1997) Where is the cost in copper tolerance in *Mimulus guttatus*? Testing the trade-off hypothesis. Funct Ecol 11:764–774
Hossain MA, Piyatida P, Teixeira da Silva JA et al (2012) Molecular mechanism of heavy metal toxicity and tolerance in plants: central role of glutathione in detoxification of reactive oxygen species and methylglyoxal and in heavy metal chelation. J Bot. doi:10.1155/2012/872875
Ibáñez S, Medina MI, Agostini E (2011) Phenol tolerance, changes of antioxidative enzymes and cellular damage in transgenic tobacco hairy roots colonized by arbuscular mycorrhizal fungi. Chemosphere 83:700–705
Ibáñez SG, Sosa Alderete LG, Medina MI et al (2012) Phytoremediation of phenol using *Vicia sativa* L. plants and its antioxidative response. Environ Sci Pollut Res Int 19:1555–1562
Javed MT, Stoltz E, Lindberg S et al (2013) Changes in pH and organic acids in mucilage of *Eriophorum angustifolium* roots after exposure to elevated concentrations of toxic elements. Environ Sci Pollut Res Int 20:1876–1880
Kopittke PM, Blamey FPC, Asher CJ et al (2010) Trace metal phytotoxicity in solution culture: a review. J Exp Bot 61:945–954
Krzesłowska M (2011) The cell wall in plant cell response to trace metals: polysaccharide remodeling and its role in defense strategy. Acta Physiol Plant 33:35–51
Kudo A, Miyahara S (1991) A case history: Minamata mercury pollution in Japan – from loss of human lives to decontamination. Water Sci Technol 23:283–290
Lassat MM (2002) Phytoextraction of toxic metals: a review of biological mechanisms. J Environ Qual 31:109–120
Leontiadou H, Mark AE, Marrink SJ (2007) Ion transport across transmembrane pores. Biophys J 92:4209–4215
Lewandowski I, Schmidt U, Londo M (2006) The economic value of the phytoremediation function – assessed by the example of cadmium remediation by willow (*Salix* ssp). Agr Syst 89: 68–89
Lindqvist O, Johansson K, Aastrup M et al (1991) Mercury in the Swedish environment: recent research on causes, consequences and corrective methods. Water Air Soil Pollut 55:1–261
López SC, Melaj MA, Tomellini G et al (2005) Rizofiltración en el tratamiento de aguas contaminadas con uranio. Ing Sanit Amb 80:78–81
Madrid L (2010) "Heavy metals": reminding a long-standing and sometimes forgotten controversy. Geoderma 155:128–129
Maine MA, Suñe N, Hadad H et al (2006) Nutrient and metal removal in a constructed wetland for wastewater treatment from a metallurgic industry. Ecol Eng 26:341–347
Maine MA, Suñe N, Hadad H et al (2009) Influence of vegetation on the removal of heavy metals and nutrients in a constructed wetland. J Environ Manage 90:355–363
Merini LJ, Bobillo C, Cuadrado V et al (2009) Phytoremediation potential of the novel atrazine tolerant *Lolium multiflorum* and studies on the mechanisms involved. Environ Pollut 157: 3059–3063

Miretzky P, Saralegui A, Fernández Cirelli A (2004) Aquatic macrophytes potential for the simultaneous removal of heavy metals (Buenos Aires, Argentina). Chemosphere 57:997–1005

Miretzky P, Saralegui A, Fernández Cirelli A (2006) Simultaneous heavy metal removal mechanism by dead macrophytes. Chemosphere 62:247–254

Monferrán MV, Pignata ML, Wunderlin DA (2012) Enhanced phytoextraction of chromium by the aquatic macrophytes *Potamogeton pusillus* in presence of copper. Environ Pollut 161:15–22

Mufarrege M, Hadad HR, Maine MA (2010) Response of *Pistia stratiotes* to heavy metals (Cr, Ni, and Zn) and phosphorous. Arch Environ Contam Toxicol 58:53–61

Nagajyoti PC, Lee KD, Sreekanth TVM (2010) Heavy metals, occurrence and toxicity for plants: a review. Environ Chem Lett 8:199–216

Orroño DI, Schindler V, Lavado RS (2012) Heavy metal availability in *Pelargonium hortorum* rhizosphere: interactions, uptake and plant accumulation. J Plant Nutr 35:1374–1386

Philippi LS, Sezerino PH, Panceri B et al (2006) Root zone system to treat wastewater in rural areas in south of Brazil. Proc 10th Int Conf Wet Sys Water Pollut Cont, Lisbon, pp 901–908

Pilon-Smits E, Freeman JL (2006) Environmental cleanup using plants: biotechnological advances and ecological considerations. Front Ecol Environ 4:203–210

Podazza G, Rosa M, González JA et al (2006) Cadmium induces changes in sucrose partitioning, invertase activities and membrane functionality in roots of Rangpur lime (*Citrus limonia* L. Osbeck). Plant Biol 8:706–714

Podazza G, Arias ME, Prado FE (2012) Cadmium accumulation and strategies to avoid its toxicity in roots of the citrus rootstock Citrumelo. J Hazard Mater 215–216:83–89

Prado C, Rodríguez-Montelongo L, González JA et al (2010a) Uptake of chromium by *Salvinia minima*: effect on plant growth, leaf respiration and carbohydrate metabolism. J Hazard Mater 177:546–553

Prado C, Rosa M, Pagano E et al (2010b) Seasonal variability of physiological and biochemical aspects of chromium accumulation in outdoor-grown *Salvinia minima*. Chemosphere 81:584–593

Prado C, Pagano E, Prado FE et al (2012) Detoxification of Cr(VI) in *Salvinia minima* is related to seasonal-induced changes of thiols, phenolics and antioxidative enzymes. J Hazard Mater 239–240:355–361

Prado C, Rosa M, Pagano E et al (2013) Metabolic interconnectivity among alternative respiration, residual respiration, carbohydrates and phenolics in leaves of *Salvinia minima* exposed to Cr(VI). Environ Exp Bot 87:32–38

Pulford ID, Watson C (2003) Phytoremediation of heavy metal contaminated land by trees: a review. Environ Int 29:529–540

Rai PK (2009) Heavy metal phytoremediation from aquatic ecosystems with special reference to macrophytes. Crit Rev Environ Sci Technol 39:697–753

Rascio N, Navari-Izzo F (2011) Heavy metal hyperaccumulating plants: how and why do they do it? And what makes them so interesting? Plant Sci 180:169–181

Rizzo PF, Bres PA, Arreghini S et al (2012) Remediation of feedlot effluents using aquatic plants. Rev Fac Cien Agric 44:47–64

Scholz RW, Hansmann R (2007) Combining experts' risk judgments on technology performance of phytoremediation: self-confidence ratings, averaging procedures, and formative consensus building. Risk Anal 27:225–240

Siedlecka A (1995) Some aspects of interactions between heavy metals and plant mineral nutrients. Acta Soc Bot Pol 64:265–272

Sosa Alderete LG, Ibáñez SG, Agostini E et al (2012) Phytoremediation of phenol at pilot scale by tobacco hairy roots. Int J Environ Sci 3:398–407

Talano MA, Busso DC, Paisio CE et al (2012) Phytoremediation of 2,4-dichlorophenol using wild type and transgenic tobacco plants. Environ Sci Pollut Res 19:2202–2211

Tamasi G, Cini R (2004) Heavy metals in drinking waters from Mount Amiata (Tuscany, Italy). Possible risks from arsenic for public health in the Province of Siena. Sci Total Environ 327:41–51

Vassilev A, Schwitzguébel JP, Thewys T et al (2004) The use of plants for remediation of metal-contaminated soils. World Sci J 4:9–34

Wevar Oller AL, Agostini E, Talano MA et al (2005) Overexpression of a basic peroxidase in transgenic tomato (*Lycopersicon esculentum* Mill. cv. Pera) hairy roots increases phytoremediation of phenol. Plant Sci 169:1102–1111

Wright JW, Stanton ML, Scherson R (2006) Local adaptation to serpentine and non-serpentine soils in *Collinsia sparsiflora*. Evol Ecol Res 8:1–21

Wu G, Kang H, Zhang X et al (2010) A critical review on the bio-removal of hazardous heavy metals from contaminated soils: issues, progress, eco-environmental concerns and opportunities. J Hazard Mater 174:1–8

Wuana RA, Okieimen FE (2011) Heavy metals in contaminated soils: a review of sources, chemistry, risks and best available strategies for remediation. ISRN Ecol. doi:10.5402/2011/402647

Zietz BP, Dieter H, Lakomek M et al (2003) Epidemiological investigation on chronic copper toxicity to children exposed via the public drinking water supply. Sci Total Environ 302: 127–144

Chapter 8
Microbial Consortia, a Viable Alternative for Cleanup of Contaminated Soils

Liliana Beatriz Villegas, María Alejandra Martínez, Analia Rodríguez, and María Julia Amoroso

Abstract The growth population and anthropogenic activity are constantly threatening the environment caused by the accumulation of different kinds of pollutants in the biosphere, especially in soils and sediments. Co-contaminated of environment with toxic organic and inorganic substance is often actually. For the remediation of soils contaminated with mainly petroleum, pesticide and heavy metals, several physical or chemical techniques have been developed inadequately. Inside the technologies "eco-friendly remediation," the bioremediation have emerged as an option using natural biological activity. Bioremediation are methods where microorganisms degrade one or various pollutants to nontoxic compounds, so working individually or coordinately inside a microbial consortium. A microbial consortium is the natural association of two or more microbial populations of different species, which act together in a complex system. The success of a bioremediation process with pure cultures is very low and restricted. Therefore, use of a microbial consortium appears to be more feasible and reliable.

The chapter aims to review of the techniques for the elimination or degradation of pollutants using microbial consortia and highlight the importance of microbial

L.B. Villegas (✉)
Instituto de Química San Luis (INQUISAL), CONICET,
Chacabuco y Pedernera 5700 San Luis, Argentina
e-mail: lbvilleg@hotmail.com

M.A. Martínez
Planta Piloto de Procesos Industriales y Microbiológicos (PROIMI), CONICET,
Av. Belgrano y Pasaje Caseros, 4000 Tucumán, Argentina

Facultad de Ciencias Exactas y Tecnología, Universidad Nacional de Tucumán,
Avda. Independencia 1800, 4000 Tucumán, Argentina

A. Rodríguez
Planta Piloto de Procesos Industriales y Microbiológicos (PROIMI), CONICET,
Av. Belgrano y Pasaje Caseros, 4000 Tucumán, Argentina

M.J. Amoroso
Planta Piloto de Procesos Industriales Microbiológicos (PROIMI), CONICET,
Av. Belgrano y Pasaje Caseros, 4000 Tucumán, Argentina

Facultad de Bioquímica, Química y Farmacia, Universidad Nacional de Tucumán,
4000 Tucumán, Argentina

Universidad del Norte Santo Tomás de Aquino (UNSTA), 4000 Tucumán, Argentina

A. Alvarez, M.A. Polti (eds.), *Bioremediation in Latin America: Current Research and Perspectives*, DOI 10.1007/978-3-319-05738-5_8

consortia assessment. Also afford a discussion on the application of these techniques to the development of strategies and remediation policies.

8.1 Introduction

Soil is the surface area of the Earth's crust, which functions as support and habitat for living beings. In general terms, it consists of 50 % of free space (half the pore space occupied by gas and the other half by liquid), 40–50 % of solid mineral (rock weathering) and 0.5–5 % organic matter (humus and derivatives, living and dead biomass) (Brock and Madigan 2000). There is a variety of different soil types, according to the proportions of sand, silt, clay and organic matter mainly humus and derivatives is determined by the weather, the kind of vegetation, organisms, soil age and bedrock and especially rainfall and temperature.

Earth is a complex habitat, with numerous microenvironments and niches. Whereas populations in soil include macrofauna, mesofauna, microfauna and microflora, it is noted that 80–90 % of the biological reactions are mediated by soil microorganisms (Nannipieri and Badalucco 2003). In soil a variety of microbial populations can be found with diverse metabolic capabilities, also conditioned by the bioavailability of different sources of carbon and energy, which in turn is determined by the physicochemical characteristics of each soil (Kästner 2000). These diverse catabolic activities are due to the presence of catabolic gene and enzymes. Moreover, the microorganisms possess adaptation strategies such as the ability to modify the cellular membrane to maintain the necessary biological functions. Microorganisms are not just an essential part of living soil and critical to soil health, but they are also excellent indicators of healthy soil, because they reflect to changes in soil ecosystems quickly (Nielsen and Winding 2002). Unfortunately, population growth and industrialization are constantly threatening the environment. This is caused by the accumulation of organic and inorganic contaminants in the biosphere, especially in soils and sediments.

For the remediation of soils contaminated with mainly petroleum, pesticide and heavy metals, several physical, chemical and biological techniques have been developed. Based on data compiled by De Souza et al. (2013), countries member of the European Environment Agency (EEA), estimated in 2007 that about 250,000 areas require remediation due to anthropogenic activities and if this tendency remains, the number of areas that require repair will increase in 50 % awaiting 2025. In these countries, approximately 35 % of the costs for remediation are public. In the USA the report of the USEPA states that, although there have been various efforts to remediate contaminated areas, a considerable amount of work is still needed. The estimate cost for the remediation of these areas is around 209 billion dollars, funded by the responsible for the contamination, private or public entity.

Resource necessary, both human and financial, to achieve successful implementation of remediation programs are high. While, developed countries have adequate mechanisms (technical and financial) to implement these programs on a large scale, developing or underdeveloped countries this is not possible at the moment, and these contaminated areas are a problem that is compounded every day.

The traditional technologies, commonly used, are expensive and deficient because they require high power or large amounts of reagents. Inside the technologies "eco-friendly remediation," the bioremediation have emerged as an interesting option using natural biological activity.

In this context, this chapter aims to provide an exploration of current concepts of bioremediation, a review of the techniques for the elimination or degradation of pollutants using microbial consortia, furthermore provide a discussion on the application of these techniques to the development of strategies and remediation policies.

8.2 Environmental Pollution and Remediation

Soils are complex systems, multiple components with a range of different types of contaminants that exist in different physical and chemical forms. Despite this complexity, organic pollutants and toxic metals and their interactions are usually studied separately, however the effects of remediation procedures which can have both types of pollutants have been neglected (Amezcua-Allieri et al. 2005).

The structure of microbial communities can be determined by defining the metabolic capabilities of microorganisms and their biogeochemical consequences (heterotrophs, nitrifiers, denitrifiers, nitrogen fixers, etc.) and/or participation in ecosystem processes (decomposition, biodegradation, oxidation, among others), delimiting functional groups and structural units of the community (Hooper et al. 2002). The metabolic diversity of microorganisms that they have, may be exploited for cleaning up of contaminated soil and recovery of ecosystem services in affected ecosystems, through bioremediation. Bioremediation is a general concept that includes all those biological processes and actions that take place in order to biotransform an environment, already altered by contaminants, to its original status (Thassitou and Arvanitoyannis 2001). Bioremediation are methods where one or various microorganisms degrade one or various pollutants to nontoxic compounds present in soil, water or air, so working individually or coordinately (through synergies) inside a microbial consortium. These practices could be defined as an assisted and controlled elimination, attenuation or transformation of polluting substances that provides a permanent solution, which differs from other technologies that only provide a temporary solution by isolation and containment of the pollutant. This new technology uses dead or alive microorganisms or their enzymes free of cellular components, in order to achieve the mineralization or partial transformation of the contaminant to less toxic or harmless substances to the environment and human health (Röling et al. 2002).

Bioremediation has become one of the most important tools for removing or reducing contamination caused by various compounds of anthropogenic origin spilled into the environment, either accidentally or deliberately. In broadly, bioremediation techniques typically are cheaper than conventional methods such as incineration washed soil, excavation, ion exchange columns, and treatments with alkalis solutions or flocculating agents.

8.3 Microbial Consortium

The success of a bioremediation process with pure cultures is very low. On the other hand, use of a microbial consortium appears to be more feasible and reliable. Yu et al. (2005) showed that the addition of bacterial consortium formed for three strains into artificially contaminated sediments, significantly enhanced the efficiency of total degradation of polycyclic aromatic hydrocarbons after 4 weeks of incubation. In other work, a microbial consortium consisting of *Pseudomonas* strains was used for biodegradation of phenolic compounds in leachate- and oil-contaminated microcosm. They observed the shift in microbial populations in various microcosm and they suggested that different pathways of metabolism of aromatic compounds dominated under defined conditions and bacteria optimized the response to contaminants present in their surrounding (Heinaru et al. 2005). Therefore, in many cases consortia were more effective than a single strain, by the fact that intermediates of a catabolic pathway of one strain may be further degraded by other strains possessing suitable catabolic pathway. Microbial consortia are ubiquitous in nature and are involved in key natural processes, such as biogeochemical cycles. They have evolved over millions of years attaining resistance to environmental fluctuations thus promoting the stability of its members over time.

In nature, most processes are performed by a plethora of microorganisms, many of which have not been yet even cultured. Indeed, microbial consortia are implicated in practices of great importance to humans, from environmental remediation and wastewater treatment to assistance in food digestion. Microbial mediated environmental protection and remediation processes involve multiple microbial consortia since such associated microorganisms can perform functions that are difficult or even impossible for individual strains or species; it is individual consortium fitness, rather than individual strain performance, that is critical to achieve an efficient process (Hamer 1997; Brenner et al. 2008).

A microbial consortium is defined as the natural association of two or more microbial populations of different species, which act together as a community in a complex system where everyone benefits from the activities of others. The synergic association reflects life styles or syntrophyc where cyclic growth and nutrient flow is conducted more effectively and efficiently than individual populations (Ochoa Carreño and Montoya Restrepo 2010). A microbial consortium comprises two features. One is the communication among its members, either by the exchange of substances or molecular signals. Each population detects and responds to the presence of others within the consortium, exerting on them a positive or negative control to their growth and/or metabolism (Keller and Surette 2006). This feature allows the second important feature, which is the division of labor. Thus, a consortium exceeds the sum of the parts that maintain ecological compatibility metabolic members and to the extent that environmental changes allow them to coexist.

Since consortia are found naturally and can perform more complicated tasks and endure changing environments than individual strains, a reasonable approach to address bioremediation is finding and/or selecting microbial consortia competent to leave residual levels below concentration limits established by regulatory authorities and able to retain their catabolic skills stable when they are exposed to intermittent concentrations of the pollutants.

8.4 Microbial Consortia Assessment

Consortia may be defined or undefined. A defined consortium, known as mixed cultures, is a combination of isolates with degradative capabilities that may be complementary to each other (Casellas et al. 1998; Foght et al. 1999). Undefined consortia are the result of enrichment procedures from environmental samples (Sugiura et al. 1997; Budzinski et al. 1998). The characterization of complex microbial communities is a requirement to properly describe and understand it as well as to evaluate its performance. While biochemical and morphological characteristics could allow the analysis of defined mixed cultures, the use of independent culture approaches is of relevance when a microbial consortium should be assessed. This is particularly necessary when microbial consortia comprise non-cultivable members.

Indeed, independent culture methods are useful for processes when all members of a community are defined or not. Among them, clonal libraries of 16S rRNA genes are useful yet laborious and time consuming method for a large number of samples. On the other hand, denaturing gradient gel electrophoresis (DGGE) of 16S rRNA gene or other functional genes, terminal restriction fragment length polymorphism (T-RFLP), and sequence analysis have been widely used for monitoring a variety of microbial consortia (Van den Abbeele et al. 2010; Zanaroli et al. 2010; Stursová et al. 2012; Jiménez et al 2013). Fluorescent in situ hybridization (FISH), quantitative real-time polymerase chain reaction (qPCR), and microarrays have also proven to be powerful tools to analyze changes in a microbial community along complex processes. However, each method has limitations mainly due to biases originated in the lack of priming with the so-called "universal primers," which target different ranges of bacteria and will thus show the diversity and dynamics of different fractions of a community. To minimize these limitations, a combination of different techniques must be used (Nettmann et al. 2010; Kieu et al. 2013).

PCR-based techniques are commonly used to characterize microbial communities, but are subject to bias that is difficult to assess. Variability of PCR-fingerprinting methods may be caused by the sampling, DNA extraction, PCR amplification, and the analysis itself (e.g., DGGE). Thus, replicates must be processed in order to reach accurate conclusions. The primer selection and potential primer bias are critical factors when analyzing microbial communities in environmental samples (Fredriksson et al. 2013). It is frequently the use of the 357F-GC and 518R, although other 16S rRNA regions could be targeted (Muyzer et al. 1993; Lin et al. 2006; Mahmood et al. 2006; Van den Abbeele et al. 2010). For archaeal or fungi, primers towards 16S rRNA or intergenic spacers of ribosomal sequences can be used, respectively (Stursová et al. 2012). In general, nested and semi-nested PCR-based approaches render higher resolution than direct PCR-based DGGE for detection and community characterization on low-copy-number targets in environmental samples that may also contain PCR inhibitors, and generate good-quality amplification products for community profile fingerprinting and sequencing (Junier et al. 2008).

Most of microorganisms are not yet cultivated or described, therefore limiting the current knowledge of a vast majority of the environmental microbiota. However, high-resolution techniques such as pyrosequencing and microarrays, have provided access to information in unprecedented detail.

Next-generation sequencing (NGS) technologies permit the identification and characterization of complex microbial populations by delivering fast, inexpensive, and accurate genome information. In this context, metagenomics and metatranscriptomics emerge as key tools to describe and disclose complex degradation pathways (Zakrzewski et al. 2012). By means of metagenomic data, the analysis of ribosomal gene sequences discloses the composition of a microbial community. Inferences over the metabolic profiles are relatively useful when it is made over the taxonomic composition. Nevertheless, additional analyses are required in order to produce accurate conclusions. Massive sequencing projects without prior amplification produces the identification of environmental gene tags (EGTs), which are short sequences from the DNA of microbial communities that contain fragments of functional genes. Each EGT from genes that are important for survival and adaptation will be present in many genomes and will therefore occur repeatedly in the EGT data. This is indicative of genes that are adaptive in that habitat.

Once the consortium of interest is defined, fingerprinting procedures regain usefulness to detect prevalent and/or active members of the community. However, NGS via amplicon sequencing approaches rivals heavily since allows tracking multiple projects, thus emerging as a feasible alternative (Martínez et al. 2013).

Beyond the description and characterization of a consortia acting over a complex environment, it could be useful to set up procedures to ensure in short time a representative number of sequences. This information would supplement bioaugmentation, i.e., feed bioaugmentation, to correct or improve a consortium whose microbial components may be modified along a complex process.

Living in community is thought both to generate robustness to environmental fluctuations and to promote stability through time for the members of a consortium. Compared with monocultures, communities might be more capable of resisting invasion by other species. Furthermore, they might be able to weather periods of nutrient limitation better because of the diversity of metabolic modes available to a mix of species combined with the ability to share metabolites within the community (Brenner et al. 2008).

8.5 Bioremediation Policies

The processes that occur during bioremediation occur naturally without any technological intervention and is called "passive bioremediation," natural attenuation or intrinsic." But when a bioremediation project is addressed the goal is to clean (more or less demanding) for shorter periods possible. This technological solution applied to initiate the "active bioremediation."

The bioremediation of a contaminant in a soil depends on several factors including the presence of a potentially active microbial community degrading, the molecular structure of the contaminant, its concentration and bioavailability, and environmental factors such as temperature, pH, soil moisture, presence electron acceptors, and the existence of inorganic nutrients (N and P source) available. Other crucial factors are soil properties and history of contamination. The results are strongly "site dependent" and should be used for each particular case, since this will determine the success or failure of the process. Most bioremediation systems are run under aerobic conditions, but running a system under anaerobic conditions may permit microbial organisms to degrade otherwise recalcitrant molecules (Shukla et al. 2010).

Similar to other remediation technologies, bioremediation can be performed ex situ treatment, where material is excavated and moved out of the polluted site or in situ, that is in the same place (Velasco and Volke Sepúlveda 2003).

8.5.1 In Situ Bioremediation

This alternative is the most desirable due some contaminants can be addressed on site, reducing exposure risks to cleanup staff, or potentially wider exposure resulting from transport accidents. However, the treatment is limited by the depth of the soil that can be effectively treated. In many soils effective oxygen diffusion for desirable rates of bioremediation extend to a range of only a few centimeters to about 30 cm into the soil, although depths of 60 cm and greater have been effectively treated in some cases (Vidali 2001).

In situ techniques can be divided into two major groups: Biostimulation in order to enhance the growth of the bacteria; Bioaugmentation, is an alternative which is considered when the autochthonous microbiota of contaminated are under treatment is scarce or inadequate to metabolize the pollutants therefore is necessary to add a pollulant-degrading bacteria.

Biostimulation is the most common in situ treatment and involves oxygen supply to stimulate naturally occurring soil microorganisms to degrade compounds in soil. This method, also known as Bioventing, employs low air flow rates and provides only the amount of oxygen necessary for the biodegradation while minimizing volatilization and release of contaminants to the atmosphere. It works for simple hydrocarbons and can be used where the contamination is deep under the surface. Bioventing is the technology of choice for remediating many petroleum wastes and may eventually be used to treat a wider variety of more recalcitrant contaminants (Vidali 2001; Shukla et al. 2010). In some cases the oxygen applied is not enough and other nutrients must be added (including alternative electron acceptors) by circulating aqueous solutions through contaminated soils to stimulate naturally occurring bacteria. This technique also called "In situ biodegradation" can be used for soil and groundwater (Vidali 2001). In such treatments, nature of the microbial ecological niches is unpredictable, and their efficient is limited by the metabolic skills of the indigenous microbiological populations. When biostimulation unsuccessful the bioaugmentation may be an option.

Bioaugmentation, is the technique for improvement of the capacity of a contaminated matrix to remove pollution by the introduction of specific competent strains or consortia of microorganisms, indigenous or exogenous. Also, genetically engineered bacterial strains can be used, but they are seldom used. The basic premise for this intervention is that the metabolic capacities of the indigenous microbial community already present in the biotope slated for cleanup will be increased by an exogenously enhanced genetic diversity, thus leading to a wider repertoire of productive biodegradation reactions (Fantroussi and Agathos 2005). According to Forsyth et al. (1995, cited by Mrozik and Piotrowska-Seget 2010) bioaugmentation should be applied in soils with low number of contaminant-degrading microorganisms, on site contaminated with substance that requiring multiprocess remediation, including processes detrimental or toxic to microorganisms and for small scale sites on which cost of nonbiological methods are too expensive. However, this approach is limited by two main factors: nonindigenous cultures rarely compete well enough with an indigenous population to develop and sustain suitable population levels and most soils with long-term exposure to biodegradable waste have indigenous microorganisms that are effective degrades if the land treatment is well managed (Vidali 2001). There are still only a few well-documented field applications as the vast majority of reports in the literature deal with laboratory-scale or demonstration-scale studies. Many reports demonstrated the utility of bioaugmentation there are also case in which it has failed. One of the most difficult issues to overcome is survival of strains introduced to soil, the number of exogenous microorganisms sometimes decrease after soil inoculation. Many studies have shown that both abiotic and biotic factors impact the effectiveness of bioaugmentation. The most important abiotic influences are temperature, moisture, pH, and organic matter content (balance in the C:N:P ratio) and biotic factors including competition between indigenous and exogenous microoganisms for limited carbon source as well as antagonistic interaction and predation by protozoa and bacteriophages. All of these factors play essential roles in the final results of bioaugmentation (Mrozik and Piotrowska-Seget 2010).

In any case, for successful soil bioremediation, it is important to know the characteristic of the contaminate site before beginning treatment and also suitable strain or their consortia microbial. For example, some author reported that the bioestimulation of soil microorganisms upon the addition of nutrients, as N and P, had no or few effect on the decontamination of soil contaminated with diesel oil (Seklemova et al. 2001; Bento et al. 2005). While Gallego et al. (2001) found that using biostimulation technique achieved 90 % of diesel oil degradation in laboratory scale. These results are clearly linked to bioavailability of nutrients, a phenomenon that is directly related to the nature of the soil to be treated. Interestingly, some work relating both methods such as Bento et al. (2005), who compared the biostimulation and bioagumentation methods for degraded of total petroleum hydrocarbons in soil and found that the last one was the best. Similarly, Tongarun et al. (2008) associated the degradation rate of 4-chloroaniline for individually methods in two agriculture soil and they no found significative differences. However, these methods were much more effective in relation with bioattenuation. Bioattenuation is the method that relies on natural processes to dissipate contaminants through biological transformation (Mrozik and Piotrowska-Seget 2010). On the other hand, Ruberto et al. (2009) showed that bioaugmentation with indigenous bacterial consortium in Antarctic soil contaminated with hydrocarbons, showed no advantage as compared with biostimulation of the indigenous microflora with N and P. The biostimulated systems showed significant increased in total heterotrophic aerobic and hydrocarbons-degrading bacterial counts. Other works reported that the greatest bioremediation of oil-contaminated soil occurred when bioaugmentation (with Antarctic bacteria capable of efficient low temperature hydrocarbon degradation) and biostimulation techniques were combined (Stallwood et al. 2005).

8.5.2 Ex Situ Bioremediation

Landfarming is a simple technique in which contaminated soil is excavated and spread over a prepared bed and periodically tilled until pollutants are degraded. The goal is to stimulate indigenous biodegradative consortium and facilitate their aerobic degradation of contaminants. In general, the practice is limited to the treatment of superficial 10–35 cm of soil. Since landfarming has the potential to reduce monitoring and maintenance costs, as well as cleanup liabilities, it has received much attention as a disposal alternative. From an engineering perspective, landfarming is a “managed treatment and ultimate disposal process that involves the controlled application of a waste to a soil or soil-vegetation system” (Loehr et al. 1985 cited by Thassitou and Arvanitoyannis 2001). Landfarming relies on the principles applied in agriculture and aims at controlling the biocycling of natural compounds. The biodegradation conditions by the natural indigenous microbial populations of soil are optimized by the dilution of contaminated soil with clean soil, tilling of the soil to reduce initial toxicity, as well as by controlling physical parameters, such as aeration, pH, soil moisture content, and temperature. It is a slow process therefore long incubation periods is required.

Composting is a process that involves anaerobic microorganisms, converting solid organic wastes into humus-like material. The presence of these organic materials supports the development of a rich microbial population and elevated temperature characteristic of composting. This process occurs typically at elevated temperatures in the range of 55–65 °C. The increased temperatures result from the heat produced by microorganisms during the degradation or the organic material in the waste. For this, contaminated soils are excavated and screened to remove large rocks and debris. Soils and amendments are layered into long piles known as windrows (Chatterjee et al. 2008; Shukla et al. 2010).

Biopiling is a hybrid technique of landfarming and composting. This method is a full-scale technology in which excavated soils are mixed with soil amendments, placed on a treatment area, and bioremediated using forced aeration. The contaminants are reduced to carbon dioxide and water (Vidali 2001). Biopiles provide a favorable environment for indigenous aerobic and anaerobic microorganisms. The basic biopile system includes a treatment bed, an aeration system, an irrigation/nutrient system, and a leachate collection system. Moisture, heat, nutrients, oxygen, and pH are controlled to enhance biodegradation. The irrigation/nutrient system is buried under the soil to pass air and nutrients either by vacuum or positive pressure. Soil piles can be up to 20 ft high and may be covered with plastic to control runoff, evaporation and volatilization, and to promote solar heating. If volatile organic compounds in the soil volatilize into the air stream, the air leaving the soil may be treated to remove or destroy them before the atmosphere liberation. Treatment time is 3–6 months usually (Wu and Crapper 2009).

Bioreactors, this performance involves treating of contaminated water or solid material (soil, sediment, sludge) through an engineered containment system. A slurry bioreactor may be defined as a containment vessel and apparatus used to create a three-phase; solid, liquid, and gas mixing condition to increase the bioremediation rate of soil bound and water-soluble pollutants as a water slurry of the contaminated soil and biomass. The biomass can be a single strain, or consortium capable of degrading target contaminants. Indigenous microorganisms are used mostly. The biodegradation ratio and magnitude are greater in a bioreactor system than in situ systems because the contained environment is more controllable and more predictable. Nevertheless the polluted soil requires pretreatment as excavation or alternatively the contaminant can be stripped from the soil via soil washing or physical extraction before being placed in a bioreactor (Vidali 2001).

8.6 Influence of Heavy Metals Presence on Bioremediation Processes

Both ex situ and in situ treatments have been applied mainly for biodegradation of organic contaminants. In recent years, it has increased the co-contamination with inorganic pollutants such as heavy metals. Forty percent of hazardous waste sites on the Environmental Protection Agency's (EPA) National Priority List (NPL) are co-contaminated with organic pollutants and heavy metals (Olaniran et al. 2013). The heavy metals cannot be degraded biologically and their action on the environment depends on their mobility and bioavailability. Besides these phenomenons depend on metal total concentration and their association with the solid phase to which they are bound. The speciation and bioavailability of metals may change with variation in the environmental factors (Villegas et al. 2013a). At high concentrations, heavy metal ions can completely inhibit the microbial population; consequently, their presence has adversely affected the degradation of organic compounds directly by interacting with enzymes involved in biodegradation or indirectly by their various metabolic activities like protein denaturation, inhibition of cell division, and cell membrane disruption (Olaniran et al. 2013). Diverse chemical species of an element possess different physics and chemistry properties that influence bioavailability and solubility, affecting the absorption and the elimination and therefore the toxicity characteristics of them (Templeton et al. 2000; Villegas et al. 2011). Thus, estimating the heavy metal toxicity effects on organic pollutant biodegradation in co-contaminated soil and water environments is challenging. Regarding this some works can be found in the literature; for example on dechlorination and biodegradation of 2-chlorophenol (2-CP) and

3-chlorobenzoate (3-CB), the Cd(II), Cu(II), Cr(VI), and Hg(II) presence increased acclimation periods, reduced dechlorination or biodegradation rates, and failed to dechlorinate or biodegrade the target compound (Kuo and Genthner 1996). It was also demonstrated that arsenic co-contamination showed an inhibitory effect on the breakdown of 1,1,1-trichloro-2,2-bis(4-chlorophenyl)ethane (DDT) (Van Zwieten et al. 2003). Similar results were obtained in presence of elevated soil copper concentrations in pip and stone fruit orchard soils in the Auckland region, reducing the ability of the indigenous soil microbial community to degrade DDT (Gaw et al. 2003). A negative correlation between the degradation of the pyrethroid pesticides by microbial consortium in soil and Cu(II) addition was observed by Liu et al. (2007). Hoffman et al. (2005) investigated the impact of cadmium on naphthalene biodegradation by an isolated bacterium in three chemically defined minimal salts media and they suggested that the type of medium determines the degrees and patterns by which metals inhibit biodegradation and emphasized the importance of coupling metal toxicity and bioavailability data. On the other hand, enhanced biological phosphorus removal system (EBPR) is a widely implemented process in wastewater treatment plants to remove phosphorus using a microbial community. Fang et al. (2012) revealed that Cr(VI) with concentrations more than 5 mg L^{-1} had significantly inhibited P removal by EBPR system; these authors proposed that the chromium presence modified the composition of bacterial population and disturbed their enzymatic activities.

Co-contamination with organic and inorganic pollutants frequently origins a synergistic effect on cytotoxicity and both types of pollutants frequently must be treated with different methods (Hoffman et al. 2005; Lin et al. 2006). For this reason, it is very important to include heavy metal-resistant microorganisms in bioremediation technology. Roane et al. (2001) demonstrated the positive effect using the dual-bioaugmentation strategy with metal-detoxifying population and an organic-degrading population that cooperatively functioned to remediate both metal and organic pollutants in a co-contaminated site. Similar results were also obtained by Fernandes et al. (2009) when bacterial strains capable of withstanding considerable concentrations of Cd(II) or Hg(II) or Pb(II) were coupled with strains that showed good performance at degrading methyl tertiary butyl ether or trichloroethane. Dual bioaugmentation demonstrated the most degradation of 1,2-dichloroethane (1,2-DCA) in water co-contaminated with heavy metals compared to biostimulated or bioaugmented microcosms. However these works did not evaluate the heavy metal removal. In relation to this subject, the biostimulation of an actinobacterium strain in culture medium with phosphate or sulfate ions, increased the speed and biological potential of Cr(VI) removal, optimizing both the time and costs involved (Villegas et al. 2013b). Obviously more studies are necessary for achieving a deeper knowledge.

8.7 Concluding Remarks

Due to the fact that many polluted area frequently contain contamination if diverse nature simultaneously, it is very important to consider remediation procedures utilizing microbial consortia formed for selected microorganisms with an enhanced or ability to remove different pollutants. On the other hand, bioremediation is clearly a field of interdisciplinary work. This is partially due to its complexity involving several different and complementary approaches; various branches of engineers, biologists (i.e., microbiologists, ecologists), chemists, geologists among other professionals have key contributions to properly design and develop of efficient remediation operations. Thus, besides microbial consortia, an interdisciplinary and collaborative human consortium is necessary to design and set up successful bioremediation projects.

References

Amezcua-Allieri MA, Lead JR, Rodríguez-Vázquez R (2005) Impact of microbial activity on copper, lead and nickel mobilization during the bioremediation of soil PAHs. Chemosphere 61:484–491

Bento FM, Camargo FAO, Okeke BC et al (2005) Compartive bioremediation of soils contaminated with diesel oil by natural attenuation, biostimulation and bioaugmentation. BioresTechnol 96:1049–1055

Brenner K, You L, Arnold FH (2008) Engineering microbial consortia: a new frontier in synthetic biology. Trends Biotechnol 9:483–489

Brock D, Madigan M (2000) Microbiología, 8th edn. Prentice Hall Hispanoamericana, Mexico

Budzinski H, Raymond N, Nadalig T et al (1998) Aerobic biodegradation of alkylated aromatic hydrocarbons by a bacterial community. Org Geochem 28:337–348

Casellas M, Grifoll M, Sabate J et al (1998) Isolation and characterization of a 9-fluorenone-degrading bacterial strain and its role in synergistic degradation of fluorene by a consortium. Can J Microbiol 44:734–742

Chatterjee S, Chattopadhyay P, Roy S et al (2008) Bioremediation: a tool for cleaning polluted environments. J App Biosci 11:594–601

De Souza RB, Maziviero TG, Christofoletti CA et al (2013) Soil contamination with heavy metals and petroleum derivates: impact on edaphic fauna and remediation strategies. In: Hernandez Soriano MC (ed) Soil processes and current trends in quality assessment. Intech, Rijeka

Fang J, Sun PD, Xu SJ et al (2012) Impact of Cr(VI) on P removal performance in enhanced biological phosphorus removal (EBPR) system based on the anaerobic and aerobic metabolism. Bioresour Technol 121:379–385

Fantroussi SE, Agathos SN (2005) Is bioaugmentation a feasible strategy for pollutant removal and site remediation? Curr Opin Microbiol 8:268–275

Fernandes VC, Albergaria JT, Oliva-Teles T et al (2009) Dual augmentation for aerobic bioremediation of MTBE and TCE pollution in heavy metal-contaminated soil. Biodegradation 20:375–382

Foght J, Semple K, Gauthier C et al (1999) Effect of nitrogen source on biodegradation of crude oil by a defined bacterial consortium incubated under cold, marine conditions. Environ Technol 20:839–849

Fredriksson NJ, Hermansson M, Wilén BM (2013) The choice of PCR primers has great impact on assessments of bacterial community diversity and dynamics in a wastewater treatment plant. PLoS One 8(10):e76431

Gallego JR, Loredo J, Llamas JF et al (2001) Bioremediation of diesel-contaminated soils: evaluation of potential in situ techniques by study of bacterial degradation. Biodegradation 12:325–335

Gaw SK, Palmera G, Kimb ND, Wilkinsa AL (2003) Preliminary evidence that copper inhibits the degradation of DDT to DDE in pip and stonefruit orchard soils in the Auckland region, New Zealand. Environ Pollut 122:1–5

Hamer G (1997) Microbial consortia for multiple pollutant biodegradation. Pure Appl Chem 69:2343–2356

Heinaru E, Merimaa M, Viggor S et al (2005) Biodegradation efficiency of functionally important populations selected for bioaugmentation in phenol- and oil-polluted area. FEMS Microbiol Ecol 51:363–373

Hoffman DR, Okan JL, Sandrin TR (2005) Medium composition affects the degree and pattern of cadmium inhibition of naphthalene biodegradation. Chemosphere 59:919–927

Hooper D, Solan M, Symstad A et al (2002) Species diversity, functional diversity and ecosystem functioning. In: Loreau M, Naeem S, Inchausti P (eds) Biodiversity and ecosystem functioning, synthesis and perspectives. Oxford University Press, Oxford, pp 195–281

Jiménez DJ, Korenblum E, van Elsas JD (2013) Novel multispecies microbial consortia involved in lignocellulose and 5-hydroxymethylfurfural bioconversion. Appl Microbiol Biotechnol. doi:10.1007/s00253-013-5253-7

Junier P, Kim OS, Hadas O et al (2008) Evaluation of PCR primer selectivity and phylogenetic specificity by using amplification of 16S rRNA genes from betaproteobacterial ammonia-oxidizing bacteria in environmental samples. Appl Environ Microbiol 74:5231–5236

Kästner M (2000) Humification: process or formation of refractory soil organic matter. In: Klein J (ed) Biotechnology. Environmental processes II. Soil decontamination. Wiley-VCH, Weinheim, pp 89–126

Keller L, Surette MG (2006) Communication in bacteria: an ecological and evolutionary perspective. Nat Rev Microbiol 4:249–258

Kieu HT, Horn H, Müller E (2013) The effect of heavy metals on microbial community structure of a sulfidogenic consortium in anaerobic semi-continuous stirred tank reactors. Bioprocess Biosyst Eng. doi:10.1007/s00449-013-1012-5

Kuo CW, Genthner BRS (1996) Effect of added heavy metal ions on biotransformation and biodegradation of 2-chlorophenol and 3-chlorobenzoate in anaerobic bacteria consortia. Appl Environ Microbiol 62:2317–2323

Lin Q, Zhaowei W, Ma S et al (2006) Evaluation of dissipation mechanisms by Lolium. perenne L, and Raphanus. sativus for pentachlorophenol (PCP) in copper co-contaminated soil. Sci. Total Environ 36:814–822

Liu TF, Sun C, Hong TJ et al (2007) Effect of copper on the degradation of pesticides cypermethrin and cyhalothrin. J Environ Sci 19:1235–1238

Mahmood S, Freitag TE, Prosser JI (2006) Comparison of PCR primer-based strategies for characterization of ammonia oxidizer communities in environmental samples. FEMS Microbiol Ecol 56:482–493

Martínez MA, Romero H, Peroti NI (2013) Two amplicon sequencing strategies revealed different facets of the prokaryotic community associated with the anaerobic treatment of vinasses from ethanol distilleries. Biores Technol. doi:10.1016/j.biortech.2013.12.030

Mrozik A, Piotrowska-Seget Z (2010) Bioaugmentation as a strategy for cleaning up of soils contaminated with aromatic compounds. Microbiol Res 165:363–375

Muyzer G, de Waal EC, Uitterlinden AG (1993) Profiling of complex microbial populations by denaturing gradient gel electrophoresis analysis of polymerase chain reaction-amplified genes coding for 16S rRNA. Appl Environ Microbiol 59:695–700

Nannipieri P, Badalucco L (2003) Processes in the soil-plant system: modelling concepts and applications. Eur J Soil Sci 54:655–670

Nettmann E, Bergmann I, Pramschüfer S et al (2010) Polyphasic analyses of methanogenic archaeal communities in agricultural biogas plants. Appl Environ Microbiol 76:2540–2548

Nielsen MN, Winding A (2002) Microorganisms as indicators of soil health, NERI Technical Report No. 388, National Environmental Research Institute, Copenhagen, Denmark

Ochoa Carreño DC, Montoya RA (2010) Microbial consortia: a biological metaphor applied to enterprise association in agricultural production chains. Rev Fac Cienc Econ 18:55–74

Olaniran AO, Balgobind A, Pillay B (2013) Bioavailability of heavy metals in soil: impact on microbial biodegradation of organic compounds and possible improvement strategies. Int J Mol Sci 14:10197–10228

Roane TM, Josephson KL, Pepper IL (2001) Dual-bioaugmentation strategy to enhance remediation of co-contaminated soil. Appl Environ Microbiol 67:3208–3215
Röling WF, Milner MG, Jones M et al (2002) Robust hydrocarbon degradation and dynamics of bacterial communities during nutrient-enhanced oil spill bioremediation. Appl Environ Microbiol 68:5537–5548
Ruberto L, Dias R, Lo Balbo A et al (2009) Influence of nutrients addition and bioaugmentation on the hydrocarbon biodegradation of a chronically contaminated Antarctic soil. J Appl Microbiol 106:1101–1110
Seklemova E, Pavlova A, Kovacheva K (2001) Biostimulation based bioremediation of diesel fuel: field demonstration. Biodegradation 12:311–316
Shukla KP, Singh NK, Sharma S (2010) Bioremediation: developments, current practices and perspective. Genet Eng Biotechnol J 2010:1–20
Stallwood B, Shears J, Williams PA et al (2005) Low temperature bioremediation of oil-contaminated soil using biostimulation and bioaugmentation with a Pseudomonas sp. from maritime Antarctica. J Appl Microbiol 99:794–802
Stursová M, Zifčáková L, Leigh MB et al (2012) Cellulose utilization in forest litter and soil: identification of bacterial and fungal decomposers. FEMS Microbiol Ecol 80:735–746
Sugiura KM, Ishihara T, Shimauchi T et al (1997) Physicochemical properties and biodegradability of crude oil. Environ Sci Technol 31:45–51
Templeton D, Ariese F, Cornelis R et al (2000) Guidelines for terms related to chemical speciation and fractionation of elements. Definitions, structural aspects, and methodological approaches. Pure Appl Chem 72:1453–1470
Thassitou PK, Arvanitoyannis IS (2001) Bioremediation: a novel approach to food waste management. Trends Food Sci Technol 12:185–196
Tongarun R, Luepromchai E, Vangnai AS (2008) Natural attenuation, biostimulation and bioaugmentation in 4-chloroaniline-contaminated soil. Curr Microbiol 56:182–188
Van den Abbeele P, Grootaert C, Marzorati M et al (2010) Microbial community development in a dynamic gut model is reproducible, colon region specific, and selective for bacteroidetes and clostridium cluster IX. Appl Environ Microbiol 76:5237–5246
Van Zwieten L, Ayres MR, Morris SG (2003) Influence of arsenic co-contamination on DDT breakdown and microbial activity. Environ Pollut 124:331–339
Velasco JA, Volke Sepúlveda TL (2003) El composteo: una alternativa ecológica para la biorremediación de suelos en México. Gaceta Ecol 66:41–53
Vidali M (2001) Bioremediation. An overview. Pure Appl Chem 7:1163–1172
Villegas LB, Amoroso MJ, Figueroa LIC (2011) Interaction of copper or chromium with yeasts: potential application on polluted environmental clean up. In: Mason AC (ed) Bioremediation: biotechnology, engineering and environmental management. Nova Science Publisher, New York
Villegas LB, Pereira CE, Colin VL et al (2013a) The effects of sulphate and phosphate ions on Cr(VI) reduction by Streptomyces sp. MC1, including studies of growth and pleomorphism. Int Biodeter Biodegr 82:149–156
Villegas LB, Rodríguez A, Pereira CE et al (2013b) Cultural factors affecting heavy metals removal by actinobacteria. In: Amoroso MJ, Benimeli CS, Cuozzo SA (eds) Actinobacteria, application in bioremediation and production of industrial enzymes. CRC PressI Llc, Boca Raton, FL
Wu T, Crapper M (2009) Simulation of biopile processes using a hydraulics approach. J Hazar Mat 171:1103–1111
Yu SH, Ke L, Wong YS et al (2005) Degradation of polycyclic aromatic (PAHs) by a bacterial consortium enriched from mangrove sediments. Environ Int 32:149–154
Zakrzewski M, Goesmann A, Jaenicke S et al (2012) Profiling of the metabolically active community from a production-scale biogas plant by means of high-throughput metatranscriptome sequencing. J Biotechnol 158:248–258
Zanaroli G, Di Toro S, Todaro D et al (2010) Characterization of two diesel fuel degrading microbial consortia enriched from a non acclimated, complex source of microorganisms. Microb Cell Fact 9:10

Chapter 9
Application of Integrated Microbial Processes for Heavy Metal Recovery from Industrial Wastes of Buenos Aires, Argentina

Josefina Plaza Cazón, Nadia Yagnentovsky, Marisa Viera, and Edgardo Donati

Abstract To allow the final disposal of galvanic sludge from a wastewater treatment plant of automotive factory, integrated microbial processes were performed to release and recover nickel and zinc contained in the solid residue. The metals were successfully leached using sulfuric acid continuously produced by *Acidithiobacillus thiooxidans* biofilm on elemental sulfur. The best condition using a sludge pulp density of 1 % w/v and a dilution rate of 0.22 h^{-1} 100 % allowed more than 90 % of nickel and zinc dissolution, respectively; higher values of dilution rate and/or pulp density decreased the percentages of metal dissolution. The recovery of metals from the leachates was performed by continuous adsorption on *Undaria pinnatifida* biomass after raising the pH up to 4. Adsorption processes allowed great recoveries of nickel and zinc from monometallic solutions at the lowest flow rate (100 and 90 %); the recoveries from the leachates were not so good and to increase them three fixed bed columns in series were used reaching approximately 50 and 80 % of nickel and zinc recovery respectively, even at high dilution rates (more than 1.5 h^{-1}). This integrated process could be applied on a higher scale.

9.1 Introduction

Industrial activity generates wastes and sludges with high heavy-metal content, and their inadequate disposal can cause serious environmental impacts whether those heavy metals are mobilized and reach soils and groundwater. The mobility of the heavy metals from the wastes in aqueous system is a cause of concern because of their intrinsic solubility at varying pH, reduction–oxidation characteristics and

---

J. Plaza Cazón • N. Yagnentovsky • M. Viera • E. Donati (✉)
Centro de Investigación y Desarrollo en Fermentaciones Industriales (CINDEFI),
Centro Científico Tecnológico (CCT-La Plata), Consejo Nacional de Investigaciones
Científicas y Técnicas (CONICET), Universidad Nacional de La Plata,
Calle 50 N° 227, 1900 La Plata, Argentina
e-mail: donati@quimica.unlp.edu.ar

A. Alvarez, M.A. Polti (eds.), *Bioremediation in Latin America: Current Research and Perspectives*, DOI 10.1007/978-3-319-05738-5_9

complex formation tendencies. Unlike organic pollutants which can be degraded to harmless chemical species, heavy metals cannot be destroyed. Application of a suitable extraction process to recover some of the toxic metals present in the wastes is an appropriate approach to avoid their dispersion into the environment. Although different technologies can be applied to removing heavy metals, biological processes have some advantages especially when dealing with large volumes of wastes and not so high content of heavy metals (Viera and Donati 2004; Li et al. 2012; D'Emilio et al. 2013; Lee et al. 2013).

One of the possible biological processes applicable to remove heavy metals from solid wastes is based on the biological methodologies to extract metals from minerals. These methodologies—called bioleaching and biooxidation—use the ability of some microorganisms to derive energy for their metabolism from the catalysis of the aerobic oxidation of ferrous iron and/or reduced sulfur compounds (sulfides, elemental sulfur). Bacteria from *Acidithiobacillus* genus are the most important microorganisms with such abilities. These cells can generate acidic (sulfuric acid is the last product during sulfur-compound oxidation) and oxidizing media (ferric iron is produced during the oxidation of ferrous iron) which are able to attack metallic sulfides present into minerals. During this attack metals are solubilized and then they can be recovered from the solution. Bioleaching processes are used at industrial scale for the recovery of copper (Cu), zinc (Zn), and uranium (U) especially from low-grade ores while biooxidation processes are used for the pre-treatment of refractory gold ores prior to cyanidation (Donati and Sand 2007).

Applying similar processes, metals can be removed from industrial solid wastes that cannot be handled with conventional techniques. Several reports about such treatments to recover metals from slag, galvanic sludges, tannery wastes, catalysts, batteries, etc., have been published in the last years (Cerrutti et al. 1998; Viera et al. 2003; Cabrera et al. 2007; Bosio et al. 2008; Coto et al. 2008; Pathak et al. 2009; Fang et al. 2011; Hoque and Philip 2011; Kaksonen et al. 2011; Asghari et al. 2013; Ilyas et al. 2013).

After metals are released from the solid wastes, they should be recovered from the liquid phase for the final disposal. Conventional treatment implies the precipitation with lime although large sludge volumes (and low stability due to high solubility of hydroxides) are obtained. There are certain disadvantages of this type of treatment but especially that the process leaves such hazardous and voluminous sludges which need to be safely disposed of; in addition, metals cannot be economically recovery from the sludges. Different biological alternatives can be used to replace chemical precipitation for recovering metals. Probably biosorption is the most promissory technology between these alternatives (Kratochvil and Volesky 1998; Mudhoo et al. 2012). Biosorption uses different biomasses to concentrate and immobilize heavy metals. Using abundant and cheap biomasses makes biosorption more attractive than other similar technologies like ion exchange treatment. Biosorbents using algae, fungi, or bacteria (usually metabolically inactive) have been studied. However, due to low costs and availability industrial waste biomass and seaweed biomass are considered the most adequate for environmental applications at commercial scale. Seaweed biomass offers excellent metal-sorbing properties (Romera et al. 2007; Plaza Cazón et al. 2012b). *Undaria pinnatifida* is a brown alga

(Phaeophyta) that grows in water along the south coast of Argentina. Since it is an invasive species, *U. pinnatifida* is producing undesirable changes into the diversity of the Argentinean coast. This seaweed biomass is frequently deposited on the beach causing unpleasant odors and that is why it can be considered an available natural waste. There are very few reports on the use of this biomass as biosorbent for heavy metal removal although the presence of alginate in its cell wall could be an important characteristic for the biosorption process (Plaza Cazón et al. 2011, 2012a, 2013).

In this chapter we describe an integrated process that involves heavy metals dissolution from solid wastes using *A. thiooxidans* action and subsequent metal adsorption on *U. pinnatifida* biomass.

9.2 Bioleaching Experiment of Solid Wastes

A galvanic sludge from a wastewater treatment plant of automotive factory was used throughout the experiments. Sludge was dried, ground and sieved and the particles with size between 100 and 200 mesh were used for the experiments. The chemical analysis of the sludge indicated that the main toxic components of the sludge were zinc (Zn) (59.2 mg Zn/g dried sludge) and nickel (Ni) (14.6 mg Ni/g dried sludge) that are 40- and 30-fold higher (for Zn and Ni, respectively) than the required levels for final disposal. Sequential chemical extractions (Ngiam and Lim 2001) were carried out for the partitioning of Zn and Ni into five fractions: exchangeable, bound to carbonates, bound to iron–manganese (Fe–Mn) oxides, bound to organic matter and residual. More than half of Zn and Ni were found in the residual fraction while the metal contents in the other fractions—in ascending order—were 0.2, 20, 23, and 4 (for Zn) and 0.5, 22, 16, and 11 % (for Ni).

The strain of *A. thiooxidans* (DSM 11478) used throughout this work was inmobilized in bioreactors able to produce sulfuric acid. For this purpose, three percolation columns were prepared adding different amount of elemental sulfur. *A. thiooxidans* culture was inoculated at 10 % v/v in an exponential stage of growth. According to the volume of medium and the amount of sulfur, 24, 54 and 94 % w/v of pulp density were reached in the different columns (Fig. 9.1). The procedure was repeated until a constant rate of sulfuric acid production was reached (this situation indicates the maximum attached bacterial population which means biofilm formation).

The performance of each reactor was evaluated through the total acid production while the pH was below 2 (needed to maintain the metals into the solution). Best performance was achieved (data not shown) in the second bioreactor (100 g elemental sulfur, 54 % w/v pulp density) at 0.22 h^{-1}, so was used in the bioleaching experiments.

For the bioleaching experiments, solid wastes were added to other columns, filled with biogenic sulfuric acid (produced into the bioreactors) at different pH values (1, 1.5, and 2) to achieve 1, 2, and 4 % w/v of pulp density and fed with the same biogenic sulfuric acid at a dilution rate of 0.22 h^{-1} (Fig. 9.2a, b). Metal dissolution was really fast using biogenic sulfuric acid at the lowest pH while the dissolution was slow and lower (pH 2).

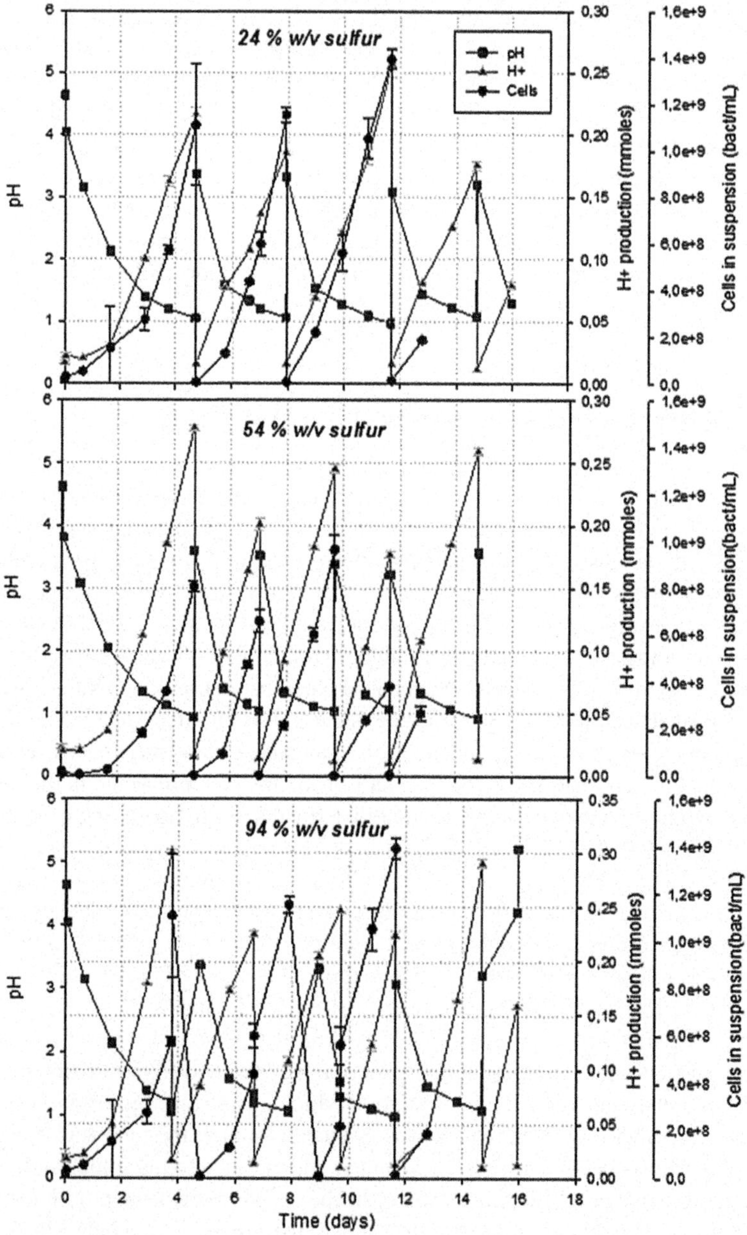

Fig. 9.1 pH evolution, acid production, and cells concentration in solution during biofilm formation in three bioreactors with different pulp density

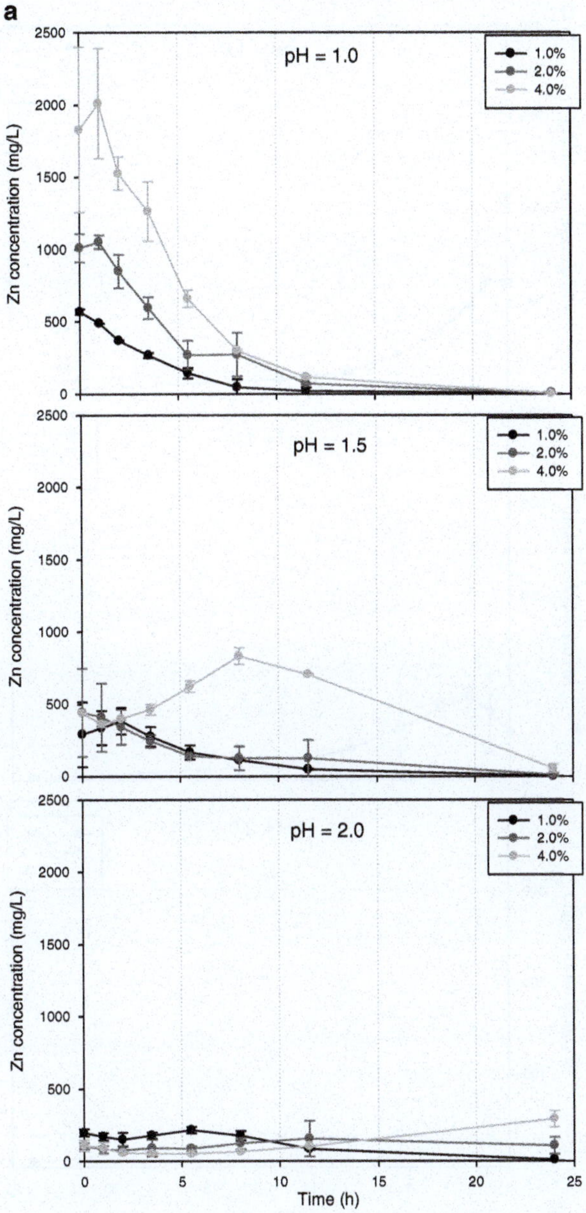

Fig. 9.2 Metal dissolution during bioleaching of solid waste using biogenic sulfuric acid at different pH values (**a**): zinc, (**b**): nickel

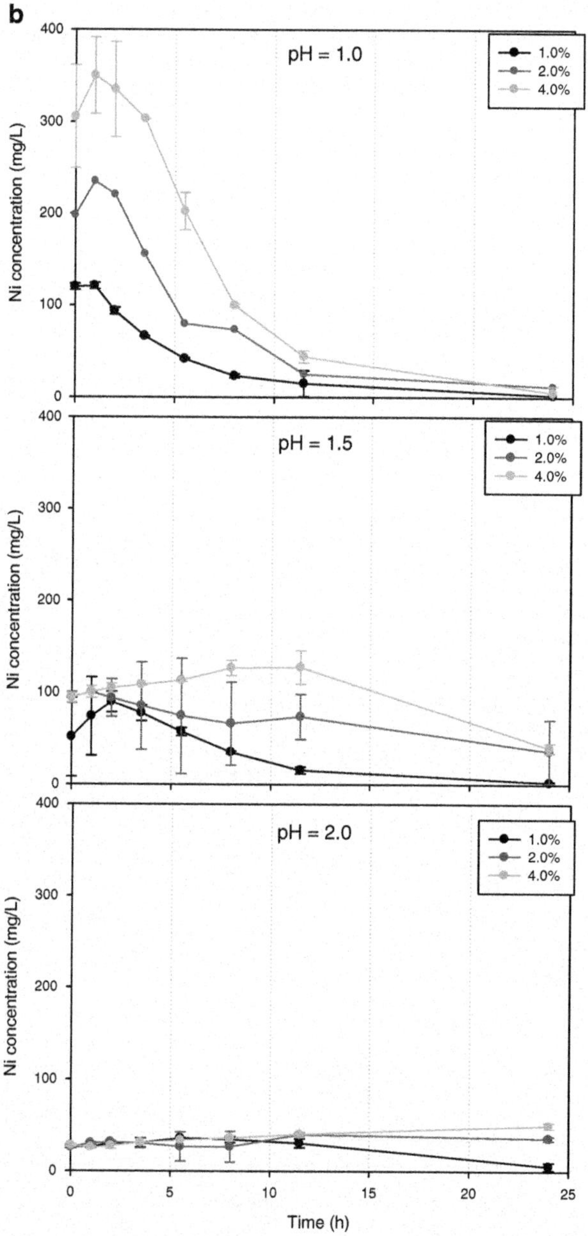

Fig. 9.2 (continued)

At the lowest pulp density (1 % w/v) Zn recoveries were almost 90 % (Fig. 9.3a) while Ni recoveries were even higher independently of the pH value (Fig. 9.3b). Higher pulp densities negative and significantly affected the metal recoveries at the highest pH (2). Analysis of variance showed that the effects of initial pH values and pulp densities on Zn(II) and Ni(II) recoveries were significantly ($P<0.05$).

Fig. 9.3 Metal recoveries during bioleaching of solid waste using biogenic sulfuric acid at different pH values (**a**): zinc, (**b**): nickel

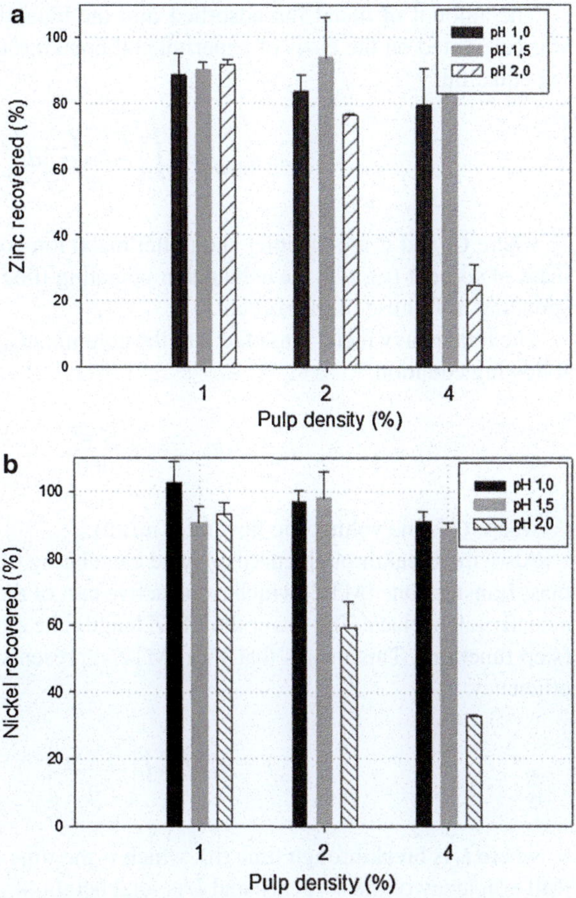

9.3 Metal Biosorption from the Leachates by Seaweed Biomass

Algae biomass was ground and sieved and the 10–16 mesh particle size was selected.

Studies were carried out using a plastic column packed with 2 g of *U. pinnatifida* biomass. At the top and the base of the column, a layer of glass wool was placed to prevent clogging of the inlet or outlet piping.

The procedure was the following: firstly distilled water was pumped to avoid a sudden initial increase in the metal initial concentration due to a rapid absorption of water by the dried seaweed. This procedure was adequate to allow the homogeneous contact between the solution and the biosorbent preventing the formation of preferential channels. After the contact with water Zn(II) solution (200, 370, and 550 mg/L) and Ni(II) solution (50, 100, and 175 mg/L) at pH 4 were fed upward through the column at 5 and 10 mL/h. Effluent was collected from the top of the column to determine metal concentration and other parameters.

The amount of metal ion adsorbed into the biosorbent in column experiments was calculated on the basis of experimental breakthrough curves using the following equation:

$$q_{\text{column}} = \int_{t=0}^{t=t_e} \left(\frac{C_0 - C}{m} \right) dt$$

where C_0 and C are the inlet and outlet metal ion concentration (mg/L), m is the mass of sorbent (g), t_e is the exhaustion operating time (h) which is the time corresponding to full use of the sorbent.

The total heavy metal mass fed into the column, m_{total} (mg) was calculated by the following equation:

$$m_{\text{total}} = \frac{C_0 - Qt_e}{1{,}000}$$

where Q is the volumetric flow rate (mL/h).

Also, the breakthrough curve allowed calculating others parameters, such as the mass transfer zone (MTZ) which is the active part of the fixed bed where adsorption actually takes place. The lower the MTZ length, the closer the system is to ideality (step function). This means that short MTZ represents systems with high removal efficiency.

$$\text{MTZ} = L\left(\frac{t_e - t_b}{t_e} \right)$$

where t_b is breakthrough time (h), which is the time for complete metal recovery (full efficiency of the sorbent), and L is total height of packed bed (cm).

At lowest flow rate and lowest inlet Zn(II) concentration no metal appeared in the outlet for more than 30 h (more than 50-fold higher than the residence time). Lower breakthrough times were found at higher inlet Zn(II) concentrations. Higher flow rates provoked much lower breakthrough times independently of the inlet metal concentration (Fig. 9.4a, b).

On the other hand, at lowest flow rate and lowest inlet Ni(II) concentration, breakthrough time was not reached up to 200 h (more than 350-fold higher than the residence time) The breakthrough times were 120- and 70-fold higher than the residence time for the other inlet Ni(II) concentrations (150 and 175 mg/L, respectively). At 10 mL/h the breakthrough times became much lower (between 20 and 40 h). (Fig. 9.5a, b). Comparing results for similar inlet metal concentration (the maximum inlet Ni(II) concentration is close to the minimum Zn(II) concentration) biosorption was more efficient for Zn(II) than for Ni(II).

The results from continuous assays were fitted employing the equation models presented in Table 9.1. Figures 9.6 and 9.7 show experimental data and fittings using the different models for two specific conditions; similar results have been found for other conditions.

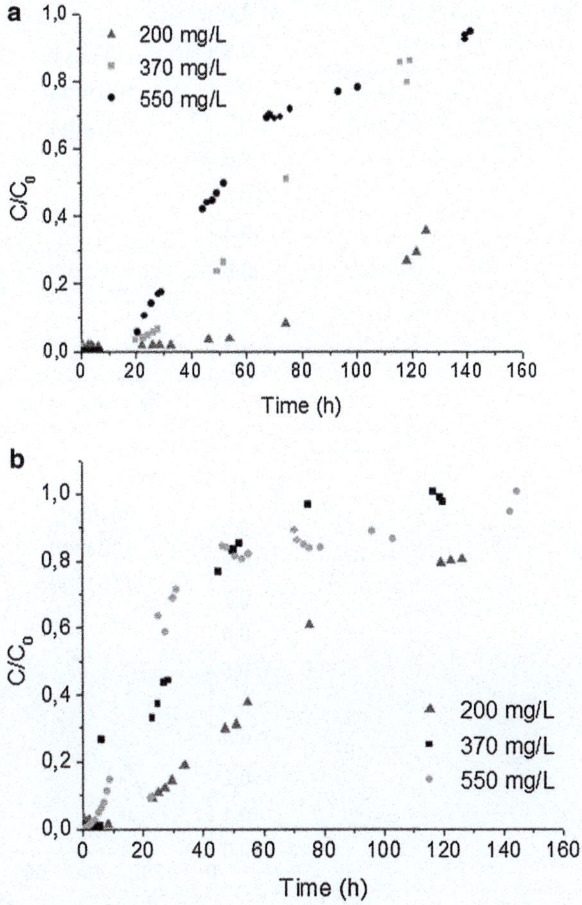

Fig. 9.4 Zinc(II) breakthrough curves in fixed bed column packed with *U. pinnatifida* biomass (**a**): 5 mL/h, (**b**): 10 mL/h

The amount of Zn retained per mass unit of biosorbent varies under the different conditions (flow rate and inlet metal concentration); probably it indicates that biomass saturation was not really obtained under all conditions used. The maximum amount of Zn adsorbed was 98.88 mg/g. As it is expected t_e and t_b decreased when flow rate and/or inlet metal concentration increased. Although the three models fit quiet well (except at the highest flow rate and metal concentration) to experimental data corresponding to Zn(II) biosorption, according to the correlation coefficients the dose-response model was the best.

As it was obtained for Zn(II) biosorption, also in Ni(II) biosorption there were differences in the amount of metal retained per gram of biosorbent. The maximum value reached was 45 mg/g (at the highest flow rate and highest inlet metal concentration). All the models fit well with the experimental data although the dose-response model was the best.

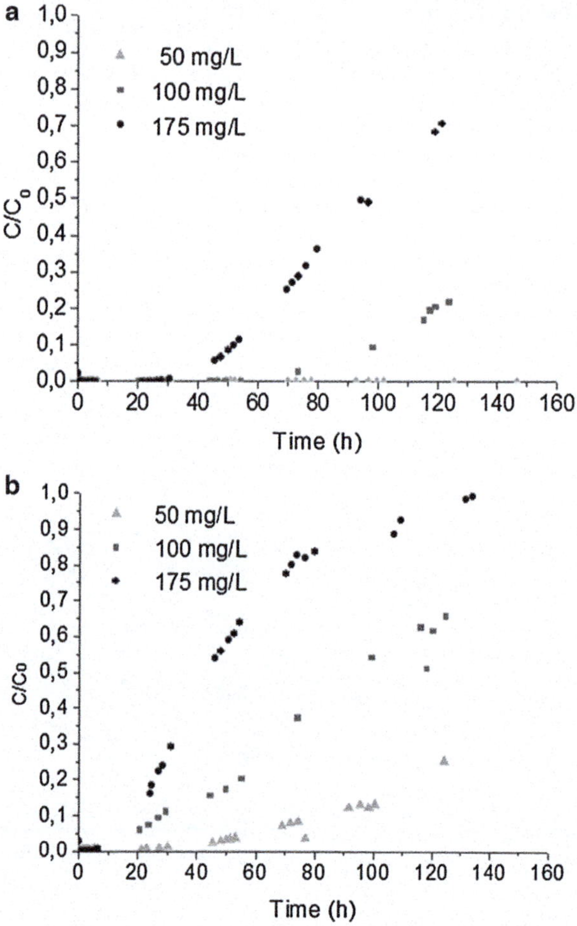

Fig. 9.5 Nickel(II) breakthrough curves in fixed bed column packed with *U. pinnatifida* biomass (**a**): 5 mL/h, (**b**): 10 mL/h

Table 9.1 Equations applied for the analysis of the experimental results from dynamic experiments

Model	Integrated equation	Reference
Thomas	$\frac{C}{C_0}=\frac{1}{1+\exp\left[\frac{k_{\mathrm{Th}}}{Q}\left(q_0 m - C_0 V_{\mathrm{eff}}\right)\right]}$	Senthilkumar et al. (2006)
Yoon and Nelson	$\ln = \left(\frac{C_0}{C}-1\right) = k_{\mathrm{YN}}\tau - k_{\mathrm{YN}}t$	
Dose-response	$\frac{C}{C_0}=1-\frac{1}{1+\left(\frac{C_0 V_{\mathrm{eff}}}{q_0 m}\right)^a}$	

k_{Th} Thomas rate constant, k_{YN} Yoon-Nelson rate constant, V_{eff} volume of metal solution passed through into column, τ time required to reach 50 % of total adsorption (on the breakthrough curve), q_0 maximum solute capacity on the solid phase

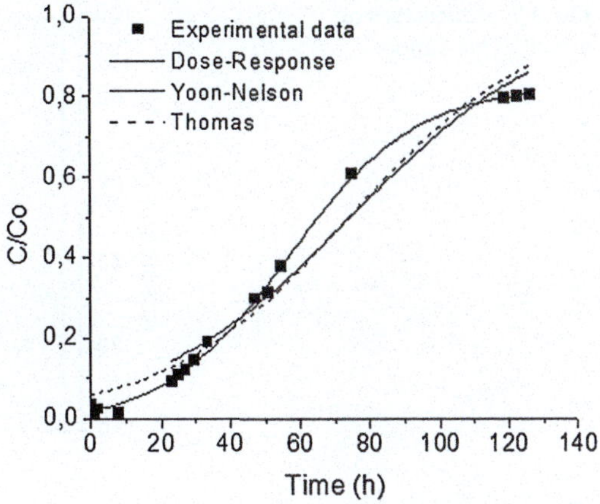

Fig. 9.6 Fitting experimental data in zinc(II) adsorption (200 mg/L; 10 mL/h) using Thomas, Yoon-Nelson, and Dose-Response models

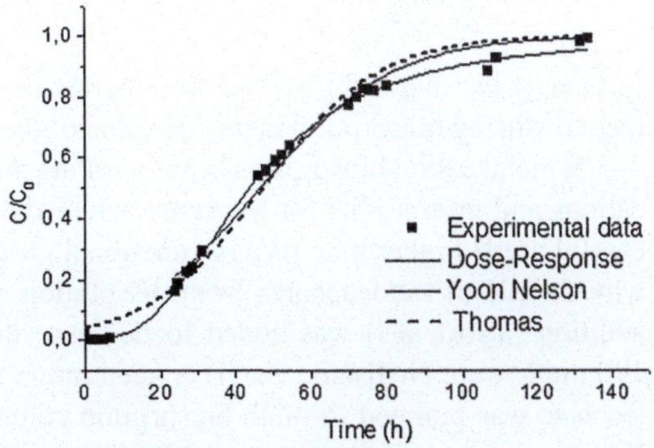

Fig. 9.7 Fitting experimental data in nickel(II) adsorption (175 mg/L; 10 mL/h) using Thomas, Yoon-Nelson, and Dose-Response models

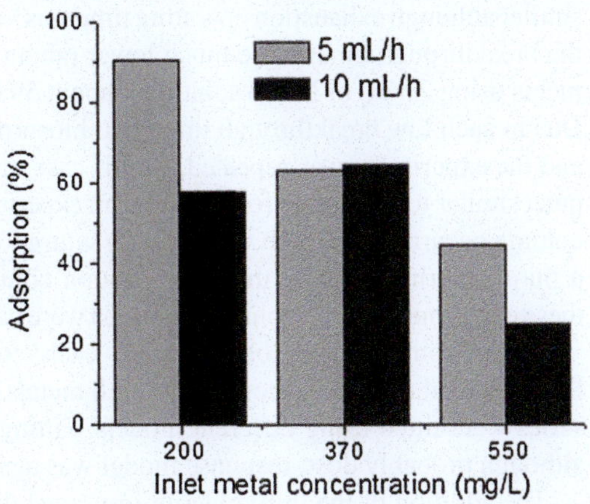

Fig. 9.8 Percentages of zinc(II) biosorption

Independently of the inlet metal concentration Zn(II) adsorption was higher at the lowest flow rate Increasing the flow rate, less volume of influent passed through the column and consequently less amount of zinc was removal (Fig. 9.8).

Similar situation was observed during Ni(II) adsorption except there are no significant differences between the adsorption percentages increasing the inlet metal concentration at the lowest flow rates (Fig. 9.9).

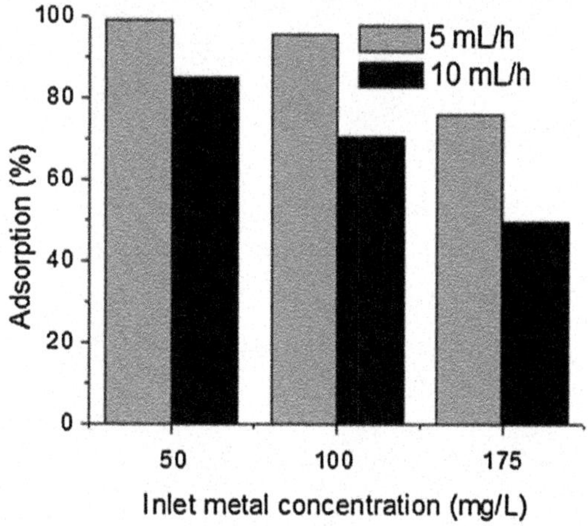

Fig. 9.9 Percentages of nickel(II) biosorption

One of the major difficulties to integrate the solubilization step with the metal recovery using biosorption is the pH value of the leachates. At pH values lower than 3–3.5, metal cation biosorption is not working due to the competence between those cations and proton ions for the same active sites. Biosorption becomes to be successful at pH higher than p*K*a of functional groups on the biomass surface. That is why the pH of the leachates from dissolution step was raised to pH 4 by NaOH addition (also $CaCl_2$ was added to decrease the effect of sodium ions). 48 and 190 mg/L were Ni(II) and Zn(II) concentration respectively into the leachate. This leachate was pumped through biosorption columns (prepared as indicated above). When only one column was used, breakthrough times for Zn(II) and Ni(II) were similar although exhaustion operating time was lower for the second ion. Surprisingly the breakthrough times were much lower (about 7 h) than those obtained in experiments using Zn(II) or Ni(II) solutions (about 59 h and more than 80 h respectively). Due to such low breakthrough time, two biosorption columns were placed in series and the experiment was repeated. In this case, the behavior of the first column was quiet similar to that found for one-column experiment; Ni(II) saturation in the second column occurred after 50 h while Zn(II) saturation had not been reached yet. Finally a third experiment using three biosorption columns in series was carried out. All these experiments using columns in series were carried out with a flow rate of 5 mL/h.

Breakthrough curves obtained for each column can be seen in Fig. 9.10. Experimental data corresponding to both metals biosorption using three columns in series were fitted using different models. Fitting was quiet good for all the models although probably dose-response model was again the best.

The addition of the third column improved the performance of biosorption process but even in such case the amount of metals adsorbed into the column was lower than those found when monometallic solutions were used. In any case, up to 41 and 86 % of total Ni(II) and Zn(II) were retained in the biomass present in the three fixed bed columns.

Fig. 9.10 Breakthrough curves in three consecutive fixed bed columns packed with *U. pinnatifida* biomass

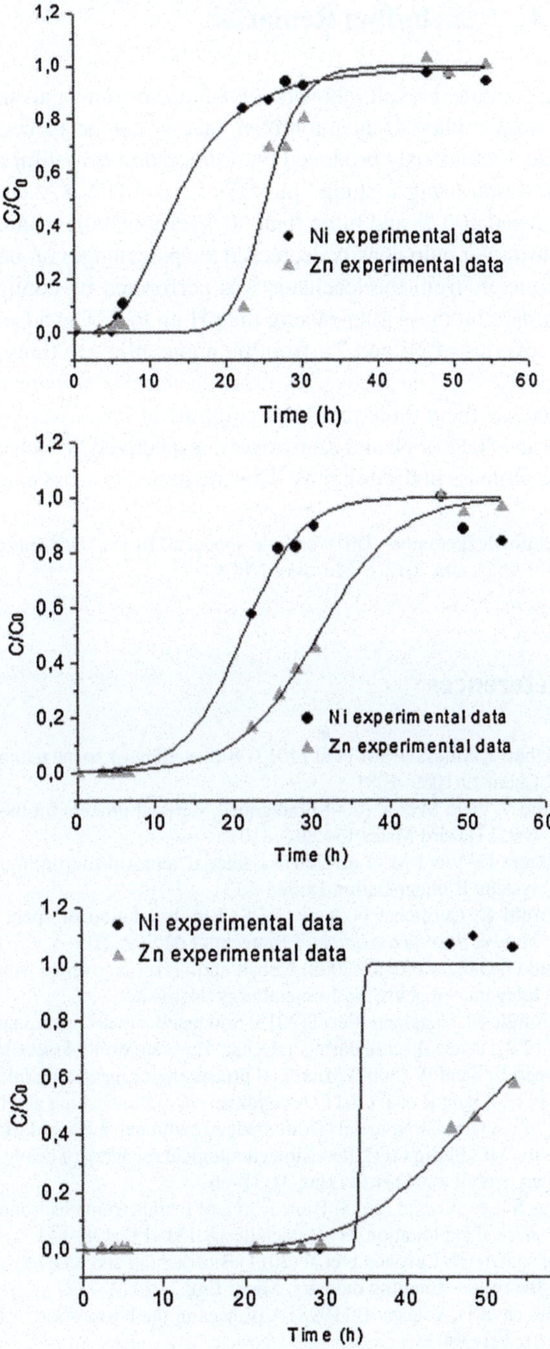

9.4 Concluding Remarks

The obtained results show that Ni and Zn from a galvanic sludge from a wastewater treatment plant of an automotive factory can be successfully leached using sulfuric acid continuously produced by *A. thiooxidans* biofilm on elemental sulfur. The best condition using a sludge pulp density of 1 % w/v and a dilution rate of 0.22 h^{-1} allowed 100 % and more than 90 % of metal dissolution. Higher values of dilution rate and/or pulp density decreased the percentages of metal dissolution. The recovery of metals from the leachates was performed by continuous adsorption on *U. pinnatifida* biomass after raising the pH up to 4. The adsorption process allowed great recoveries of Ni and Zn from monometallic solutions at the lowest flow rate (100 and 90 %). The recoveries from the leachates were not so good and in order to increase them three fixed bed columns in series were used reaching approximately 50 and 80 % of Ni and Zn recovery, respectively, which could be improved increasing the biomass in the columns. This integrated process could be applied at higher scale.

Acknowledgements This work was supported by PIP 0368 from CONICET and PICT 2010-0749, 2012-0623, and 2012-1428 from ANPCyT

References

Asghari I, Mousavi SM et al (2013) Bioleaching of spent refinery catalysts: a review. J Ind Eng Chem 19:1069–1081
Bosio V, Viera M et al (2008) Integrated bacterial process for the treatment of a spent nickel catalyst. J Hazard Mater 154:804–810
Cabrera G, Viera M et al (2007) Bacterial removal of chromium(VI) and (III) in a continuous system. Biodegradation 18:505–513
Cerrutti C, Curutchet G et al (1998) Bio-dissolution of spent nickel-cadmium batteries using *Thiobacillus ferrooxidans*. J Biotechnol 62:209–219
Coto O, Galizia F et al (2008) Cobalt and nickel recoveries from laterite tailings by organic and inorganic bioacids. Hydrometallurgy 94:18–22
D'Emilio M, Caggiano R et al (2013) Soil heavy metal contamination in an industrial area: analysis of the data collected during a decade. Environ Monit Assess 185:5951–5964
Donati E, Sand W (2007) Microbial processing of metal sulphides. Springer, Drodrecht, p 314
Fang D, Zhang R et al (2011) A combination of bioleaching and bioprecipitation for deep removal of contaminating metals from dredged sediment. J Hazard Mater 192:226–233
Hoque ME, Philip OJ (2011) Biotechnological recovery of heavy metals from secondary sources – an overview. Mater Sci Eng 31:57–66
Ilyas S, Lee JC et al (2013) Bioleaching of metals from electronic scrap and its potential for commercial exploitation. Hydrometallurgy 131–132:138–143
Kaksonen AH, Lavonen L et al (2011) Bioleaching and recovery of metals from final slag waste of the copper smelting industry. Miner Eng 24:1113–1121
Kratochvil D, Volesky B (1998) Advances in the biosorption of heavy metals. Trends Biotechnol 16:291–300
Lee PK, Youm SJ et al (2013) Heavy metal concentrations and contamination levels from Asian dust and identification of sources: a case-study. Chemosphere 91:1018–1025

Li X, Liu L, et al (2012) Integrated Assessment of Heavy Metal Contamination in Sediments from a Coastal Industrial Basin, NE China. PLoS ONE 7:p.e39690

Mudhoo A, Garg VK et al (2012) Removal of heavy metals by biosorption. Environ Chem Lett 10:109–117

Ngiam LS, Lim PE (2001) Speciation patterns of heavy metals in tropical estuarine anoxic and oxidized sediments by different sequential extraction schemes. Sci Total Environ 275:53–61

Pathak A, Dastidar MG et al (2009) Bioleaching of heavy metals from sewage sludge: a review. J Environ Manage 90:2343–2353

Plaza Cazón J, Viera M et al (2011) Biosorption of mercury by *Macrocystis pyrifera* and *Undaria pinnatifida*: influence of zinc, cadmium and nickel. J Environ Sci 23:1778–1786

Plaza Cazón J, Viera M et al (2012a) Equilibrium and kinetic study of the biosorption of chromium(III) by two brown algae: *Macrocystis pyrifera* and *Undaria pinnatifida*. Eng Life Sci 12:95–103

Plaza Cazón J, Bernardelli C et al (2012b) Zinc and cadmium biosorption by untreated and calcium-treated *Macrocystis pyrifera* in a batch system. Bioresource Technol 116:195–203

Plaza Cazón J, Viera M et al (2013) Zinc and cadmium removal by biosorption on *Undaria pinnatifida* in batch and continuous processes. J Environ Manage 129:423–434

Romera E, González F et al (2007) Comparative study of biosorption of heavy metals using different types of algae. Bioresource Technol 98:3344–3353

Senthilkumar R, Vijayaraghavan K et al (2006) Seaweeds for the remediation of wastewaters contaminated with zinc(II) ions. J Hazard Mater 136:791–799

Viera M, Donati E (2004) Microbial processes to metal recovery from waste products. Curr Top Biotechnol 1:117–127

Viera M, Curutchet G et al (2003) A combined bacterial process for the reduction and immobilization of chromium. Int Biodeter Biodegr 52:31–34

Chapter 10
Microbial Generation of Acid Mine Drainage: Its Bioremediation in Buenos Aires, Argentina

Leonardo Benítez, Josefina Plaza Cazón, Graciana Willis Poratti, and Edgardo Donati

Abstract Acid mine drainages are highly acidic toxic solutions with elevated heavy metal concentrations. These acidic solutions are generated when rocks containing pyrite and other metal sulfides are brought and exposed to oxygen and water. The activity of iron- and sulfur-oxidizing microorganisms which thrive at a pH range of 1–4 can significantly catalyze that process and the resulting drainages produce a strong impact on the mining region causing contamination to streams and rivers and negative effects on the biodiversity. To reduce its impact, different physical, chemical, and biological treatments are used together in natural and constructed wetlands. Biosorption and bioprecipitation are two of the biological processes within the wetlands. In the first case nonviable biomasses can be used to retain heavy metals while in the second case the metals are precipitated as sulfides through the action of sulfate-reducing microorganisms. In this chapter we describe the situation in an abandoned silver–zinc–lead mine including acid mine drainage generation by microbial action and the evaluation of its treatment using biosorption on plant biomasses and bioprecipitation by sulfate-reducing bacteria.

10.1 Introduction

Waters draining active and abandoned mines and mine wastes are often extremely acidic. They usually contain high concentrations of metals, mainly iron (Fe) and aluminum (Al) but also other heavy metals [copper (Cu), zinc (Zn), nickel (Ni), cadmium Cd)] and frequently some metalloids like arsenic (As). Although its generation is of minor importance when a mine is in active production and water levels are kept artificially low by pumping, when mines are closed and the pumps turned

L. Benítez • J.P. Cazón • G.W. Poratti • E. Donati (✉)
Centro de Investigación y Desarrollo en Fermentaciones Industriales (CINDEFI), Centro Científico Tecnológico (CCT-La Plata), Consejo Nacional de Investigaciones Científicas y Técnicas (CONICET), Facultad de Ciencias Exactas, Universidad Nacional de La Plata, Calle 50 N° 227, 1900 La Plata, Argentina
e-mail: donati@quimica.unlp.edu.ar

A. Alvarez, M.A. Polti (eds.), *Bioremediation in Latin America: Current Research and Perspectives*, DOI 10.1007/978-3-319-05738-5_10

off, the rebound of the water table can contaminate the groundwater and discharged to the environment. These acidic metal-rich mine drainage waters, denominated acid mine drainages (AMD), produce serious damages including some catastrophic events on the biodiversity around the mine but also far away from the origin due to AMD can be transported through streams and rivers (Ayora et al. 2013; Nieto et al. 2013; Sun et al. 2013).

Acid mine drainages is produced during the oxidation of sulfide minerals [mainly pyrite (FeS_2)] when the mine wastes are exposed to oxygen and water. Acidophilic microorganisms, ubiquitous in mining areas, can strongly catalyze that process. Most of these microorganisms are Fe oxidizers which can get energy from the catalysis of the following chemical process:

$$4Fe^{2+} + 4H^{+} + O_2 \rightarrow 4Fe^{3+} + 2H_2O \qquad (10.1)$$

Some Fe oxidizing microorganisms (like *Gallionella* sp.) can mediate the same oxidation at pH above 4 although chemical oxidation is dominant at such pH values. Below 4 the chemical oxidation is negligible and the ferric iron production (and the subsequent generation of AMD as it can be seen below) is microbiologically controlled. That is why the presence of these microorganisms has been verified in most AMD with pH values lower than 4 (Mohapatra et al. 2011). *Acidithiobacillus ferrooxidans*, *Leptospirillum ferrooxidans*, and other Fe oxidizing acidophiles have been found in those environments and they are the main responsible of keeping high ferric iron concentrations. *A. ferrooxidans* is an obligate autotrophic mesophilic bacterium able to derive energy from the oxidation of ferrous iron and also of several reduced sulfur compounds. *L. ferrooxidans*, belonging to the phylum Nitrospira, is also an obligate acidophile able to catalyze only the oxidation of ferrous iron. Other bacteria and archaea capable of oxidizing iron at different ranges of temperature (mainly mesophilic but also moderately and extremely thermophilic microorganisms) have also been found in AMD environments (Hallberg 2010).

Other microorganisms also detected in AMD are sulfur-oxidizers and they can catalyze the following process:

$$S_8 + 12O_2 + 8H_2O \rightarrow 8SO_4^{2-} + 16H^{+} \qquad (10.2)$$

Although Eq. (10.2) only shows the oxidation of sulfur, this process also works for other sulfur reduced compounds like sulfides, polythionates, sulfites, among others. Between sulfur oxidizing microorganisms, some species of *Thiobacillus* and *Thiomonas* can grow under neutrophilic conditions. These species are responsible to lower the pH of the AMD to 4.5–4. After that, other sulfur-oxidizing acidophilic bacteria and archaea (mainly species of *Acidithiobacillus* and *Acidiphilium*) can continue lowering the pH value by oxidizing the elemental sulfur to sulfuric acid. Between mesophiles and moderate thermophiles, *A. thiooxidans* and *A. caldus* seem to be dominant in these environments. Although some eukaryotes (fungi, yeasts) can be abundant in acidic environments usually prokaryotic organisms are more important (Hallberg 2010).

The main oxidant agent involved in metal sulfides dissolution in most natural situations is ferric iron rather oxygen (O_2). Equation (10.3) shows the dissolution of pyrite by the oxidizing action of ferric iron which can be continually regenerated according to Eq. (10.1).

$$FeS_2 + 14Fe^{3+} + 8H_2O \rightarrow 15Fe^{2+} + 16H^+ + 2SO_4^{2-} \quad (10.3)$$

As it has been described elsewhere this dissolution is a multistep process involving the production of different reduced sulfur compounds which can be oxidized by the microorganisms to sulfuric acid as the last product, keeping a very low pH in the water [Eq. (10.3)].

10.2 Acid Mine Drainages: A Case of Study

The region of Argentina adjacent to the Andes is possibly one of the most metal-rich areas in the world. Important reserves of copper (Cu), gold (Au), silver (Ag), lithium (Li), and Zn, among others, have been detected and just a little amount of them are currently under exploitation; also vast undiscovered mineral deposits are been predicted through geologic assessments and they could place Argentina into the top seven countries in the world according to its metalliferous reserves. Due to different political and economical reasons, mining has been long periods of low or even null activity. This situation and the neglect of environmental regulations have generated lots of abandoned mines without adequate closure constituting passive mines.

In the north of Argentina there are several passive mines; one of them is Pan de Azúcar Mine, located in the northwest of Argentina (Jujuy province), at 3,700 m above sea level, at 22° 32–22°38 S and 66°01 66°08 O. The main mineral is a vein-type with a high sulfide content (main mineralogical species are pyrite, sphalerite, and galena) and also rich in Ag. Extractive activities for lead (Pb) and Ag ceased in 1986 and the site was abandoned (Fig. 10.1).

Fig. 10.1 Pan de Azúcar mine. View of the tailings (*left*). Grinding plant (*right*)

Fig. 10.2 Laguna de Pozuelos

This abandoned mine was especially chosen as the study site because it is situated at 25 km south of Laguna Pozuelos (Fig. 10.2), National Natural Monument, Biosphere Reserve (UNESCO 1990) and Ramsar site (1982). The site with neutral to alkaline water (pH 8–9) hosts 44 species of birds and is under a serious risk because Cincel River, the mainly tributary of the lake, receives direct input from acid mine drainage that drained during the summer season.

Huge amounts of tailings are accumulated in the place and even dispersed along several kilometers from the mine; also some traces of AMD could be detected. In the grinding plant a hole with concentrated AMD was found; this AMD presented pH lower than 0.6, ferric iron concentration around 100 g/L, Zn concentration higher than 25 g/L, and Pb concentration higher than 1 mg/L (metal concentrations were determined using atomic absorption spectrophotometry).

Different studies are been carried out in the zone although in this chapter we focus on those related with the eventual isolation of microorganisms feasible to generate AMD and their activity on the tailings.

Samples from small streams of AMD and from the tailings at different depths were collected and utilized in the experiences. Samples were enriched for iron- and/or sulfur oxidizing microorganisms using 0 K medium (KCl 0.1 g/L; $MgSO_4 \cdot 7H_2O$ 0.5 g/L; K_2HPO_4 0.5 g/L; $Ca(NO_3)_2$ 0.01 g/L; $(NH_4)_2SO_4$ 3 g/L) supplemented with sulfur powder (10 % w/v) or ferrous iron (9 g/L) as $FeSO_4$. Positive enrichments were obtained in many of the experimental samples taken either from tailings or AMD. Those cultures were analyzed using Fluorescence In Situ Hybridization (FISH) with specific probes for *Acidithiobacillus ferrooxidans*, *Acidithiobacillus thiooxidans*, and *Leptospirillum ferrooxidans*. For that, aliquots of 1 mL of cultures were fixed with paraformaldehyde 4 % and then filtered through 0.25 Millipore membranes (0.22 μm). Filters were washed and neutralized with buffer and air dried. Hybridizations were done following Amman's protocol (Amann 1995) using probes TF539 (specific for *Acidithiobacillus ferrooxidans*), ATT223 (specific for *Acidithiobacillus thiooxidans*), and LF665 (specific for *Leptospirillum ferrooxidans*). 4′,6′-diamidino-2-phenylindole (DAPI) stain was used in all hybridizations to evaluate total cells number.

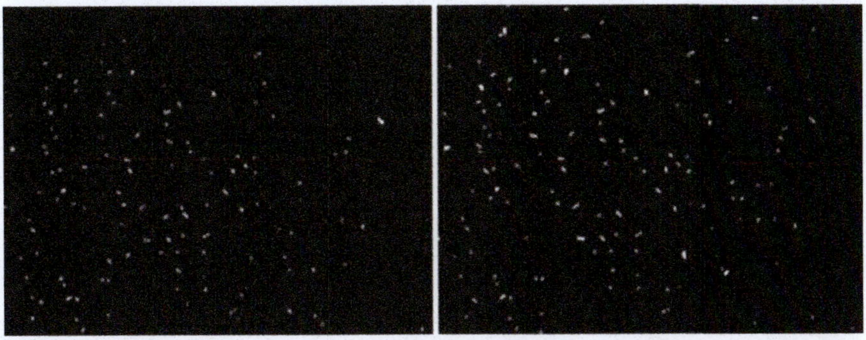

Fig. 10.3 Enrichment using ferrous iron as energy source. 4′,6′-diamidino-2-phenylindole (DAPI) stain (*left*) and hybridization with probe TF539 (*right*) specific for *Acidithiobacillus ferrooxidans*

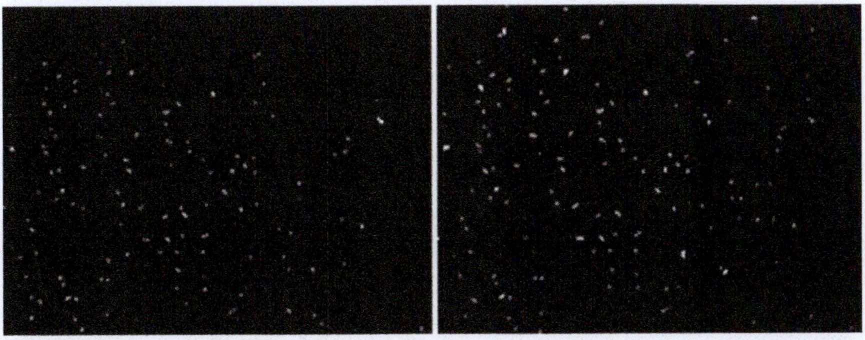

Fig. 10.4 Enrichment using sulfur as energy source. 4′,6′-diamidino-2-phenylindole (DAPI) stain DAPI (*left*) and hybridization with probe ATT223 (*right*) specific for *Acidithiobacillus thiooxidans*

Hybridization with TF539 was positive for most of cells stained by DAPI in the enrichment using ferrous iron as the energy source (Fig. 10.3), while most cells in the enrichment with sulfur as energy source showed positive hybridization with ATT223 (Fig. 10.4).

Similar procedure on the original samples (fixed in situ) showed the presence of the same microorganisms. According to these results, sulfur oxidizers (*A. thiooxidans* and *A. ferrooxidans*) and iron oxidizers (*A. ferrooxidans*) are ubiquitous not only where the effects are already visible (AMD) but also into the tailings.

In order to evaluate the risk associated to AMD generation boreholes from tailing surface up to 1 m depth were carried out (Fig. 10.5) and tailing samples were taken for bioleaching studies.

The experiments were performed at two different pulp densities (2 and 5 % w/v) by adding the samples to 150 mL of 0 K medium, inoculating with the enrichments (5 % v/v of each culture) where the presence of *A. ferrooxidans* and *A. thiooxidans* had been confirmed, and incubating:

Fig. 10.5 *Left*: Tailing surface (the *arrow* shows the place where boreholes were done). *Right*: Samples taken from different depths

Fig. 10.6 Bioleaching of tailing sample (the deepest one, 70–100 cm depth) by a mixed culture composed by *A. ferrooxidans* and *A. thiooxidans*

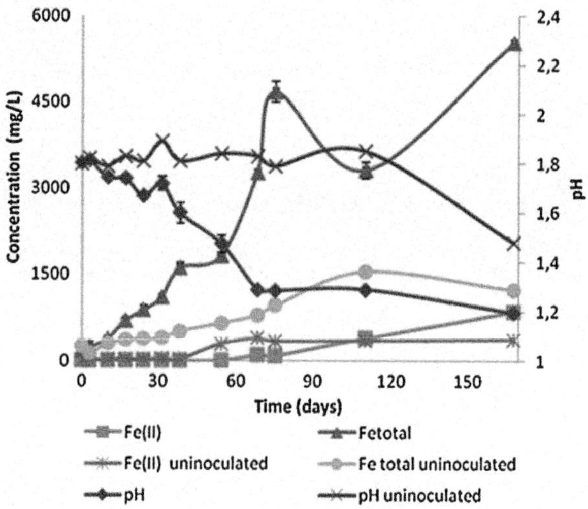

Samples but mainly the deepest were attacked by the microorganisms generating a strong decrease of pH values and high iron dissolution (which are indicators of AMD generation). Figure 10.6 shows the results for deepest sample at 5 % w/v. It is possible to see a high solubilization of iron (reaching more than 5,500 mg/L after 150 days); at the final pH value (lower than 1.2). So, iron oxidation by *A. ferrooxidans* was inhibited and that is the reason because ferrous iron concentration increased toward the end of the experiment. Uninoculated control showed signs of microbial contamination after the first 100 days.

The microorganisms detected and isolated in the tailings showed to be able to leach them generating AMD with low pH value and high contents of Fe and Zn. This means a serious environmental risk due to these AMD could reach Cincel River, especially during summer season where rains are abundant, and subsequently Laguna Pozuelos affecting the biodiversity in the place.

10.2.1 Acid Mine Drainages Treatment

Acid mine drainages usually present low pH water with elevated concentrations of Fe, sulfates and heavy metals but usually metal load is of greater concern than the acidity according to their possible environmental damage. Acid mine drainages is often neutralized using chemicals like lime, calcium carbonate, caustic soda, among others, producing huge amounts of sludge and its disposal represents an additional environmental problem (Johnson and Hallberg 2005). Wetlands have emerged as an alternative low cost technology (low investment and operating costs, no external energy input) capable of removing heavy metals from AMD besides domestic, commercial and industrial waste water. Wetlands are mainly filters with several bio-physicochemical processes working together. Although natural wetlands have been used for centuries, constructed wetlands are even better because they can be controlled and manipulated allowing great load rates and removal efficiencies of most metals (Sheoran and Sheoran 2006). Between several physical, chemical, and biological processes within the wetlands, we have selected some biological processes which are relevant for metal immobilization.

Microorganisms can contribute to immobilization of heavy metals by different processes: (1) sorption to cell components or exopolymers, (2) transport and intracellular sequestration, and (3) precipitation as insoluble compounds, either organic as oxalate or inorganic as sulfides, phosphates, carbonates, and co-precipitation with Fe compounds. Depending upon the metal chemistry, the reduction of high-valency species could lead to immobilization [e.g., Cr (VI) to Cr (III)] or mobilization [Mn(IV) to Mn (II)]. Immobilization processes are applicable to removing metals from mobile phases such as groundwater, leachates, and liquid effluents.

10.3 Biosorption and Bioaccumulation

Metal biosorption can occur either by metabolism-independent passive sorption or by intracellular, metabolism-dependent active uptake. Both processes may occur in the same organism. The terms sorption and adsorption are usually used when passive accumulation is considered while uptake and accumulation are used when metabolism-dependent intracellular transport is implied. The non metabolic biosorption processes comprise different mechanisms: chemisorption, ion exchange, complexation, coordination, chelation; and physical adsorption (Davis et al. 2003; Viera and Donati 2004).

High metal-binding capacities of different biological materials have been studied. In recent years biosorption studies involving low-cost easily available biomass were published: marine algae, yeast, dried leaves, biopolymers such as alginate, chitosan, etc. (Ghaedi et al. 2013; Kiplagat and Kituyi 2013). Most of these studies evaluated the uptake capacities of biosorbents using isotherms curves and the effect of several parameters such as pH, sorbent loading, pretreatments, metal initial concentrations. Mechanistic models for Cu and Cd biosorption have been developed taking into account the acidic properties of the cell wall constituents derived from biomass characterization. These models reveal the complexity of the metal biosorption phenomenon and the need to consider different chemico-physical mechanisms operating simultaneously (Volesky 2001).

There are several reports on the higher biosorption capacity of live biomass compared to dead biomass (Volesky 2001). Growing cells might be better option due to their ability of self-replenishment, continuous metabolic uptake of metals after physical adsorption, and the potential for optimization through development of resistant species and cell surface modification. Further the metals diffused into the cells during detoxification get bound to intracellular proteins before being incorporated into vacuoles and other intracellular sites. These processes are often irreversible and ensure less risk of metal releasing back to the environment. Apart from this, using growing cultures in bioremoval could avoid the need for separate biomass production processes, e.g., cultivation, harvesting drying processing and storage prior to the use, in contrast to conventional chemo-physical and biosorptive methods. The use of active microorganisms may allow development of a single stage process for removal of most of the pollutants present in industrial effluents. However, biosorption utilizing nonliving biomass to accumulate heavy metals from wastewaters seems to be a more competitive, effective, and economically attractive treatment method than bioaccumulation, because maintaining a viable biomass during the metal removal process can be rather difficult. Other disadvantages of using living biomass can be the higher difficulty in biomass separation can be encountered, mass loss after regeneration, and insufficient understanding of the process (Volesky 2001). Algae are among the most promising biosorbents; due to presence of alginate in their cell wall, brown algae are probably the best. Recently we have published the results of metal biosorption using nonviable biomasses from two brown algae, *Macrocystis pyrifera* and *Undaria pinnatifida* (Plaza Cazón et al. 2011, 2012a, b, 2013). *M. pyrifera* grows in water along the south coast of Argentina and it is frequently deposited on the beach of Bahía de Camarones causing unpleasant odors and a negative impact on local tourism and can be considered an easily available natural waste. On the other hand, *U. pinnatifida* is not a native species and only appeared recently (i.e., since 2005) on the Argentinean coast.

10.3.1 Study of Biosorption and Bioaccumulation by Different Plants Species

In order to determine the impact of Pan de Azucar mine into the environment, several sampling sites were selected based on the distance from mine, topography, vegetation, etc. Soil samples were sieved before treatment with concentrated nitric and sulfuric acids. Acid extracts were filtered and then analyzed using atomic absorption spectrophotometry. Zinc concentrations higher than 100 mg/kg were

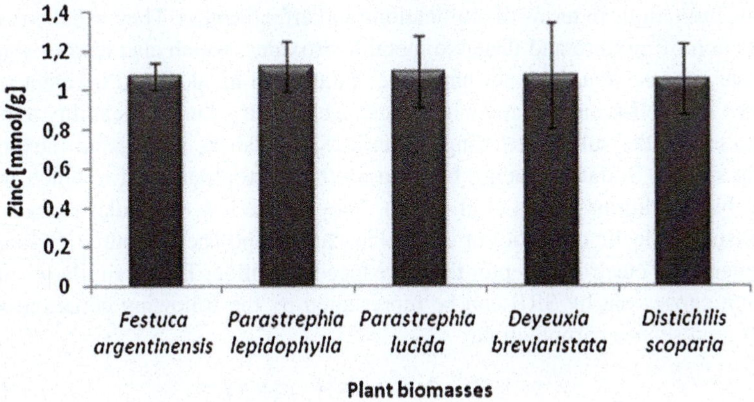

Fig. 10.7 Zinc sorption by plant biomasses

measured in all sampling sites while Pb was found only in the closer places with concentrations higher than 50 mg/kg.

Plant samples were also collected from the sites and identified using Darwiniana dichotomous key and classified making a herbarium. *Parastrephia lepidophylla*, *Parastrephia lucida*, *Festuca argentinensis*, *Adesmia horrida*, *Distichlis scoparia*, and *Deyeuxia breviaristata* were found in the samples being the last which was in the soil with highest Zn and Pb concentrations (500 and 400 mg/kg, respectively). Zinc and Pb contents were analyzed in roots and aerial parts. All the plants contained Zn (up to 0.6 mg/g) while the maximum lead concentration was 0.2 mg/g. *Parastrephia lucida*, *Adesmia horrida*, and *Deyeuxia breviaristata* showed higher Zn contents, while the last contained the highest Pb concentration.

Since these plants showed certain capacity to retain metals into their biomasses and there are no other biosorbents available in the zone, the sorption capacity of nonviable plant biomasses was evaluated. For that the aerial part of the plants was ground and sieved. 0.1 g biomasses were added to 100 mL of 50 mg/L Zn(II) prepared from $ZnSO_4$. Flasks were maintained under agitation for 3 h. Samples were collected and filtered. Zinc concentrations were measured using atomic absorption spectrophotometry.

Results from Zn sorption experiments can be seen in Fig. 10.7. Sorption values higher to 1 mmol Zn/g (more than 50 mg/g) were achieved being the maximum for *Parastrephia lepidophylla* biomass. These results suggest the possibility of using the plants from the zone for metal removal in constructed wetlands.

10.4 Bioprecipitation

As it was indicated above, when selecting the method of choice to carry out precipitation is important that it is efficient and it can be operated at reasonable costs. Conventional treatments using chemical products to precipitate metals present some serious limitations in terms of applications and effectiveness. They usually result in production of mixture and unstable metal hydroxides, which also lead to a greater disposal expense. On the other hand, precipitation of divalent and trivalent ions as sulfides is a well-known strategy in inorganic chemistry. Due to its ability to reduce sulfate to sulfide, sulfate-reducing bacteria (SRB) can be used as an inexpensive sulfide source. Sulfate-reducing bacteria are heterotrophic bacteria which require anaerobic conditions (Kikot et al. 2010). They utilize low molecular mass organic acids (such as lactic or acetic) or ethanol as carbon and energy source. Sulfate acts as the terminal electron acceptor that is reduced to sulfide. The dissimilatory sulfate reaction carries out by SRB can be represented by the following equation, where CH_2O represents a carbohydrate.

$$2CH_2O + SO_4^{=} \rightarrow SH_2 + HCO_3^{-} \tag{10.4}$$

From a biotechnological point of view, this reaction implies important applications: not only the precipitation of metal as sulfides but also the decrease of sulfate concentration and the increase of pH (through bicarbonate formation). These facts are especially important if SRB are used to remediate metal-rich effluents like acid mine drainage (Neculita et al. 2007).

The major limitation of active bioremediation of AMD is the sensitivity of SRB to acidity and high metal concentration. To avoid inhibition of SRB growth by high acidic conditions or by heavy metals, several strategies can be applied: (1) the adaptation of bacterial population to low pH values; (2) a previous step of soft neutralization; or (3) the biological sulfate reduction can take place in a reactor separate from the metal precipitation stage with sulfhydric acid [SH_2 (coming from the SRB reactor)] facilitating metal separation and recovery. There are few examples of processes utilizing SRB in the treatment of wastewater at industrial scale. The best known is the groundwater treatment process installed at the Budelco Zinc Refinery (The Netherlands) since 1992. A 1,800 m^3 SRB reactor (capable of treating 7,000 m^3/day of extracted groundwater) used a consortium of SRB with ethanol as carbon and energy source. Methanogenic bacteria present in the consortium removed the acetate produced by the SRB, leaving an effluent that is environmentally acceptable and satisfies the regulatory authorities (Barnes et al. 1994).

Although the use of SRB to precipitate heavy metals has been widely studied, this is no the only possibility for metal precipitation (Viera and Donati 2004). Microbial phosphatases, as those presents in *Citrobacter* sp., can hydrolyse phosphate from glycerol 2-phosphate. The phosphate accumulates at high concentration at the cell surface where it reacts with divalent cations yielding inorganic crystals trapped in the exocellular lipopolysaccharide. *Acinetobacter johnsonii* can accumulate phosphate as polyphosphate aerobically. Exposure to anaerobic conditions promotes polyphosphate degradation with concomitant release of phosphate into the medium.

This process could be used for the treatment of wastewaters contaminated with phosphate and heavy metals at concentrations too low to exceed its solubility product. Experiments done with *Acinetobacter johnsonii* in the presence of 0.2–0.5 mM of

Cd^{+2} or uranyl (UO_2^{+2}) showed that the presence of uranium (U) has a stimulatory effect on polyphosphate degradation. Conversely, Cd has a negative effect on the process. Metals like Zn, Cu, Ca, Mn, and strontium (Sr) can be immobilized as oxalates by oxalic acid produced by fungi such as *Aspergillus* and *Penicillum*. Iron can be precipitated by previous oxidation to ferric iron. As it has been indicated above, this process [represented by Eq. (10.1)] can be catalyzed by *A. ferrooxidans* among other microorganisms. Even at very low pH values (2–2.5) ferric iron yields insoluble compounds such as jarosites [Fe(III) basic sulfates] which may incorporate toxic elements within their structure or adsorb them at their surface. In that way, this precipitation process can be used for adsorption of heavy metals from wastewater.

10.4.1 Bioprecipitation: Case of Study

Copahue-Caviahue geothermal area is located in the Cordillera Norpatagónica in the north west of Neuquén province, Argentina. The area is crowned by Copahue volcano (37° 51′ S and 71° 10′ O, 2,965 m.a.s.l.), a predominantly andesitic stratovolcano which contains a small, very acidic, volcanic crater lake (pH 0.2–1.1). The area is characterized by many fumaroles, ponds, and hot springs. They are geographically grouped in different zones: Copahue thermal centre, Las Máquinas, Las Maquinitas, Anfiteatro, and Chancho-Co (situated in the Chilean side of the cordillera). Las Máquinas (Fig. 10.8), with a moderate temperature pool (38 °C), low pH value (3), and reducing conditions (Eh about −70 mV), was chosen because it has been less affected by human activity. Samples from Las Máquinas were enriched in media with glycerol 3 mM, yeast extract 0.01 % (w/v), Zn 4 mM, $FeSO_4$ 100 μM, K_2SO_4 0.87 g/L, and trace elements. Nitrogen was bubbled through

Fig. 10.8 Las Máquinas

the media to displace O_2 to create an environment for the growth of anaerobic microorganisms. Flasks were incubated in anaerobic jars at 30 and 40 °C. Sulfate and glycerol were measured using ionic chromatography and the microbial community were analyzed by Terminal restriction fragment length polymorphisms (T-RFLP). One microorganism close to *Desulfobacillus acidavidus* was found in both enrichments.

Zinc bioprecipitation was analyzed as one of the possibility to treat AMD in the case of Pan de Azucar mine. Bioprecipitation experiments were carried out in sealed flasks. Postgate medium with different Zn concentration (1, 2, 4, and 8 mg/L) was inoculated with cells coming from the enrichments at 30 and 40 °C. Zinc concentration was measured using atomic absorption spectrophotometry. Abiotic controls and controls without the addition of Zn were also carried out. All systems were placed in anaerobic jars at 30 or 40 °C.

Results showed that up to 4 mM the precipitation was complete at 40 °C. At 30 °C, no significant precipitation compared with abiotic controls was found (Figs. 10.9 and 10.10). In addition, results showed a close relationship between Zn bioprecipitation (associated to sulfide production) and glycerol consumption. Although higher Zn concentrations inhibited the microbial activity, sulfate-reducing activity at this low pH value is not so common and they could be used for metal removal from ADM in the anaerobic zone of wetlands.

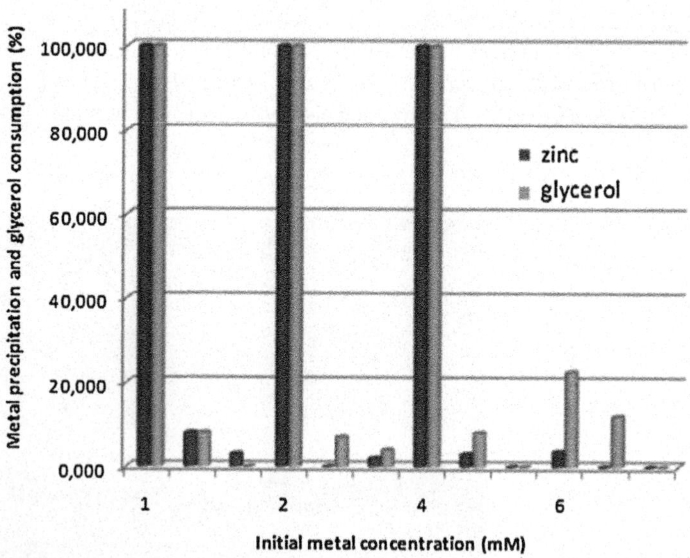

Fig. 10.9 Zinc bioprecipitation at 40 °C

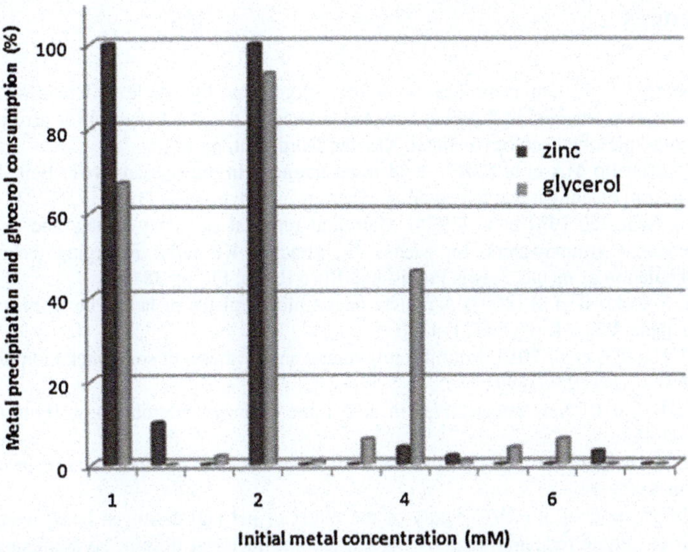

Fig. 10.10 Zinc bioprecipitation at 30 °C

10.5 Concluding Remarks

Acid mine drainages are highly acidic toxic solutions with elevated heavy metal concentrations generated when rocks containing pyrite and other metal sulfides are brought and exposed to O_2 and water catalized by the activity of Fe- and sulfur-oxidizing microorganisms. Sampling tailings in an abandoned mine (Pan de Azucar, Jujuy province, Argentine) allowed the detection and isolation of some species of microorganisms belonging to *Acidithiobacillus* and *Leptospirillum* genus which were able to leach the tailings generating AMD with low pH value and high contents of Fe and Zn. This means a serious environmental risk due to these AMD could reach Cincel River, especially during summer season where rains are abundant, and subsequently Pozuelos Lake which is a natural reserve just some kilometers from the mine. Soil and plant samples around the mine showed high levels of Zn and Pb which is a signal of the environmental impact on the zone but also suggested the possibility of using the plants from the zone for metal removal in constructed wetlands. In this way biomasses of *Parastrephia lepidophylla* (collected from the zone) showed capacity of Zn sorption even higher than 1 mmol Zn/g. Finally acidic sulfate-reducing communities (isolated in Caviahue-Copahue geothermal zone) showed complete Zn bioprecipitation up to 4 mM at 30 and 40 °C. Both biological processes, biosorption and bioprecipitation, could be used in artificial wetlands to treat acid mine drainages generated in the tailings.

Acknowledgements This work was supported by PIP 0368 from CONICET and PICT 2010-0749, 2012-0623, and 2012-1428 from ANPCyT.

References

Amann RI (1995) In situ identification of microorganisms by whole cell hybridization with rRNA-targeted nucleic acid probes. In: Akkermans ADL, van Elsas JD, de Bruijn FJ (eds) Molecular microbial ecology manual. Kluwer, Dordrecht, pp 1–15

Ayora C, Caraballo MA et al (2013) Acid mine drainage in the Iberian Pyrite Belt: 2. Lessons learned from recent passive remediation experiences. Environ Sci Pollut R 20:7837–7853

Barnes LJ, Scheeren PJM et al (1994) Microbial removal of heavy metals and sulfate from contaminated groundwaters. In: Means JL, Hinchee RE (eds) Emerging technology for bioremediation of metals. Lewis Publishers, Boca Raton, FL, pp 38–40

Davis TA, Volesky B et al (2003) A review of the biochemistry of heavy metal biosorption by brown algae. Water Res 37:4311–4330

Ghaedi M, Hajati S et al (2013) Equilibrium, kinetic and isotherm of some metal ion biosorption. J Ind Eng Chem 19:987–992

Hallberg KB (2010) New perspectives in acid mine drainage microbiology. Hydrometallurgy 104:448–453

Johnson DB, Hallberg KB (2005) Acid mine drainage remediation options: a review. Sci Total Environ 338:3–14

Kikot P, Mignone C et al (2010) Study of the effect of pH and dissolved heavy metals on the growth of sulfate-reducing bacteria by a fractional factorial design. Hydrometallurgy 104: 334–341

Kiplagat A, Kituyi L (2013) Comparative study of rates of biosorption for selected single and mixed metal ions using natural products. Chem Mater Res 3:1–6

Mohapatra BR, Douglas Gould W et al (2011) Tracking the prokaryotic diversity in acid mine drainage-contaminated environments: a review of molecular methods. Miner Eng 24:709–718

Neculita CM, Zagury GJ et al (2007) Passive treatment of acid mine drainage in bioreactors using sulfate-reducing bacteria: critical review and research needs. J Environ Qual 36:1–16

Nieto JM, Sarmiento AM et al (2013) Acid mine drainage in the Iberian Pyrite Belt: 1. Hydrochemical characteristics and pollutant load of the Tinto and Odiel rivers. Environ Sci Pollut R 20:7509–7519

Plaza Cazón J, Viera M et al (2011) Biosorption of mercury by *Macrocystis pyrifera* and *Undaria pinnatifida*: influence of zinc, cadmium and nickel. J Environ Sci 23:1778–1786

Plaza Cazón J, Bernardelli C et al (2012a) Zinc and cadmium biosorption by untreated and calcium-treated *Macrocystis pyrifera* in a batch system. Bioresour Technol 116:195–203

Plaza Cazón J, Viera M et al (2012b) Equilibrium and kinetic study of the biosorption of chromium(III) by two brown algae: *Macrocystis pyrifera* and *Undaria pinnatifida*. Eng Life Sci 12:95–103

Plaza Cazón J, Viera M et al (2013) Zinc and cadmium removal by biosorption on *Undaria pinnatifida* in batch and continuous processes. J Environ Manage 129:423–434

Sheoran AS, Sheoran V (2006) Heavy metal removal mechanism of acid mine drainage in wetlands: a critical review. Miner Eng 19:105–116

Sun J, Tang C et al (2013) Hydrogeochemical characteristics of streams with and without acid mine drainage impacts: a paired catchment study in karst geology, SW China. J Hydrol 504:115–124

Viera M, Donati E (2004) Microbial processes to metal recovery from waste products. Curr Top Biotechnol 1:117–127

Volesky B (2001) Detoxification of metal-bearing effluents: biosorption for the next century. Hydrometallurgy 59:203–216

Chapter 11
Co-contaminated Soils Bioremediation by Actinobacteria

Juan D. Aparicio, María Z. Simón Solá, Mariana C. Atjián, Claudia S. Benimeli, and María Julia Amoroso

Abstract More than one-third of contaminated areas are found to have more than one type of pollutant. Co-contaminated environments with metals and organic compounds are difficult to remediate because of the mixed nature of the pollulants. Actinobacteria is an important group of microorganisms found in soils, with high metabolic versatility and abilities to bioremediation. Actinobacteria are currently studied for bioremediation of soils contaminated by pesticides and heavy metals. In this chapter we review the potential of actinobacteria isolated from contaminated environments for simultaneous soil bioremediation of Cr(VI) and the organochlorine pesticide lindane. Four actinobacteria, tolerant to Cr(VI) and lindane mixture were used: *Streptomyces* spp. A5, M7, MC1 and *Amycolatopsis tucumanensis* DSM 45259. Sterilized soil samples were inoculated with actinobacteria strains, either individually or as a consortium (formed by all selected actinobacteria) then contaminated with Cr(VI) and lindane, and incubated at 30 °C for 14 days. All actinobacteria

J.D. Aparicio (✉) • M.Z. Simón Solá
Planta Piloto de Procesos Industriales Microbiológicos (PROIMI), CONICET, Av. Belgrano y Pasaje Caseros, 4000 Tucumán, Argentina

Facultad de Bioquímica, Química y Farmacia, Universidad Nacional de Tucumán, 4000 Tucumán, Argentina
e-mail: dani_aparicio@hotmail.com.ar

M.C. Atjián
Planta Piloto de Procesos Industriales Microbiológicos (PROIMI), CONICET, Av. Belgrano y Pasaje Caseros, 4000 Tucumán, Argentina

C.S. Benimeli
Planta Piloto de Procesos Industriales Microbiológicos (PROIMI), CONICET, Av. Belgrano y Pasaje Caseros, 4000 Tucumán, Argentina

Universidad del Norte Santo Tomás de Aquino (UNSTA), 4000 Tucumán, Argentina

M.J. Amoroso
Planta Piloto de Procesos Industriales Microbiológicos (PROIMI), CONICET, Av. Belgrano y Pasaje Caseros, 4000 Tucumán, Argentina

Facultad de Bioquímica, Química y Farmacia, Universidad Nacional de Tucumán, 4000 Tucumán, Argentina

Universidad del Norte Santo Tomás de Aquino (UNSTA), 4000 Tucumán, Argentina

A. Alvarez, M.A. Polti (eds.), *Bioremediation in Latin America: Current Research and Perspectives*, DOI 10.1007/978-3-319-05738-5_11

were able to grow and remove both contaminants, the consortium formed by *Streptomyces* spp. A5, M7, MC1 and *A. tucumanensis* showed the highest Cr(VI) removal, while *Streptomyces* sp. M7 produced the maximum lindane removal. In non-sterile soil samples, *Streptomyces* sp. M7 and the consortium removed more than 40 % of the lindane, while *Streptomyces* sp. M7 demonstrated the greatest Cr(VI) removal. According to these results, it could be concluded that the use of *Streptomyces* sp. M7 is the strategy more appropriate for the bioremediation of soils contaminated with Cr(VI) and lindane.

11.1 Introduction

The great expansion of industrial activity has resulted in an increase in scenarios of serious and complex environmental contamination by both organic compounds (herbicides, plastics, tannins, polyphenols, pesticides, etc.) and inorganic compounds (As, Cd, Cu, Pb, Cr, Hg, etc.) (Volke Sepúlveda and Velasco Trejo 2002). Co-pollution is a very important issue because more than one third of contaminated areas are found to have more than one type of pollutant (Mansour 2012; Tang et al. 2010). Moreover, environments co-contaminated with metals and organic compounds are difficult to remediate because of the mixed nature of these pollutants.

11.1.1 Chromium (VI)

Chromium (VI) (Cr(VI)) compounds have several uses in industry (Bhadra and Mahananda 2013; Polti et al. 2007) and contamination by these compounds has been detected in soil and water and around a wide variety of industrial sites (Benimeli et al. 2003; Nie et al. 2010; Srinivasa Gowd et al. 2010). Cr(VI) is a harmful pollutant characterized by its chronic toxicity, neurotoxicity, dermatotoxicity, genotoxicity, carcinogenicity, and immunotoxicity (Bagchi et al. 2002), and it is approximately 1,000 times more toxic and mutagenic than Cr(III) (Dana Devi et al. 2001; USEPA 1998).

11.1.2 Lindane

The systematic use of pesticides has led to great improvements in terms of agricultural production levels. However, massive and indiscriminate application of pesticide products has also led to adverse effects on human health, the environment, and even the effectiveness of the products themselves (Johri et al. 2000; Phillips et al. 2005). The gamma isomer of hexachlorocyclohexane (γ-HCH), commercially known as lindane, is a highly chlorinated, recalcitrant organochlorine pesticide (OP). Lindane residues persist in the environment and have been reported in soils, water, air, plants, agricultural products, animals, foods, and microbial environments, as well as in the human body. Since the toxicity of γ-HCH has been established, it is now imperative to develop methods to remove lindane from the contaminated environments (Fuentes et al. 2011).

11.2 Bioremediation

The intense search for a solution to co-contamination has led to the development of remediation technologies that can simultaneously deal with multiple contaminants (Ma et al. 2010; Srivastava et al. 2007; Wasi et al. 2011).

In the last 10 years a stronger emphasis has come to be placed on the study of the physiological, biochemical, and molecular approaches to microbial bioremediation of environments co-contaminated with heavy metals and pesticides. Soils with long-term exposure to mixed contamination with organic compounds and heavy metals have been shown to have structural and functional microbial communities with the ability to adapt and grow under these conditions. This suggests that bioremediation based on microorganisms is feasible for recovery of such sites by microbial transformation of both organic compounds and heavy metals into nontoxic products. These strategies depend mainly upon the catabolic biological activities of the microorganisms, and therefore their ability to utilize the contaminants as nutrients and energy sources (Atlas and Unterman 1999; Boopathy 2000).

The impacts that metals have on biodegradation are complex and are influenced by the matrix structure, which determines the bioavailable metal concentrations. Metals inhibit biodegradation using different mechanisms and patterns, which depend upon the biological and physicochemical characteristics of each system. A variety of approaches to bioremediation of co-contaminated sites are under development, and they include addition of metal-resistant microorganisms as well as additives that reduce metal bioavailability (Sandrin and Hoffman 2007).

Several authors have evaluated bioremediation in media co-contaminated with metals and persistent organic compounds. Olaniran et al. (2009) investigated the impact of lead and mercury on biodegradation of 1,2-dichloroethane in soils, and they concluded that heavy metals have a negative impact on this bioprocess. These authors also found that biostimulation can have a positive influence on 1,2-dichloroethane degradation.

Another emerging approach is bioaugmentation. Sprocati et al. (2012) used this strategy to remediate soils co-contaminated with diesel oil and heavy metals. The bioaugmentation was performed by introducing a consortium composed of 12 allochthonous bacterial strains, previously isolated from a site with long-term pollution. This strategy showed high efficiency in the bioremediation process.

11.3 Actinobacteria

It is important to consider that when allochthonous microorganisms are incorporated into a soil, they usually cannot fully participate in the community activity in a meaningful way. This is why the use of indigenous microorganisms in bioremediation processes is so important. The actinobacteria are a group of bacteria that is found in high concentrations in soils. They play an important ecological role in recycling substances in the natural world, using humic acids for their growth as well as organic matter, which is difficult to degrade (Kieser et al. 2000). The physiological diversity of actinobacteria allows the production of a large number of metabolites with biotechnological importance included antibiotics, which are synthesized and excreted into a medium (Ensign 1990; Genilloud et al. 2011; Goodfellow et al. 1988). The important role played by actinobacteria in the environment is also demonstrated by their ability to remove oil, rubber, plastics, pesticides, and heavy metals, among other substances (Albarracín et al. 2005, 2010b; Benimeli et al. 2003, 2006, 2007; Goodfellow et al. 1988; Polti et al. 2009, 2011; Vobis 1997).

There have been previous studies focused on biotransformation of OPs by actinobacteria, particularly in relation to lindane degradation (Benimeli et al. 2006, 2007; Fuentes et al. 2011; Saez et al. 2012). *Streptomyces* spp. M7, A2, A5, and A11, isolated from sediments and soils contaminated with OPs, were found to be able to degrade lindane, as revealed by the release of chloride ions when the microorganisms were grown on media containing this pesticide as a sole carbon source (Benimeli et al. 2003, 2006; Cuozzo et al. 2009; Fuentes et al. 2010). Biotransformation of heavy metals Cu(II), Cd(II) and Cr(VI) by actinobacteria, particularly in terms of uptake and/or reduction to less toxic forms, has also been studied (Albarracín et al. 2008a; Polti et al. 2007; Siñeriz et al. 2009). *Streptomyces* sp. MC1, isolated from contaminated sugar cane, has shown the ability to reduce Cr(VI) to Cr(III) in both liquid and solid culture media (Polti et al. 2009, 2010). *Amycolatopsis tucumanensis* DSM 45259, isolated from sediments contaminated with heavy metals has also shown resistance to copper and chromium under a variety of culture conditions (Albarracín et al. 2005, 2008b, 2010a).

In this chapter we could see how actinobacteria, as pure or mixed culture, could be able to remediate soil co-contaminated with lindane and Cr(VI).

11.4 Cr(VI)-Lindane Tolerant Actinobacteria

Studies of tolerance to Cr(VI) and lindane were performed using Minimal Medium (containing in g L^{-1}: glucose, 10.0; L-asparagine, 0.5; K_2HPO_4, 0.5; $MgSO_4 \times 7H_2O$, 0.20; $FeSO_4 \times 7H_2O$, 0.01) agar plates because the toxic elements do not interact with the medium components and they therefore remain bioavailable to the actinobacteria (Amoroso et al. 2001, 2002; Rathnayake et al. 2013). Rectangular troughs were cut in the center of the plate and then filled with 500 mg L^{-1} of Cr(VI) and/or 250 μg L^{-1} of lindane.

Six previously isolated actinobacteria were assayed: three isolated from environments contaminated with pesticides and heavy metals (*Streptomyces* sp. M7, *Streptomyces* sp. MC1, and *Amycolatopsis tucumanensis* DSM 45259) (Albarracín et al. 2005; Benimeli et al. 2003; Polti et al. 2007), and three isolated from a lindane-contaminated environment in Santiago del Estero, Argentina, where in 1994 about 30 tons of organochlorine pesticides were spilled: *Streptomyces* sp. A2, *Streptomyces* sp. A5, and *Streptomyces* sp. A11 (M. S. Fuentes et al. 2010). The strains were inoculated by streaking perpendicular to the troughs, and the Petri dishes were incubated at 30 °C for 7 days. Microbial growth was used as a qualitative parameter of toxicity tolerance. Control samples for growth were also created using a medium without the addition of toxics (Fuentes et al. 2013). When the individual toxic elements were assayed, *Streptomyces* spp. A5, A11, M7, MC1, and *Amycolatopsis tucumanensis* all showed similar growth to that observed in the uncontaminated control, while *Streptomyces* sp. A2 showed little growth and was thus considered to have low tolerance to Cr(VI) and lindane. Previously, Polti et al. (2007) used Cr(VI) 260 mg L^{-1}, and Benimeli et al. (2003) used 10 μg L^{-1} of lindane to select Cr(VI) or lindane tolerant actinobacteria, respectively. Therefore, used toxic concentrations ensure selection of bacteria with high tolerance to such compounds.

Degradation of organic contaminants by microorganisms generally corresponds to an inducible system. However, in co-contaminated environments the presence of heavy metals inhibits the degrading metabolism, so it is necessary to evaluate the toxicity of both pollutants in combination in order to select the most suitable microorganisms for bioremediation processes (Alisi et al. 2009; Moreira et al. 2013; Thavamani et al. 2012). Therefore, as described above the two types of contaminants were mixed to evaluate their combined effect on the six evaluated actinobacteria.

No inhibition of growth was seen in *Streptomyces* spp. A5, A11, M7, MC1, or *Amycolatopsis tucumanensis*. However, little growth was seen in *Streptomyces* sp. A2 and it was thus considered to be a strain that with low tolerance to this mixture of contaminants, probably because the combination of Cr(VI) and lindane enhanced their inhibitory activity against this strain. This effect has in fact already been observed by other authors studying co-contaminated systems (Alisi et al. 2009; Sandrin and Hoffman 2007; Sandrin and Maier 2003). However, these results indicate that the contaminant concentrations used were not inhibitory for the growth of five of the six actinobacteria under the experimental testing conditions. The contaminant concentrations tested were selected based upon previous studies, while also taking international standards for permissible levels in soils into consideration (9 mg kg^{-1} for Cr(VI) and 10 μg kg^{-1} for lindane), in order to ensure that microorganisms with high toxicity resistance could be obtained (Benimeli et al. 2006). The concentrations used are also consistent with those observed by a variety of authors in co-contaminated environments (El Deeb and Altalhi 2009; Olaniran et al. 2009, 2013; Roane et al. 2001; Shi et al. 2013), who have reported lindane and Cr(VI) concentrations in the order of μg L^{-1} and mg L^{-1}, respectively, in different environmental compartments such as soil, groundwater, rainwater, etc. Such contamination levels produce acute toxicity in animals (Harris et al. 2011; Srivastava et al. 2007). Based upon their tolerance to the individual toxic elements and the mixture, the strains *Streptomyces* spp. A5, A11, M7, MC1, and *Amycolatopsis tucumanensis* were initially selected for use in the further studies discussed below.

The use of a single population involves many metabolic limitations, which could be avoided by using a mixed community of microorganisms. In nature, microorganisms exist as elements of microbial consortia, made up of multiple populations that coexist and carry out complex chemical processes and physiological functions in order to enable survival of the community. Microbial consortia can combine the catalytic specialties of different species to metabolize new substrates, including pesticides (Dejonghe et al. 2003; Fuentes et al. 2011; Shong et al. 2012; Smith et al. 2005; Yang et al. 2010).

A microbial consortium formed by resistant actinobacteria could thus enhance the potential to simultaneously remove Cr(VI) and lindane, however, the absence of antagonism between the consortium members is a major issue. The presence of potential antagonistic effects among the isolated strains was evaluated (Fuentes et al. 2011), Petri dishes with solid MM were sown as follows: one of the strains was spread in the center of the plate and faced transversely with the other microorganisms to be assayed. It was considered a strain to be antagonistic to the other evaluated strains if a growth inhibition was observed. In this way, the presence of antagonism among the strains studied was assessed by considering all possible combinations. However, when these individual strains confronted each other on solid MM, it was observed that *Streptomyces* sp. A11 had an inhibitory effect on the growth of *Streptomyces* sp. MC1 and *Streptomyces* sp. M7. These results suggested that it would be best to develop a consortium with *Streptomyces* spp. A5, MC1, M7, and *Amycolatopsis tucumanensis* for removal of lindane and Cr(VI).

11.5 Soil Bioremediation Performance

11.5.1 Sterilized Soil Samples Co-contaminated with Cr(VI) and Lindane

This study was conducted in order to determine the ability of the selected actinobacteria to grow and to remove Cr(VI) and lindane in sterilized SS.

Non-polluted soil samples (SS) were collected from near the surface (5–15 cm deep) and stored in the dark at 10–15 °C until being utilized. Glass pots were filled with 200 g of soil and kept at 20 % humidity using distilled water. The SS were steam-sterilized (three successive sterilizations at 24 h intervals, at 100 °C for 1 h each) (Polti et al. 2009). The sterilized soil samples (SSS) were each inoculated with either an individually selected actinobacterium or with the mixed culture (the four actinobacteria selected after the resistance assay) to a final inoculum concentration of 2 g kg^{-1} of soil (wet weight).

The inoculated SSS were then contaminated with 25 μg $kg^{-1}$of lindane and 50 mg kg^{-1} of Cr(VI). Also, inoculated SSS without toxics and non-inoculated SSS with both toxics were used as controls.

Table 11.1 Microbial growth in SSS after 14 days at 30°C

Strain	Non contaminated SS	Contaminated SS
A. tucumanensis	2×10^8 (a)[a,b]	4×10^7 (ab)
Streptomyces sp. MC1	2×10^8 (a)	5×10^7 (a)
Streptomyces sp. M7	2×10^8 (a)	4×10^8 (a)
Streptomyces sp. A5	2×10^8 (a)	2×10^8 (a)
Consortium	1×10^9 (c)	2×10^8 (c)

[a]Different letters indicate significant differences ($p<0.05$)
[b]CFU L^{-1}

After 14 days at 30 °C, samples were taken at the end of each assay to determine the lindane and chromium concentrations, also microbial growth was evaluated.

Microbial growth was determined as CFU g^{-1} by transferring 1 g of soil from each pot into a sterile flask, containing 9 ml of a sterile sodium hexametaphosphate solution (1.66 g L^{-1}, pH 7), Samples were then vortexed during 10 min and tenfold serial dilutions were made in NaH_2PO_4 (0.05 M, pH 7) and plated out onto solid MM in triplicate. Plates were incubated at 30 °C for 72 h (Polti et al. 2009).

After 14 days, the growth of the individual strains *Streptomyces* spp. A5, M7, and *Amycolatopsis tucumanensis* as well as the growth of the consortium was significantly inhibited by the contaminants ($p<0.05$). However, *Streptomyces* sp. MC1 showed similar growth levels in the presence or absence of both contaminants (Table 11.1). This result agrees with those of previous studies carried out in sterilized SS contaminated with 50 mg $kg^{-1}$of Cr(VI) and inoculated with *Streptomyces* sp. MC1 (Polti et al. 2009). On the other hand, for *Streptomyces* sp. M7, Benimeli et al. (2008) found no growth inhibition in SS contaminated with 100 μg $kg^{-1}$of lindane, and it would thus appear that the combined presence of the two types of contaminants probably caused the observed growth inhibition in this strain.

Potentially bioavailable chromium in the soil was measured by a physical method: 100 g of soil were centrifuged at 5,050 × *g* during 60 min, to reproduce the maximal plant suction (soil water potential: −1,500 kPa, conventional wilting point) (Csillag et al. 1999). After centrifugation, the supernatant was recovered, filtered at 0.45 μm and analyzed by AAS for Cr content (APHA 1989). After 14 days of incubation, bioavailable chromium levels were determined. In control flasks, bioavailable chromium was reduced from 50 to 12 mg kg^{-1}. This result agree with previously reported by other authors (Kotas and Stasicka 2000; Mandiwana et al. 2007; Polti et al. 2011; Stewart et al. 2003), where a fraction of chromium was adsorbed by soil compounds. Over time, this concentration was kept constant. Bioavailable chromium concentration detected in control flasks (12 mg kg^{-1}) was considered as 100 % to further calculations. *Streptomyces* spp. MC1, M7, A5, and the consortium were able to completely remove the bioavailable chromium, while *Amycolatopsis*

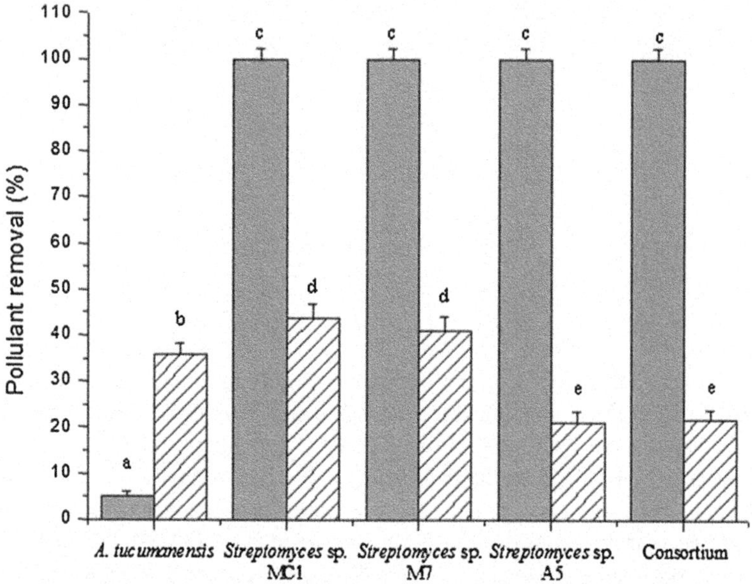

Fig. 11.1 Bioavailable chromium (*grey shaded box*) and Lindane (*stripped box*) removal in sterilized SS, after 14 days at 30 °C. Means with *different letters* are significantly different ($p<0.05$)

tucumanensis removed only 5 % (Fig. 11.1). Previous studies have demonstrated that the bioavailable chromium fraction is exclusively formed by Cr(VI) (Polti et al. 2011), and it can therefore be inferred that the reduction of bioavailable chromium is due to either Cr(VI) reduction to Cr(III) or to bioaccumulation of chromium by *Streptomyces* spp. MC1, M7, and A5. However, since *Amycolatopsis tucumanensis* showed little ability to reduce the bioavailable chromium, its Cr(VI) tolerance must reflect a different mechanism, such as metal exclusion by a permeability barrier or active transport of the metal away from the cell (Bruins et al. 2000). In this case, the metal resistance mechanism does not have relevance in terms of bioremediation processes. It can also be mentioned that Albarracín et al. (2008b) demonstrated copper accumulation by *Amycolatopsis tucumanensis*, with electron microscopy studies demonstrating the presence of copper binding-proteins inside the cell. This mechanism would thus seem to be specific to copper, or at least it is not utilized with chromium.

Polti et al. (2009) previously reported that *Streptomyces* sp. MC1 removed more than 90 % of bioavailable chromium after 14 days of incubation in SSS contaminated with 50 mg kg^{-1} of Cr(VI). In the present work, this strain maintained this ability despite the presence of a second pollutant.

The extraction and determination procedure for γ-HCH residues in soil was performed according to Fuentes et al. (2011). The changes in lindane concentration in controls were also evaluated. No variations of lindane concentrations in both control series were observed (data not shown), so, there was no evidence of noticeable contribution of abiotic processes to pesticide removal. *Streptomyces* spp. MC1, M7, A5, *Amycolatopsis tucumanensis*, and the consortium were all able to remove significant amounts of lindane ($p<0.05$). *Streptomyces* sp. MC1 and *Streptomyces* sp. M7 showed the highest removal levels (44 and 41 % respectively), while *Streptomyces* sp. A5 and the consortium removed 22 and 21 %, respectively. *Amycolatopsis tucumanensis* removed 36 % (Fig. 11.1).

The actinobacteria showing better performance in the sterilized SS were selected to carry out studies in non-sterilized SS (NSSS), in order to evaluate the influence of the native microbial flora on their ability to remove Cr(VI) and lindane.

11.5.2 Nonsterilized Soil Samples Co-contaminated with Cr(VI) and Lindane

Non-sterilized soil samples (NSSS) were inoculated with the selected actinobacteria, then contaminated with 25 μg $kg^{-1}$of lindane and 50 mg kg^{-1} of Cr(VI). Flasks were incubated at 30 °C during 14 days. Similarly, inoculated NSSS without toxics and non-inoculated NSSS with both toxics were used as controls.

In control flasks, bioavailable chromium was reduced from 50 to 18 mg kg^{-1}. This result agree with previously found in non-sterilized soils (Polti et al. 2011). The bioavailable chromium fraction was lower in SSS than in NSSS, sterilization process modifies adsorption properties of soil, probably exposing or activating reactive groups of soil, and also the adsorbing surfaces that control the heavy metals solubility (Egli et al. 2006). Similarly to that observed in SSS, over time, this concentration was kept constant. Bioavailable chromium concentration detected in control flasks (18 mg kg^{-1}) was considered as 100 % to further calculations.

Bioavailable chromium removal of by *Streptomyces* sp. M7 (28 %) was significantly higher ($p<0.05$) than by the consortium (14 %) (Fig. 11.2). It is noticeable that the bioavailable chromium removal produced by the consortium decreased significantly in the NSSS in comparison with the SSS. The main barrier to the use of communities in bioprocesses is the need for simultaneous control of both individual organisms as the ecosystem as a whole. It is possible that the different consortium members had different behaviors in relation to the native flora, resulting in a decrease in the overall performance of the consortium (Shong et al. 2012).

The changes in lindane concentration in controls were also evaluated. No variations of lindane concentrations in both control series were observed, so, there was no evidence of noticeable contribution of autochthonous microflora on the pesticide removal.

Lindane removal by the consortium and Streptomyces sp. M7 was higher than 50 % (Fig. 11.2).

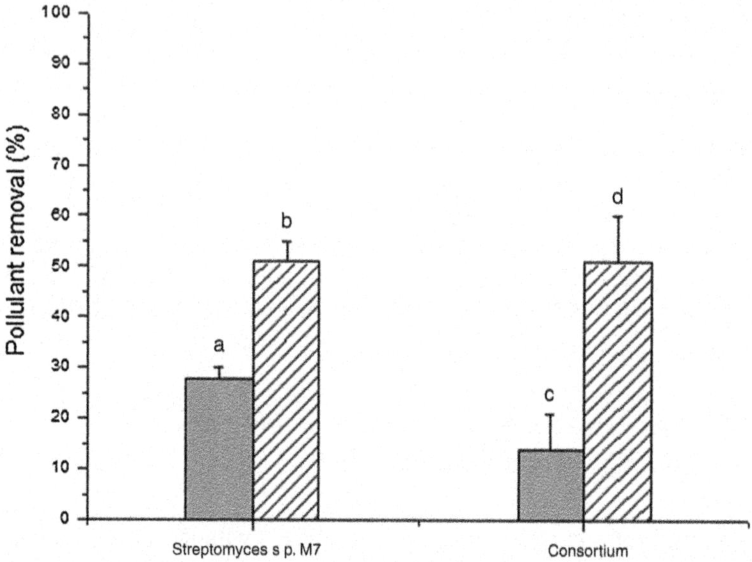

Fig. 11.2 Bioavailable chromium (*grey shaded box*) and Lindane (*stripped box*) removal in non sterilized SS, after 14 days at 30 °C. Means with *different letters* are significantly different ($p < 0.05$)

11.6 Concluding Remarks

Based upon these results, it appears that *Streptomyces* sp. M7 and the consortium makeup of the four actinobacterial strains tested could be useful for bioremediation of soils co-contaminated with Cr(VI) and lindane. However, taking into account the importance of cost–benefit ratios in biotechnological processes and the fact that the use of a consortium is more complex and time consuming and also carries higher risks of contamination, the use of *Streptomyces* sp. M7 alone would seem to be most suitable for these types of processes.

In a next step, laboratory testing must be scaled up to field. It is mandatory to transfer the acquired knowledge to benefit the affected population.

References

Albarracín VH, Amoroso MJ, Abate CM (2005) Isolation and characterization of indigenous copper-resistant actinomycete strains. Chem Erde Geochem 65:145–156

Albarracín VH, Ávila AL, Amoroso MJ et al (2008a) Copper removal ability by *Streptomyces* strains with dissimilar growth patterns and endowed with cupric reductase activity. FEMS Microbiol Lett 288:141–148

Albarracín VH, Winik B, Kothe E et al (2008b) Copper bioaccumulation by the actinobacterium *Amycolatopsis* sp. AB0. J Basic Microbiol 48:323–330

Albarracín VH, Alonso-Vega P, Trujillo ME et al (2010a) *Amycolatopsis tucumanensis* sp. nov., a copper-resistant actinobacterium isolated from polluted sediments. Int J Syst Evol Microbiol 60:397–401

Albarracín VH, Amoroso MJ, Abate CM (2010b) Bioaugmentation of copper polluted soil microcosms with *Amycolatopsis tucumanensis* to diminish phytoavailable copper for Zea mays plants. Chemosphere 79:131–137

Alisi C, Musella R, Tasso F et al (2009) Bioremediation of diesel oil in a co-contaminated soil by bioaugmentation with a microbial formula tailored with native strains selected for heavy metals resistance. Sci Total Environ 407:3024–3032

Amoroso MJ, Castro GR, Durán A et al (2001) Chromium accumulation by two *Streptomyces* spp. isolated from riverine sediments. J Ind Microbiol Biotechnol 26:210–215

Amoroso MJ, Oliver G, Castro GR (2002) Estimation of growth inhibition by copper and cadmium in heavy metal tolerant actinomycetes. J Basic Microbiol 42:231

APHA (1989) Standard methods for the examination of water and wastewater, 17th edn. American Public Health Association, Washington, DC

Atlas R, Unterman R (1999) Bioremediation. In: Demain AC, Davis JE (eds) Industrial microbiology and biotechnology. ASM Press, Washington, DC

Bagchi D, Stohs SJ, Downs BW et al (2002) Cytotoxicity and oxidative mechanisms of different forms of chromium. Toxicology 180:5–22

Benimeli CS, Amoroso MJ, Chaile AP et al (2003) Isolation of four aquatic streptomycetes strains capable of growth on organochlorine pesticides. Bioresour Technol 89:133–138

Benimeli CS, Castro GR, Chaile AP et al (2006) Lindane removal induction by *Streptomyces* sp. M7. J Basic Microbiol 46:348–357

Benimeli CS, Castro GR, Chaile AP et al (2007) Lindane uptake and degradation by aquatic *Streptomyces* sp. strain M7. Int Biodeter Biodegr 59:148–155

Benimeli CS, Fuentes MS, Abate CM, Amoroso MJ (2008) Bioremediation of lindane-contaminated soil by Streptomyces sp. strain M7 and its effects on Zea mays growth. Int Biodeterior Biodegrad 61:233–239

Bhadra A, Mahananda M (2013) Bioaccumulation of hexavalent chromium in rice (*Oryza sativa* L.) grown in paddy field soil of Basundhara coal mine area, Sundargarh, Odisha, India. Development 25:27

Boopathy R (2000) Factors limiting bioremediation technologies. Bioresour Technol 74:63–67

Bruins MR, Kapil S, Oehme FW (2000) Microbial resistance to metals in the environment. Ecotoxicol Environ Saf 45:198–207

Csillag J, Partay G, Lukacs A et al (1999) Extraction of soil solution for environmental analysis. Int J Environ Anal Chem 74:305–324

Cuozzo S, Rollán G, Abate C et al (2009) Specific dechlorinase activity in lindane degradation by *Streptomyces* sp. M7. World J Microbiol Biotechnol 25:1539–1546

Dana Devi K, Rozati R, Saleha Banu B et al (2001) In vivo genotoxic effect of potassium dichromate in mice leukocytes using comet assay. Food Chem Toxicol 39:859–865

Dejonghe W, Berteloot E, Goris J et al (2003) Synergistic degradation of linuron by a bacterial consortium and isolation of a single linuron-degrading variovorax strain. Appl Environ Microbiol 69:1532–1541

Egli M, Mirabella A, Kägi B et al (2006) Influence of steam sterilisation on soil chemical characteristics, trace metals and clay mineralogy. Geoderma 131:123–142

El Deeb B, Altalhi AD (2009) Degradative plasmid and heavy metal resistance plasmid naturally coexist in phenol and cyanide assimilating bacteria. Am J Biochem Biotechnol 5:84–93

Ensign J (1990) Introduction to the Actinomycetes. In: Balows A, Trüper H, Dworkin M, Harder W, Schleifer K (eds) The prokaryotes. A handbook on the biology of bacteria: ecophysiology, isolation, identification, applications. Springer, New York, pp 811–815

Fuentes MS, Benimeli CS, Cuozzo SA et al (2010) Isolation of pesticide-degrading actinomycetes from a contaminated site: Bacterial growth, removal and dechlorination of organochlorine pesticides. Int Biodeter Biodegr 64:434–441

Fuentes M, Sáez J, Benimeli C et al (2011) Lindane biodegradation by defined consortia of indigenous *Streptomyces* strains. Water Air Soil Pollut 222:217–231
Fuentes MS, Briceño G, Saez JM et al (2013) Enhanced removal of a pesticides mixture by single cultures and consortia of free and immobilized *Streptomyces* strains. Biomed Res Int 2013:9
Genilloud O, González I, Salazar O et al (2011) Current approaches to exploit actinomycetes as a source of novel natural products. J Ind Microbiol Biotechnol 38:375–389
Goodfellow M, Williams S, Mordarski M (1988) Actinomycetes in biotechnology. Academic Press, San Diego
Harris ESJ, Cao S, Littlefield BA et al (2011) Heavy metal and pesticide content in commonly prescribed individual raw Chinese herbal medicines. Sci Total Environ 409:4297–4305
Johri AK, Dua M, Saxena DM et al (2000) Enhanced degradation of hexachlorocyclohexane isomers by *Sphingomonas paucimobilis*. Curr Microbiol 41:309–311
Kieser T, Bibb M, Buttner M et al (2000) Practical *Streptomyces* genetics. The John Innes Foundation, Colney
Kotas J, Stasicka Z (2000) Chromium occurrence in the environment and methods of its speciation. Environ Pollut 107:263–283
Ma JW, Wang FY, Huang ZH et al (2010) Simultaneous removal of 2,4-dichlorophenol and Cd from soils by electrokinetic remediation combined with activated bamboo charcoal. J Hazard Mater 176:715–720
Mandiwana KL, Panichev N, Kataeva M et al (2007) The solubility of Cr(III) and Cr(VI) compounds in soil and their availability to plants. J Hazard Mater 147:540–545
Mansour S (2012) Evaluation of Residual Pesticides and Heavy Metals Levels in Conventionally and Organically Farmed Potato Tubers in Egypt. In: He Z, Larkin R, Honeycutt W (eds) Sustainable potato production: global case studies. Springer, Netherlands, pp 493–506
Moreira I, Amorim C, Carvalho M et al (2013) Effect of the metals iron, copper and silver on fluorobenzene biodegradation by Labrys portucalensis. Biodegradation 24:245–255
Nie M, Xian N, Fu X et al (2010) The interactive effects of petroleum-hydrocarbon spillage and plant rhizosphere on concentrations and distribution of heavy metals in sediments in the Yellow River Delta, China. J Hazard Mater 174:156–161
Olaniran AO, Balgobind A, Pillay B (2009) Impacts of heavy metals on 1,2-dichloroethane biodegradation in co-contaminated soil. J Environ Sci (China) 21:661–666
Orton TG, Saby NPA, Arrouays D et al (2013) Spatial distribution of lindane concentration in topsoil across France. Sci Total Environ 443:338–350
Phillips TM, Seech AG, Lee H et al (2005) Biodegradation of hexachlorocyclohexane (HCH) by microorganisms. Biodegradation 16:363–392
Polti MA, Amoroso MJ, Abate CM (2007) Chromium(VI) resistance and removal by actinomycete strains isolated from sediments. Chemosphere 67:660–667
Polti MA, García RO, Amoroso MJ et al (2009) Bioremediation of chromium(VI) contaminated soil by *Streptomyces* sp. MC1. J Basic Microbiol 49:285–292
Polti MA, Amoroso MJ, Abate CM (2010) Chromate reductase activity in *Streptomyces* sp. MC1. J Gen Appl Microbiol 56:11–18
Polti MA, Atjian MC, Amoroso MJ et al (2011) Soil chromium bioremediation: synergic activity of actinobacteria and plants. Int Biodeter Biodegr 65:1175–1181
Rathnayake IVN, Megharaj M, Krishnamurti GSR et al (2013) Heavy metal toxicity to bacteria – are the existing growth media accurate enough to determine heavy metal toxicity? Chemosphere 90:1195–1200
Roane TM, Josephson KL, Pepper IL (2001) Dual-bioaugmentation Strategy To Enhance Remediation of Cocontaminated Soil. Appl Environ Microbiol 67:3208–3215
Saez JM, Benimeli CS, Amoroso MJ (2012) Lindane removal by pure and mixed cultures of immobilized actinobacteria. Chemosphere 89:982–987
Sandrin TR, Hoffman DR (2007) Bioremediation of Organic and Metal Co-contaminated Environments: Effects of Metal Toxicity, Speciation, and Bioavailability on Biodegradation. In: Singh S, Tripathi R (eds) Environmental bioremediation technologies. Springer, Berlin Heidelberg, pp 1–34

Sandrin TR, Maier RM (2003) Impact of metals on the biodegradation of organic pollutants, vol 111. ETATS-UNIS, US Department of Health and Human Services, Research Triangle Park, NC

Shi Y, Lu Y, Meng F et al (2013) Occurrence of organic chlorinated pesticides and their ecological effects on soil protozoa in the agricultural soils of North Western Beijing, China. Ecotoxicol Environ Saf 92:123–128

Shong J, Jimenez Diaz MR, Collins CH (2012) Towards synthetic microbial consortia for bioprocessing. Curr Opin Biotechnol 23:798–802

Siñeriz ML, Kothe E, Abate CM (2009) Cadmium biosorption by *Streptomyces* sp. F4 isolated from former uranium mine. J Basic Microbiol 49:S55–S62

Smith D, Alvey S, Crowley DE (2005) Cooperative catabolic pathways within an atrazine-degrading enrichment culture isolated from soil. FEMS Microbiol Ecol 53:265–275

Sprocati AR, Alisi C, Tasso F, Marconi P, Sciullo A, Pinto V, Chiavarini S, Ubaldi C, Cremisini C (2012) Effectiveness of a microbial formula, as a bioaugmentation agent, tailored for bioremediation of diesel oil and heavy metal co-contaminated soil. Process Biochem 47:1649–1655

Srinivasa Gowd S, Ramakrishna Reddy M, Govil PK (2010) Assessment of heavy metal contamination in soils at Jajmau (Kanpur) and Unnao industrial areas of the Ganga Plain, Uttar Pradesh, India. J Hazard Mater 174:113–121

Srivastava S, Ahmad AH, Thakur IS (2007) Removal of chromium and pentachlorophenol from tannery effluents. Bioresour Technol 98:1128–1132

Stewart M, Jardine P, Brandt C et al (2003) Effects of contaminant concentration, aging, and soil properties on the bioaccessibility of Cr (III) and Cr (VI) in soil. Soil Sediment Contam 12:1–21

Tang X, Shen C, Shi D et al (2010) Heavy metal and persistent organic compound contamination in soil from Wenling: An emerging e-waste recycling city in Taizhou area, China. J Hazard Mater 173:653–660

Thavamani P, Megharaj M, Naidu R (2012) Bioremediation of high molecular weight polyaromatic hydrocarbons co-contaminated with metals in liquid and soil slurries by metal tolerant PAHs degrading bacterial consortium. Biodegradation 23:823–835

USEPA (1998) Toxicological review for hexavalent chromium. USEPA

Vobis G (1997) Morphology of actinomycetes. In: Miyadoh S (ed) Atlas of actinomycetes. Asakura Publishing Co., Japón, pp 180–191

Volke Sepúlveda T, Velasco Trejo JA (2002) Tecnologías de remediación para suelos contaminados. INE-SEMARNAT, Mexico

Wasi S, Tabrez S, Ahmad M (2011) Suitability of immobilized pseudomonas fluorescens SM1 strain for remediation of phenols, heavy metals, and pesticides from water. Water Air Soil Pollut 220:89–99

Yang S-C, Lei M, Chen T-B et al (2010) Application of zerovalent iron (Fe0) to enhance degradation of HCHs and DDX in soil from a former organochlorine pesticides manufacturing plant. Chemosphere 79:727–732

Chapter 12
Molecular Markers in Hydrocarbon Degradation: State of the Art and Prospective in South America

Mariana P. Lanfranconi and Héctor M. Alvarez

Abstract In South America petroleum production is one of the principal economic resources for various countries. Although petroleum and its derivatives fuel daily life, the major concern to their use is related to the high environmental risk of oil spills. Whether it is a large and acute or small and frequent exposure, the ecosystem must adapt to the scenario and rely on those microorganisms capable of regenerating the system. Bacteria have a central role in bioremediation and different catabolic pathways involved in saturated and aromatic hydrocarbons removal have been identified from isolated strains. The traditional isolation and pure cultivation approach allows a thorough physiological study of the process occurring in these strains. However, the accuracy in linking composition to function using this approach is debatable. Culture-independent techniques applied to the metabolism of components of petroleum can be used to discover novel genes and catabolic routes directly from the environment. This chapter presents an updated overview of different approaches and molecular markers used in hydrocarbon degradation in coasts from South America as well as those methods already used outside the geographic scope of this work that still need to be acknowledged in this region.

12.1 Introduction

Petroleum is a very complex mixture of hydrocarbons with different chemical structures and physical properties depending on the extraction site. For example, in a South American crude oil more than 11,000 compounds were detected, although the authors suggest that the real composition may be three times larger (Hughey et al. 2002). Despite its complexity, crude oil components can be divided in aromatic

M.P. Lanfranconi (✉) • H.M. Alvarez
Centro Regional de Investigación y Desarrollo Científico Tecnológico (CRIDECIT), Universidad Nacional de Patagonia San Juan Bosco, Ruta Provincial N°1 Km 4, Ciudad Universitaria, 9000 Comodoro Rivadavia, Chubut, Argentina
e-mail: mlanfranconi@unpata.edu.ar

A. Alvarez, M.A. Polti (eds.), *Bioremediation in Latin America: Current Research and Perspectives*, DOI 10.1007/978-3-319-05738-5_12

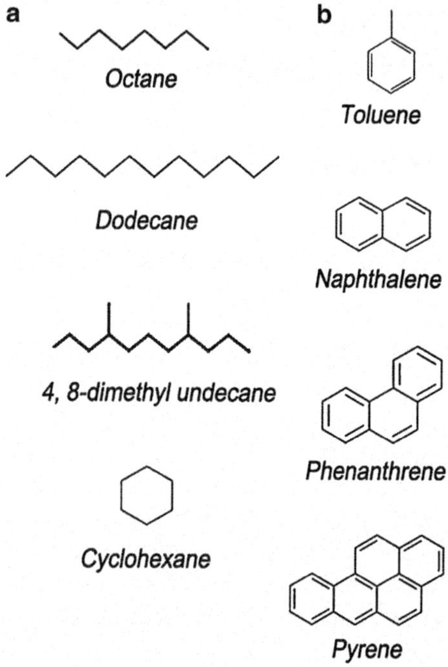

Fig. 12.1 Examples of abundant crude oil components grouped according to their structures. (**a**) Alkanes, (**b**) Aromatic hydrocarbons

(mono and polyaromatic), saturated (linear, ramified, or cyclic *n*-alkanes), and the more recalcitrant asphaltenes and resins (Aske et al. 2001). Saturated hydrocarbons and polycyclic aromatic hydrocarbon (PAH) represent the largest fraction of light crude oil (Fig. 12.1). In contrast, heavy crude oil contains a higher proportion of the polar compounds. Regarding the biodegradation dynamics when it is released in the environment, the former group is the first to be attacked followed by the latter (Head et al. 2006) (Fig. 12.2).

Although it remains unrecognized by our society, a great proportion of oil enters the marine environment as a result of daily activities related to the maritime industry or with routine hydrocarbon extraction, manipulation and consumption in land. In fact, land-based runoff and illegal discharges are two of the most important causes of pollution events in the marine environment (Committee on Oil in the Sea 2003). There are 27 oil-containing basins in South America and the most productive are located near the coasts or offshore (USGA 2012) (Fig. 12.3). Furthermore, petroleum obtained from terrestrial fields frequently travels by water to oil refineries and refined hydrocarbons return via maritime transportation (Estevez 2009). Thus, the relevance of petroleum and its derivatives in marine environments as major contaminants has not been ignored. There is a strong regional effort directed to the understanding and evaluation of the biological removal of hydrocarbons as a step in the development of in situ bioremediation strategies. In this sense, bacterial and archaeal diversity in polluted coastal areas of oil-producing countries in South America have been reported (Vieira et al.

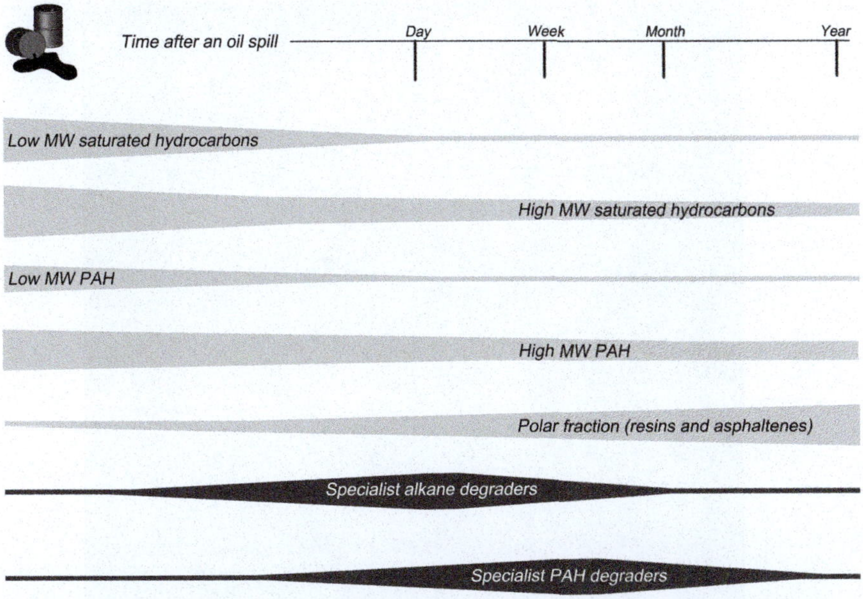

Fig. 12.2 Dynamics of different petroleum components and degradation players after an oil spill. MW: molecular weight. *PAH* polyaromatic hydrocarbons (modified from Head et al. 2006)

2007; Artigas et al. 2008) as well as their role in bioremediation (Gomes et al. 2008; reviewed in Dionisi et al. 2012). The presence of typical hydrocarbon-degrading bacteria inhabiting marine environments is not an exception in coasts of South America (Brito et al. 2006; dos Santos et al. 2011; Isaac et al. 2013). Genera like *Pseudomonas*, *Marinobacterium*, *Marinobacter*, and *Cycloclasticus*, capable of degrading PAH and *Alcanivorax*, a specialist in alkane degradation, have been detected or isolated. Other bacterial groups, which were previously not characterized as hydrocarbon consumers, have also been found (Guibert et al. 2012). Therefore, the real bacterial diversity with the ability to remove petroleum components goes beyond the classic degrader groups. Studies of microbial communities through 16S rRNA analysis can help in identifying metabolically active hydrocarbon degraders, but it provides indirect information in relation to the biodegradation potential of the community. Targeting functional markers helps in the in situ characterization of the process because it allows a detailed analysis of groups which would be undetected or in low proportion in ribosomal-based surveys. It is important to mention that hydrocarbon degradation is not exclusively found in bacteria; fungi for example are also able to degrade components of petroleum. However, they are not the scope of this chapter; as a consequence, the term microorganisms will be used when referring to bacteria.

Biodegradation of hydrocarbons can occur under aerobic or anaerobic conditions depending on the environment under study. Thus, the catabolic routes and the microorganisms involved in the process vary according to O_2 availability.

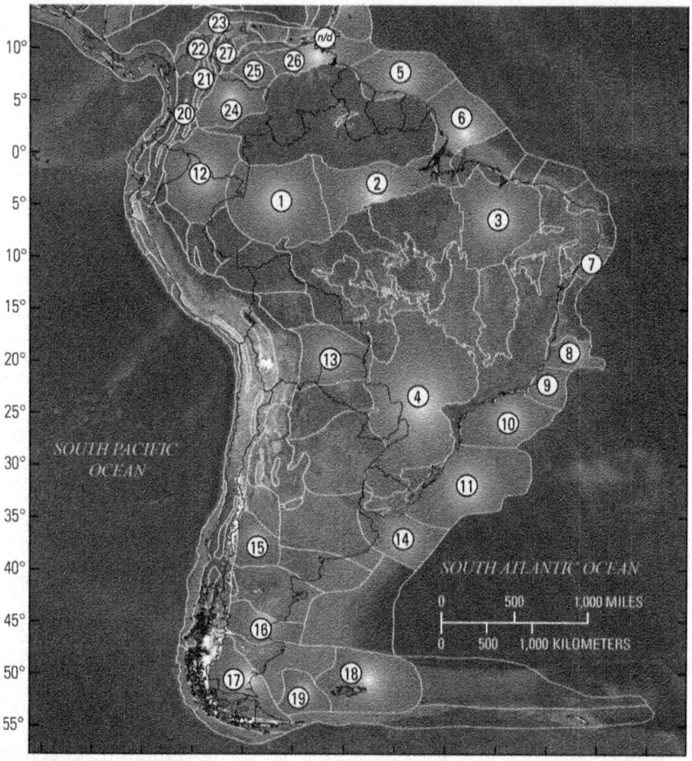

Fig. 12.3 Oil basins located in South America. *1* Solimoes, *2* Amazonas, *3* Parnaiba, *4* Parana, *5* Guyana-Suriname, *6* Foz do Amazonas, 7 Sergipe-Alagoas, *8* Espirito Santo, *9* Campos, *10* Santos, *11* Pelotas, *12* Putamayo-Oriente-Maranon, *13* Santa Cruz-Tarija, *14* Salado-Punta del Este, *15* Neuquen, *16* San Jorge, *17* Magallanes, *18* Falklands Plateau, *19* Malvinas Basin, *20* Upper Magdelena, *21* Middle Magdelena, *22* Lower Magdelena, *23* Guarija, *24* Llanos, *25* Barinas-Apure, 26 East Venezuela, 27 Maracaibo, *n/d* not detected (modified from USGA 2012)

This chapter aims at presenting the natural degradation potential found in polluted marine ecosystem studied in South America. The second section includes the unexplored metabolic capacities present in the environment and powerful *–omics* tools initiatives are considered as future prospects to be acknowledged in the region.

12.2 Aerobic Hydrocarbon Degradation in Marine Environment

Considering the high risk that threatens coastal ecosystems due to oil exploitation and the ecological value of marine microorganisms, understanding and preserving their diversity is mandatory. Particularly in South America, Brazil possesses the highest number of priority areas for marine conservation. However, this scenario is difficult to maintain and 9 % of protected areas has been ceded for offshore exploitation (Turra et al. 2013). To diagnose how healthy a specific region or habitat is, most studies consider the aerobic degradation of hydrocarbons, which is better known and understood. Moreover, the aerobic degradation process usually occurs faster in the environment. Both PAH and saturated hydrocarbons are degraded when O_2 is available in marine ecosystems. In this section, regional studies reporting the natural degradation analysis of both groups using functional gene markers will be presented.

12.2.1 Linear, Branched Alkanes and Cycloalkanes

In order to analyze the environmental capacity for xenobiotic degradation, the first approach is the selection of genes to be used as biomarkers. Marine catabolism of saturated hydrocarbons under aerobic conditions has begun to be studied recently in South America via genes encoding alkane-hydroxylases (Guibert et al. 2012). This class of enzymes is divided in three categories according to the length of the alkanes used as substrate (reviewed in van Beilen and Funhoff 2007). Among them, a membrane monooxygenase (AlkB) that mediates the conversion of the alkane into the respective alcohol is the selected target in environmental studies around the world. Several microorganisms are able to degrade aliphatic hydrocarbons, including the paradigm in alkane metabolism represented by *Pseudomonas putida* GPo1. Furthermore, AlkB is proposed to be also active in hydrocarbonoclastic marine bacteria. Transcriptomic analyses in *Alcanivorax borkumensis*, a well-known hydrocarbonoclastic marine microorganism, have confirmed the significant upregulation of *alkB* when cells were grown on hexadecane (Naether et al. 2013). Also, Kube et al. (2013) demonstrated using Real-Time PCR an increase in alkane monooxygenase gene expression by *Oleispira antarctica* during growth of cells on tetradecane. Sediments are the final sink of pollutants in marine environments. Therefore, the presence and diversity of alkane monooxygenase genes (*alkB*) in chronic polluted intertidal sediments from Ushuaia Bay in Argentina were recently analyzed using a set of degenerate primers (Guibert et al. 2012). Although alkanes are immediately attacked after being released, their concentration in the sampled sites indicated moderate polluted sediments. The authors indicated that translated functional gene sequences were less than 70 % identical to already known AlkB from isolated bacterial strains. This finding confirmed the notion that environmental degrading community goes beyond cultivated representatives with the capacity to degrade alkanes. As a result, there is a common underestimation of the real degradative potential offered by natural communities. Most sequences affiliated with gram-negative bacteria and clustered together with AlkB sequences from *Alcanivorax*, *Pseudomonas*, *Marinobacter*, *Thalassolituus*, and *Oleiphilus*. Also, sequences affiliated to gram-positive bacteria (*Actinobacteria* family) were abundant and related to *Rhodococcus* and *Arthrobacter* genera or formed a unique cluster with an uncultured clone retrieved from Artic soil however, no cultured representative closely related was found. Identifying regional alkane-degrading population by molecular methods is fundamental to understand the process occurring at low temperatures. Guibert and collaborators also combined a molecular analysis at phylogenetic (ribosomal-based) and functional level in amended slurry with crude oil. This approach could be useful to confirm results obtained from targeting metabolic genes. For example, the important role of *Oleispira* genus in cold environments evidenced by their data from 16S rRNA gene pyrosequencing analysis was confirmed recently with experiments carried out in *Oleispira antarctica* grown at low temperatures and analyzed at the expression level of *alkB* (Kube et al. 2013). However, trying to link function using a phylogenetic approach could lead to unreal conclusions. *Thalassospira* genus, belonging to the α-*Proteobacteria*, was numerous in phylogenetic-based analysis using a mixture of hydrocarbons (crude oil) but its alkane-degrading capabilities have never been reported. Furthermore, no alkane could be used as carbon and energy source and no genes related to alkane-monooxygenase were found after whole-genome analysis of *Thalassospira xiamenensis* M-5 isolated from an oil storage pool in China (Lai and Shao 2012). Thus, its role in environmental degrading consortia might not be directly connected to alkane catabolism.

Up to our knowledge, the research article portrayed above is the first and only report combining in situ alkane biodegradation and functional biomarker genes in polluted coastal environments of South America. Thus, the biodegradation potential in this region is far from being characterized. Considering the lack of attention in coastal areas in this region and the potential offered by high-throughput *–omic* techniques, their use could be helpful to explore the natural degrading capabilities.

12.2.2 Polycyclic Aromatic Hydrocarbon

The mutagenic and carcinogenic nature of aromatic hydrocarbons makes them to be considered as priority pollutants and their degradation have received the greatest attention. Culture-independent techniques based on PCR analysis to determine functional degradation diversity in polluted coastal areas are a robust approach when there is no previous knowledge to fill this gap. Aerobic catabolism of low molecular PAH begins after the incorporation of two atoms of oxygen into the substrate by a multicomponent aromatic ring dioxygenase. *nahAc* encodes the large subunit and contains the catalytic domain of the oxygenase component of Rieske nonheme dioxygenases. Polycyclic aromatic hydrocarbon degradation occurs in different bacterial genera, including the highly characterized pseudomonads (Diaz et al. 2013). Within this group, the best studied naphthalene dioxygenase is plasmid encoded in *Pseudomonas putida* NCIB 9816-4 (Dennis and Zylstra 2004). Gram-positive bacteria are also able to degrade PAH (von der Weid et al. 2007; Kim et al. 2010; Alvarez and Silva 2013), but little is known about the genes involved in the process (Kosono et al. 1997; Kim et al. 2007). Other distantly related genes involved in PAH degradation have been described in marine bacteria. For example,

Cycloclasticus sp. strains characterized on its ability to degrade aromatic hydrocarbons possess *phn*-genes that, while encoding enzymes with a similar activity, are distantly related to *nah*-genes and its ecological relevance in the environment should also be considered (Geiselbrecht et al. 1998; Kasai et al. 2003). Marine representatives of other genera such as *Neptunomonas*, *Pseudoalteromonas*, *Marinomonas*, *Halomonas*, *Sphingomonas*, and *Burkholderia* also display the ability to degrade PAH (Hedlund et al. 1999; Melcher et al. 2002; Huang et al. 2008; Pinyakong et al. 2012). Whether these genera are ecologically relevant for the removal of PAH could be assessed when working with environmental samples and analyzing the functional diversity of a specific metabolic route. Regarding PAH degradation, *nahAc* has been targeted as biomarker gene to detect bacterial functional diversity in polluted and pristine environments. One aspect that could be considered as a disadvantage is the low *nahAc* sequence similarity among PAH degrading bacteria belonging to different gene families. Therefore, designing subgroup specific primers is mandatory when attempting a thorough analysis to cover the natural diversity. It should be noted that while PCR is an extremely valuable technique to apply in environmental studies, the obtained results depend on the primer used and at the same time, primer design is conditioned by the available information in databases that usually comes from bacterial isolates. However, data derived from annotated genomes and environmental metagenomes have improved primer selection. The first regional report evaluating the diversity of genes involved in aromatic hydrocarbon degradation appeared almost 5 years ago (Gomes et al. 2007). Three mangroves located in Rio de Janeiro (Brazil) showing different levels of PAH contamination were analyzed in relation to composition and relative abundances of *nahAc*. The authors used a nested PCR-DGGE approach to compare the different sites and based on the classification system proposed by Nam et al. (2001), they focused on group III genes. Within this group, *nahAc* and *phnAc* genes encoding the a priori environmentally most relevant types of dioxygenases were found. The distribution of bands was site specific and surprisingly, no *nahAc or phnAc* was detected in the most contaminated mangrove. Instead, the bands obtained affiliated with genes of members belonging to the family *Comamonadaceae*. This result was confirmed by Southern Blot showing hybridization only when using a specific probe for the previously mentioned family.

Much more knowledge has been accumulated from coastal polluted environments located in Patagonia Argentina (Fig 12.4). Different strategies were used to detect bacterial PAH degradation, including the amplification and cloning of functional biomarker genes. This methodology was useful to address the high diversity of bacterial groups able to degrade PAH from seven intertidal sediments located along the eastern coast of Patagonia and five areas located in Tierra del Fuego, Argentina (Lozada et al. 2008) (Fig. 12.4). Levels and composition of PAH varied greatly depending on the sampled site. Those samples showing no PAH pollution did not produced amplification in PCR experiments using specific dioxygenases primers. In contrast, PCR products obtained from polluted sediments were cloned and analyzed phylogenetically. Seven novel dioxygenases gene variants were detected, all of them loosely or strongly affiliated with genes from gram-negative bacteria.

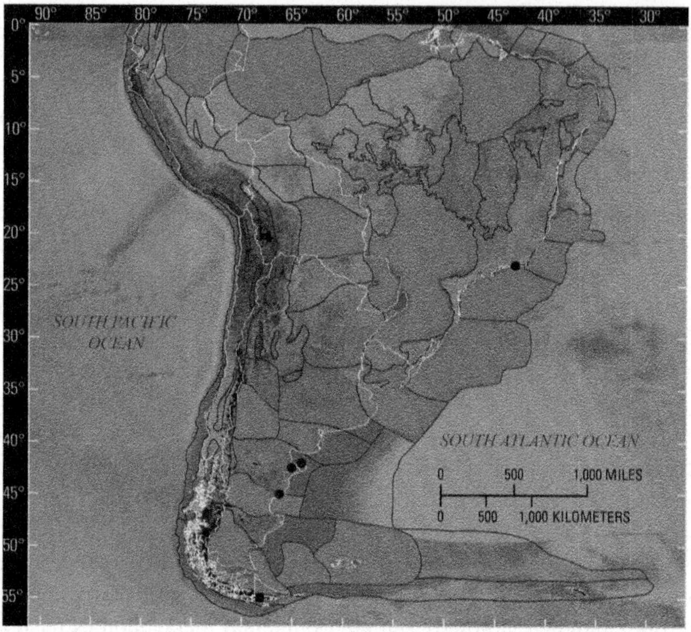

Fig. 12.4 Sampling sites and functional biomarkers analyzed along the coastline of South America. *Black square*: genes involved in polyaromatic and alkane hydrocarbons. *Black circle*: genes involved in alkane hydrocarbons (modified from USGA 2012)

The former, named from A to E, were found in sediments from colder environments while the latter showing ≥95 % identity values at the amino acid level with dioxygenases from *Alcaligenes faecalis* and *Pseudomonas* spp. dominated in samples from patagonic coastal sediments. To cover the diversity of PAH-dioxygenases related to *Cycloclasticus* genus, the authors designed a specific set of primers and tested them on chosen samples from the original 12. A product of the expected size was obtained in all samples and was almost similar to sequences belonging to *Cycloclasticus* isolates. The presence of sequences related to *Neptunomonas*, *Pseudoalteromonas*, *Burkholderia*, and *Ralstonia* sp. was also tested but no amplification was obtained when specific PAH-dioxygenases primers were used. These works broaden the knowledge regarding degrading populations in marine environments and set the bases for future studies going beyond the scope of diversity analysis.

In a recent work, the abundances of seven of the eight dioxygenases variants identified previously by Lozada et al. (2008) were analyzed using quantitative PCR (qPCR) (Marcos et al. 2012). These variants included the *phnA1* from *Cycloclasticus*, *phnAc* from *Alcaligenes faecalis* and *nahAc* from *Pseudomonas* spp. and also the gene types A to D. Due to the low recovery of variant E in clone libraries, it was not considered in this study. Samples covered two polluted sites, Ushuaia Bay and eastern Patagonia (Argentina) along a temporal span. Also, one site located in North

Patagonia without detectable levels of PAH was included (Fig 12.4). Three gene variants (B, C and D) were detected in all sediment samples from Ushuaia Bay highlighting their relevance in the community involved in PAH degradation. However, their abundances showed a reduction with time and no increase in other studied variants was observed. Thus, other PAH-degrading populations not targeted in this study might be in charge of PAH removal. In accordance with clone libraries results (Lozada et al. 2008), variants A–D were important in colder habitats and not in northern areas of Patagonia where *Cycloclasticus* could have a central role in PAH degradation.

Although the contribution of gram-positive bacteria as PAH-degraders in soil has been reported, there is a gap regarding their role in marine environments. This lack of information in the region was addressed using a primer pair specific to amplify gram-positive bacteria (Marcos et al. 2009). A total of 14 dioxygenase groups were identified and matched with PAH degradation genes from *Rhodococcus*, *Nocardioides*, *Mycobacterium*, *Terrabacter*, and *Bacillus* genus within the *Actinobacteria* family. However, the identities with them were low, varying from 33 to 62 %. Thus, the sequences could be considered as novel variants of genes encoding new dioxygenases enzymes. The authors used a new classification system (Kweon et al. 2008) to assign the oxygenase component of Rieske nonheme dioxygenases into five groups. Considering the sequence and the interactions among different enzymatic components, dioxygenases are assigned into one group or another. Particularly, the authors could relate only 4 out of 14 groups to the five types proposed, indicating a bias in the designation of the five types classification towards data obtained from the Northern Hemisphere. Interestingly, type V seems to be characteristic of gram-positive strains. The authors also used specific primers for *Rhodococcus* spp. dioxygenases-gene amplification but no product was obtained suggesting that genes belonging to this genus might not be abundant in these coastal sediments (Dionisi et al. 2011). Up to our knowledge this is the first report showing the contribution of gram-positive bacteria involved in PAH degradation by studying the diversity of a functional gene directly from the environment. Expanding the use of the classification system explained above, the authors also analyzed the sequences obtained in the same area investigated previously (Lozada et al. 2008). All of them affiliated with sequences from gram-negative bacteria and belonged to type III. Thus, in coastal sediments collected in Patagonia, at least three out of five types are present. These results show the great unexplored potential present in cold environments located in South America in relation to PAH degradation.

The identification of hydrocarbon-degrading population at functional level is critical to understand the mechanisms that drive degradation of these toxic compounds in coastal environments. Culture independent approaches have proved to be excellent tools to analyze diversity of key genes involved in the degradation process. Data from South America is scarce and locally focused but the first reports using two classic molecular techniques: PCR-DGGE and PCR-cloning allowed the identification of genes involved in alkanes and PAH degradation. Furthermore, using qPCR it was possible to monitor the dynamics of different key players in PAH degradation. Besides, this information set the bases for future regional metagenomic studies aiming at obtaining a deeper knowledge of biodegradation pathways and the role of horizontal gene transfer in shaping the community involved in hydrocarbon removal. Considering the extensive unexplored coastal area of South America and its central role in petroleum transportation and extraction, much work should be done in the field. Some studying areas that still need to be covered will be included in next sections.

12.3 Anaerobic Hydrocarbon Degradation in the Environment

Mangroves are coastal ecosystems occurring in all maritime countries of South America except Argentina, Chile and Uruguay. They are particularly important in Brazil, covering a large area of coastline. Oxygen (O_2) rarely penetrates the surface thus, the sediments remain anoxic. The impact of petroleum and its derivatives on mangroves has been addressed and the effects studied (Brito et al. 2009; dos Santos et al. 2011). However, anaerobic degradation in situ still needs to be acknowledged in this region as well as key players and genes involved remains to be elucidated. Petroleum degradation under aerobic conditions occurs faster than under anaerobic conditions however, the absence of O_2 characterizes many polluted environments. Among them, petroleum reservoirs are the most studied (Head et al. 2010) but the scope of this section includes nearer habitats such as mangroves and coastal subsurface sediments. Anaerobic degradation is carried out by bacteria, either in consortia or individually, showing a narrow versatility of compounds to be degraded. Due to their volatility and preferable removal under aerobic conditions when enter the water column, the presence of short-chain aliphatic hydrocarbon in anoxic sediments is not expected. The most studied PAH in anaerobic conditions is toluene but others such as xylene, ethylbenzene, naphthalene are also degradable (Widdel et al. 2010). Based on well-studied catabolic pathways of bacteria considered a paradigm of anaerobic degradation and the access to complete genomes, the selection of marker genes involved in hydrocarbon removal has improved lately. However, it is still poorly understood in comparison to the process under aerobic conditions. When oxygen is not available, the first step in hydrocarbon degradation pathway is the addition of fumarate to the methyl group of toluene (or other compound with a similar moiety) (Meckenstock and Mouttaki 2011). The reaction occurs via a multicomponent benzyl or alkylsuccinate synthases and the gene encoding its *r*-subunit (*bssA*) could be used as functional marker in environmental surveys. Most studies targeting *bssA* aim at characterizing a degrading consortium or identifying genes involved in anaerobic degradation in bacterial strains or enrichments (Beller et al. 2002; Botton et al. 2007; Washer and Edwards 2007). However, *bssA* sequence similarity considering different gene families is low and no single primer pair could be used to amplify the whole degradative community (Callaghan et al. 2010). Though scarce, studies evaluating direct environmental potential for anaerobic degradation through *bssA* analysis allowed the analysis of functional diversity in oil-polluted aquifers (Winderl et al. 2007) and subtidal polluted coastal sediments (Acosta-González et al. 2013) in Northern hemisphere places. In contrast, it failed to identify the local environmental potential for anaerobic hydrocarbon degradation in mangroves with a long history of petroleum pollution (Andrade et al. 2012). This could be due to the incompatibility of in situ diversity and the set of primers used. Another functional marker for PAH degradation under anoxic conditions is *bamA*, this gene encodes a hydrolase involved in the last step of the route involved in dearomatizing the ring. It is found in facultative and obligate anaerobes and has been proved useful in enrichment experiments (Kuntze et al. 2011). However, it failed to cover the gram-positive contribution to aromatic degradation. Targeting *bamA* in samples from mangroves with a long history of petroleum pollution indicated that the structure of hydrocarbon-degrading communities changed with depth (Andrade et al. 2012), although the catabolic pathway in which it is involved is not exclusive of petroleum derivatives.

We consider that much effort should be made to analyze functional diversity from natural habitats. In any database search, it is clear that many *bamA* and *bssA* sequences were originated from cultures, enrichments, or microcosms. This bias could result in overlooked environmental sequences when an in situ study is proposed. Next generation sequencing offers an alternative that might reveal mechanisms and key players in anaerobic degradation process.

12.4 Metagenomics in Hydrocarbon Polluted Coasts in South America

The application of metagenomic approaches on environmental samples has evolved from the study of microbial diversity through 16S rRNA analysis (Pace et al. 1985) to the combination with other *–omics* analyses resulting in the understanding of metabolic pathways or community responses (Mason et al. 2012; Yu and Zhang 2012). Furthermore, metagenomes analyses could be helpful to link composition and functions of microbial communities (Silva et al. 2013) and even complete genomes from uncultured microorganisms (Blainey et al. 2011; Albertsen et al. 2013). The reduced costs in next generation sequencing may possibly end in the metagenomic approach replacing the largely used 16S rRNA gene fingerprinting. However, data interpretation generated from metagenome analysis is a challenge so the advantage offered by previous information obtained from classic culture-independent techniques should not be ignored. In case of petroleum polluted areas, it is a powerful tool for identifying metabolic clusters associated with degradation of hydrocarbons. Although coastal areas with a high risk of hydrocarbon pollution remain unexplored in South America, the outcomes of metagenomic studies could be used as baselines for future studies in neighboring areas, especially along the Atlantic coastline considered a hotspot of petroleum extraction and transportation. Also, when there is a previous historical record of the polluted site (alkane and PAH degraders, functional gene marker diversity, etc.), constructing an environmental metagenome turns into an attractive strategy for obtaining a panoramic view of players and mechanisms used for hydrocarbon removal. Metagenomic libraries can be screened by gene-specific PCR to detect the presence of genes of interest within a community. However, this is a time-consuming procedure and could add a bias towards largely known degraders. The identification of novel genes and metabolic pathways could be achieved through an alternative method. It involves the selection (and further sequencing) of positive clones showing a specific phenotype. This strategy also has drawbacks, for example the heterologous in-frame expression of genes together with linked enzymatic activities is not easy to achieve in one clone. When both screening strategies are compared, only a positive clone obtained with the functional approach will contain the complete route for expressing the expected phenotype. Beyond the screening procedure used to identify biodegradation routes, the implementation of metagenomic approaches in hydrocarbon polluted coastal areas in this region is in development (Loviso et al. 2010, 2013) but to the best of our knowledge, there are no published reports in international journals or books related to functional metagenomic analysis of polluted sediments from South America yet. This evidences that diversity of hydrocarbon-degrading bacteria and their catabolic pathways is a topic that still needs to be addressed.

12.5 Concluding Remarks

South America is placed between the Atlantic and the Pacific Ocean, and as a consequence the coastal areas extend largely especially in oil-producing countries. One of the main pollution problems in the marine environment is contamination with hydrocarbons. Particularly, it heavily affects coastal waters and mainly sediments where they accumulate. The presence of contaminants exerts a selective pressure towards favoring microorganisms which are either able to use these pollutants, or to withstand their presence. The existence of recent research articles focusing on this region indicates that a large part of the coastline remains unexplored and there are only local and isolated data focusing on bacterial composition, metabolic capacities, and ecological relationships in polluted and pristine coastal areas. New generation sequencing techniques offer an excellent alternative that combined with functional marker genes could contribute to the understanding of impacted environments. Ushuaia and Guanabara Bays as well as Central Patagonia of Argentina are potential areas for metagenomic approaches due to available previous data based on library construction. Furthermore, other genes or approaches might be applied when the environmental potential in relation to hydrocarbon degradation is tested. As shown above, aerobic degradation of low molecular weight PAH is usually studied through genes encoding naphthalene dioxygenases or alkane monooxygenase. There are certainly other enzymes or metabolic pathways to be discovered that could possibly be used as molecular markers. Their detection depends heavily on metagenomic analysis and data interpretation. Regarding anaerobic degradation of saturated and aromatic hydrocarbons in coastal areas, the scenario is even worse.

There are no significant advances made in the field and there is a world of bacterial players and natural functional diversity to be untapped. Although much research effort is required in this region, the information gathered until now contributes to improve our understanding of bacterial endemism and natural biodegradation capabilities and will surely be indispensable in future research studies.

References

Acosta-González A, Rosselló-Móra R, Marqués S (2013) Diversity of benzylsuccinate synthase-like *bssA* genes in hydrocarbon-polluted marine sediments suggests substrate-dependent clustering. Appl Environ Microbiol 79:3667–3676

Albertsen M, Hugenholtz P, Skarshewski A et al (2013) Genome sequences of rare, uncultured bacteria obtained by differential coverage binning of multiple metagenomes. Nat Biotechnol 31:533–538

Alvarez HM, Silva R (2013) Metabolic diversity and flexibility for hydrocarbon biodegradation by *Rhodococcus*. In: Benimelli CS, Cuozzo SA, Amoroso MJ (eds) Actinobacteria. Application in bioremediation and production of industrial enzymes. CRC Press, Boca Raton, FL, pp 241–273

Andrade LL, Leite DCA, Ferreira EM et al (2012) Microbial diversity and anaerobic hydrocarbon degradation potential in an oil-contaminated mangrove sediment. BMC Microbiol 12:186

Artigas LF, Otero E, Paranhos R et al (2008) Towards a Latin American and Caribbean international census of marine microbes LACar-ICoMM: overview and discussion on some current research directions. Rev Biol Trop 56:183–214

Aske N, Kallevik H, Sjöblom J (2001) Determination of saturate, aromatic, resin, and asphaltenic SARA components in crude oils by means of infrared and near-infrared spectroscopy. Energy Fuels 15:1304–1312

Beller HR, Kane SR, Legler TC et al (2002) A real-time polymerase chain reaction method for monitoring anaerobic, hydrocarbon-degrading bacteria based on a catabolic gene. Environ Sci Technol 36:3977–3984

Blainey PC, Mosier AC, Potanina A et al (2011) Genome of a low-salinity ammonia-oxidizing archaeon determined by single-cell and metagenomic analysis. PLoS One 6:e16626

Botton S, van Harmelen M, Braster M et al (2007) Dominance of Geobacteraceae in BTX-degrading enrichments from an iron-reducing aquifer. FEMS Microbiol Ecol 62:118–130

Brito EM, Guyoneaud R, Goni-Urriza M et al (2006) Characterization of hydrocarbonoclastic bacterial communities from mangrove sediments in Guanabara Bay Brazil. Res Microbiol 57: 752–762

Brito EM, Duran R, Guyoneaud R et al (2009) A case study of in situ oil contamination in a mangrove swamp Rio De Janeiro, Brazil. Mar Poll Bull 58:418–423

Callaghan AV, Davidova IA, Savage-Ashlock K et al (2010) Diversity of benzyl- and alkylsuccinate synthase genes in hydrocarbon-impacted environments and enrichment cultures. Environ Sci Technol 44:7287–7294

Committee on Oil in the Sea: Inputs, Fates, and Effects. Ocean Studies Board and Marine Board, Divisions of Earth and Life Studies and Transportation Research Board, National Research Council (2003) Oil in the sea III: inputs, fates and effects. The National Academy Press, Washington, DC

Dennis JJ, Zylstra GJ (2004) Complete sequence and genetic organization of pDTG1, the 83 kilobase naphthalene degradation plasmid from *Pseudomonas putida* strain NCIB 9816-4. J Mol Biol 341:753–768

Diaz E, Jimenez JI, Nogales J (2013) Aerobic degradation of aromatic compounds. Curr Opin Biotechnol 24:431–442

Dionisi HM, Lozada M, Marcos MS et al (2011) Aromatic hydrocarbon degradation genes from chronically polluted subantarctic marine sediments. In: de Bruijn FJ (ed) Handbook of molecular microbial ecology, vol 2: metagenomics in different habitats, 1st edn. Wiley-Blackwell, New York

Dionisi HM, Lozada M, Olivera NL (2012) Bioprospection of marine microorganisms: biotechnological applications and methods. Rev Argent Microbiol 44:49–60

dos Santos HF, Cury JC, do Carmo FL et al (2011) Mangrove bacterial diversity and the impact of oil contamination revealed by pyrosequencing: bacterial proxies for oil pollution. PLoS ONE 6(3):e16943

Estevez JL (2009) Contaminación costera marina. In: Atlas de Sensibilidad Ambiental de la Costa y el Mar Argentino, Boltovskoy D (ed). http://atlas.ambiente.gov.ar. Accessed Nov 2010

Geiselbrecht AD, Hedlund BP, Tichi MA et al (1998) Isolation of marine polycyclic aromatic hydrocarbon PAH-degrading *Cycloclasticus* strains from the Gulf of Mexico and comparison of their PAH degradation ability with that of Puget Sound *Cycloclasticus* strains. Appl Environ Microbiol 64:4703–4710

Gomes NC, Borges LR, Paranhos R et al (2007) Diversity of *ndo* genes in mangrove sediments exposed to different sources of polycyclic aromatic hydrocarbon pollution. Appl Environ Microbiol 73:7392–7399

Gomes NC, Borges LR, Paranhos R et al (2008) Exploring the diversity of bacterial communities in sediments of urban mangrove forests. FEMS Microbiol Ecol 66:96–109

Guibert LM, Loviso CL, Marcos MS et al (2012) Alkane biodegradation genes from chronically polluted subantarctic coastal sediments and their shifts in response to oil exposure. Microb Ecol 64(3):605–616

Head IM, Jones DM, Röling WFM (2006) Marine microorganisms make a meal of oil. Nat Rev Microbiol 4:173–182

Head IM, Larter SR, Gray N et al (2010) Hydrocarbon biodegradation in petroleum reservoirs. In: Timmis KN (ed) Handbook of hydrocarbon and lipid microbiology. Springer, Braunschweig, Germany, pp 3098–3109

Hedlund BP, Geiselbrecht AD, Bair TJ et al (1999) Polycyclic aromatic hydrocarbon degradation by a new marine bacterium, *Neptunomonas naphthovorans* gen. nov., sp. nov. Appl Environ Microbiol 65:251–259

Huang X, Tian Y, Luo YR et al (2008) Modified sublimation to isolate phenanthrene-degrading bacteria of the genera Sphingomonas and Burkholderia from Xiamen oil port. Mar Poll Bull 57:538–543

Hughey CA, Rodgers RP, Marshall AG (2002) Resolution of 11,000 compositionally distinct components in a single electrospray ionization Fourier transform ion cyclotron resonance mass spectrum of crude oil. Anal Chem 74:4145–4149

Isaac P, Sánchez LA, Bourguignon N et al (2013) Indigenous PAH-degrading bacteria from oil-polluted sediments in Caleta Cordova, Patagonia Argentina. Int Biodeter Biodegr 82:207–214

Kasai Y, Shindo K, Harayama S et al (2003) Molecular characterization and substrate preference of a polycyclic aromatic hydrocarbon dioxygenase from *Cycloclasticus* sp. strain A5. Appl Environ Microbiol 69:6688–6697

Kim SJ, Kweon O, Jones RC et al (2007) Complete and integrated pyrene degradation pathway in *Mycobacterium vanbaalenii* PYR-1 based on systems biology. J Bacteriol 189:464–472

Kim SJ, Kweon O, Cerniglia CE (2010) Degradation of polycyclic aromatic hydrocarbons by *Mycobacterium* strains. In: Timmis KN (ed) Handbook of hydrocarbon and lipid microbiology. Springer, Braunschweig, Germany, pp 1865–1880

Kosono S, Maeda M, Fuji F et al (1997) Three of the seven *bphC* genes of *Rhodococcus erythropolis* TA421, isolated from a termite ecosystem, are located on an indigenous plasmid associated with biphenyl degradation. Appl Environ Microbiol 63:3282–3285

Kube M, Chernikova TN, Al-Ramahi Y et al (2013) Genome sequence and functional genomic analysis of the oil-degrading bacterium *Oleispira antarctica*. Nat Commun 4:2156. doi:10.1038/ncomms3156

Kuntze K, Vogt C, Richnow HH et al (2011) Combined application of PCR-based functional assays for the detection of aromatic compound-degrading anaerobes. Appl Environ Microbiol 77:5056–5061

Kweon O, Kim SJ, Baek S et al (2008) A new classification system for bacterial Rieske non-heme iron aromatic ring-hydroxylating oxygenases. BMC Biochem 9:11

Lai Q, Shao Z (2012) Genome sequence of *Thalassospira xiamenensis* type strain M-5. J Bacteriol 194:6957

Loviso CL, Guibert LM, Marcos MS et al (2010) Metagenomic analysis of aromatic ring hydroxylating dioxygenases from cold marine environments. Biocell 34:108

Loviso CL, Guibert LM, Sarango S et al (2013) Screening of a metagenomic library from polluted subantarctic sediments for the identification of gene clusters involved in hydrocarbon biodegradation, American Society for Microbiology 113th general meeting, Denver, CO, 18–21 May

Lozada M, Riva Mercadal JP, Guerrero LD et al (2008) Novel aromatic ring-hydroxylating dioxygenase genes from coastal marine sediments of Patagonia. BMC Microbiol 8:50

Marcos MS, Lozada M, Dionisi HM (2009) Aromatic hydrocarbon degradation genes from chronically polluted Subantarctic marine sediments. Lett Appl Microbiol 49:602–608

Marcos MS, Lozada M, Di Marzio WD et al (2012) Abundance, dynamics, and biogeographic distribution of seven polycyclic aromatic hydrocarbon dioxygenase gene variants in coastal sediments of Patagonia. Appl Environ Microbiol 78:1589–1592

Mason OU, Hazen TC, Borglin S et al (2012) Metagenome, metatranscriptome and single-cell sequencing reveal microbial response to Deepwater Horizon oil spill. ISME J 6:1715–1727

Meckenstock RU, Mouttaki H (2011) Anaerobic degradation of nonsubstituted aromatic hydrocarbons. Curr Opin Biotechnol 22:406–414

Melcher RJ, Apitz SE, Hemmingsen BB (2002) Impact of irradiation and polycyclic aromatic hydrocarbon spiking on microbial populations in marine sediment for future aging and biodegradability studies. Appl Environ Microbiol 68:2858–2868

Naether DJ, Slawtschew S, Stasik S et al (2013) Adaptation of the hydrocarbonoclastic bacterium *Alcanivorax borkumensis* SK2 to alkanes and toxic organic compounds: a physiological and transcriptomic approach. Appl Environ Microbiol 79:4282–4293

Nam JW, Nojiri H, Yoshida T et al (2001) New classification system for oxygenase components involved in ring-hydroxylating oxygenations. Biosci Biotechnol Biochem 65:254–263

Pace NR, Stahl DJ, Lane DJ et al (1985) Analyzing natural microbial populations by rRNA sequences. ASM News 51:4–12

Pinyakong O, Tiangda K, Iwata K et al (2012) Isolation of novel phenanthrene-degrading bacteria from seawater and the influence of its physical factors on the degradation of phenanthrene. Science Asia 38:36–43

Silva CC, Hayden H, Sawbridge T et al (2013) Identification of genes and pathways related to phenol degradation in metagenomic libraries from petroleum refinery wastewater. PLoS ONE 8:e61811

Turra A, Cróquer A, Carranza A et al (2013) Global environmental changes: setting priorities for Latin American coastal habitats. Global Change Biol 19:1965–1969

US Geological Survey World Energy Assessment Team (2012) Assessment of undiscovered conventional oil and gas resources of South America and the Caribbean. US Geological Survey, Washington, DC

van Beilen JB, Funhoff EG (2007) Alkane hydroxylases involved in microbial alkane degradation. Appl Microbiol Biotechnol 74:13–21

Vieira RP, Clementino MM, Cardoso AM et al (2007) Archaeal communities in a tropical estuarine ecosystem: Guanabara Bay, Brazil. Microb Ecol 54:460–468

von der Weid I, Marques JM, Cunha CD et al (2007) Identification and biodegradation potential of a novel strain of *Dietzia cinnamea* isolated from a petroleum-contaminated tropical soil. Syst Appl Microbiol 30:331–339

Washer CE, Edwards EA (2007) Identification and expression of benzylsuccinate synthase genes in a toluene-degrading methanogenic consortium. Appl Environ Microbiol 73:1367–1369

Widdel F, Knittel K, Galushko A (2010) Anaerobic hydrocarbon-degrading microorganisms – an overview. In: Timmis KN (ed) Handbook of hydrocarbon and lipid microbiology. Springer, Heidelberg, pp 1997–2021

Winderl C, Schaefer S, Lueders T (2007) Detection of anaerobic toluene and hydrocarbon degraders in contaminated aquifers using benzylsuccinate synthase *bssA* genes as a functional marker. Environ Microbiol 9:1035–1046

Yu K, Zhang T (2012) Metagenomic and metatranscriptomic analysis of microbial community structure and gene expression of activated sludge. PLoS One 7(5):38183

Chapter 13
Perspective in Bioremediation: Enhancing the Hexavalent Chromium Removal Using Native Yeasts from Tucumán, Argentina

Pablo M. Fernández, Elías L. Cruz, and Lucía I.C. de Figueroa

Abstract The occurrence of indigenous Cr(VI)-reducing eukaryotic microorganisms, including those with no history of Cr(VI) contamination, has provided important non-conventional yeasts with significant biological relevance and biotechnological applications. Based on physiological/biochemical characterization and molecular taxonomy analysis, these isolates were identified as *Cyberlindnera jadinii* M9 and *Wickerhamomyces anomalus* M10. *Cy. jadinii* M9 and *W. anomalus* M10 were grown in medium plus 1 mM Cr(VI) at 25 °C, causing complete chromium removal before reaching 48 h of cultivation. Flame Atomic Absorption Spectroscopy (FAAS) assays suggested that Cr(VI) disappearance was coupled to the Cr(III) concomitant production. These results indicated that reducing capacity of chromate-resistant yeasts would be the main detoxification mechanism. Crude chromate reductase (CChRs) of strains M9 and M10, were characterized based on optimal temperature, pH, use of electron donors, metal ions and initial Cr(VI) concentration in the reaction mixture. Both CChRs showed an increase in Cr(VI) reductase activity with addition of NAD(P)H as electron donor and were highly inhibited by Hg^{2+} and Mn^{2+}. The CChR from *Cy. jadinii* M9 showed the highest chromate reductase activity at 60 °C and pH 6.0 in the presence of Cu^{2+} or Na^{+}, while *W. anomalus* M10 CChR had the maximum activity at 50 °C and pH 7.0 in

P.M. Fernández (✉)
Planta Piloto de Procesos Industriales Microbiológicos PROIMI-CONICET,
Av. Belgrano y Caseros, T4001MVB Tucumán, Argentina

Calidad de Productos II, Gestión de Empresas Agroindustriales,
Universidad de San Pablo – T, Tucumán, Argentina
e-mail: pmfernandez@argentina.com; pfernandez@proimi.org.ar

E.L. Cruz
Planta Piloto de Procesos Industriales Microbiológicos PROIMI-CONICET,
Av. Belgrano y Caseros, T4001MVB Tucumán, Argentina

L.I.C. de Figueroa
Planta Piloto de Procesos Industriales Microbiológicos PROIMI-CONICET,
Av. Belgrano y Caseros, T4001MVB Tucumán, Argentina

Microbiología Superior, Facultad de Bioquímica, Química y Farmacia,
Universidad Nacional de Tucumán, Tucumán, Argentina

A. Alvarez, M.A. Polti (eds.), *Bioremediation in Latin America: Current Research and Perspectives*, DOI 10.1007/978-3-319-05738-5_13

presence of Cu^{2+}. Initial Cr(VI) concentrations of 1.3 and 1.7 mM for CChRs of *Cy. jadinii* M9 and *W. anomalus* M10 respectively were inhibitory. This chapter presents evidence of the significant potential of native selected yeasts for chromium bioremediation, thus being promising candidates for alleviating this polluting metal from environment.

13.1 Introduction

13.1.1 Chromium as a Metal, Its Uses and Health Hazards

Chromium is the sixth most abundant element in the Earth's crust. This metal always exist in combination with others elements, displaying a wide variety of colors. Although the oxidation states of chromium range from −II to +VI, the important valences are 0, II, III, and VI (Molokwane et al. 2008). Elemental chromium does not occur naturally on Earth, most chromium compounds exist as halides, oxides, or sulfides. The divalent state is a strong reductant, but rapidly decomposes on air or water to form the trivalent chromium (relatively inert). The trivalent state of chromium is thermodynamically the most stable and is commonly found in the living systems. The hexavalent (chromate) form, the second most stable state, is a strong oxidizing agent (especially in acid media). Chromate compounds usually exist as oxides or as oxohalides, and occurs only rarely in nature in the mineral, crocoite ($PbCrO_4$); therefore, most chromate results from anthropogenic origin. Cr(IV) and Cr(V) compounds are relatively unstable and exist infrequently in nature (Barceloux 1999; Kamaludeen et al. 2003).

In pH media (4–11), trivalent chromium compounds are poorly soluble in water, except for the acetate, chloride (hexahydrate), and nitrate salts. But, in alkaline media, trivalent chromium easily oxidizes to hexavalent chromium. Hexavalent chromium compounds (e.g., chromic acid, chromic trioxide, ammonium salt, sodium, or potassium alkaline metal salts) are soluble in water and traverse cell membranes and are reduced to trivalent compounds intracellularly (Cheung and Gu 2007).

Most chromium released into the environments results from human activities at stationary point sources. The primary users of chromium compounds are chemical, metallurgical, and refractory industries. Commercial applications include tanning, corrosion inhibition, plating, glassware-cleaning solutions, wood preservatives, safety match manufacture, metal finishing, and pigments (Sultan and Hasnain 2005; Thacker et al. 2006). Due to improper disposal, leakage, and poor storage, chromate has become one of the most frequently detected contaminant at the waste sites (Thacker et al. 2006). Not only that chromate is dangerously toxic, it is also difficult to contain and spreads rapidly through aquatic systems and subterranean waterways (Gonzalez et al. 2003). Thus, chromium (VI) has been designated a priority pollutant in many countries and by US EPA (Juvera-Espinosa et al. 2006; Thacker et al. 2006).

Regarding environment we should be very careful since everything what was released in soil, water or air from technology activities comes back to food chain and consequently—directly or indirectly—to human diet (Pawlisz et al. 1997). The link to environmental biochemistry of chromium is evident what is crucial for tracing chromium flux in environmental niches and understanding oxidation–reduction dynamics in chemical as well as in biological world.

Depending on its form and concentration, chromium can be toxic and even carcinogenic. Trivalent chromium is more stable and it constitutes a trace element necessary for the formation of glucose tolerance factor, metabolism of insulin, and is related to cell-membrane stability, synthesis and stability of nucleic acids and proteins (Zetic et al. 2001). However, at high concentrations it can complex with organic compounds interfering with metallo-enzyme systems (Jamnik and Raspor 2003). Hexavalent chromium ion is not essential and therefore, is considered toxic in all concentrations. Epidemiological studies of chromate workers indicate an increased risk of death from lung cancer. Hexavalent chromium compounds are both skin and pulmonary sensitizer, producing a generalized irritation of conjunctiva and mucous membranes, nasal perforations, and contact dermatitis (Poljsak et al. 2010). Its toxicity to biological systems is mainly based on the strong oxidizing potential which can lead to cellular damage (Villegas et al. 2008; Reynolds et al. 2009).

13.1.2 Microbial Resistance to Cr(VI) and Microbial Cr(VI) Removal

Biological treatment of heavy metal containing wastewater by using microorganisms is one of the most active research fields in recent years (Cheung and Gu 2007). Occurrence of indigenous Cr(VI) reducing eukaryotic microorganisms, including those no related with Cr(VI) contamination, has emerged as an important non-conventional yeasts-based bioremediation method with significant biological relevance and biotechnological applications. Several microorganisms have the exceptional ability to adapt to and colonize the noxious metal polluted environments. Microbial Cr(VI)-tolerance and Cr(VI)-reduction are independent events. However, for the Cr(VI)-reduction cells must tolerate Cr(VI), otherwise the cell growth is inhibited. These microorganisms have the capabilities to protect themselves from heavy metal toxicity by various mechanisms such as adsorption, uptake, methylation, oxidation, and reduction (Ramirez-Díaz et al. 2008). Compared to conventional chemical treatment methods (chemical reduction plus precipitation, ion exchange, adsorption on activated coal, etc.), biological treatment methods have many advantages, wich include (1) low operation cost, (2) steady performance, and (3) easy recovery of some valuable metals (Juvera-Espinosa et al. 2006; Villegas et al. 2008; Poljsak et al. 2011).

Among eukaryotic microorganisms, *Candida utilis*, *Schizosaccharomyces pombe*, and *Candida intermedia*, have been found to be effective in accumulating aggressive Cr-compounds (Paš et al. 2004; Poljsak et al. 2010), whilst others (*Candida maltosa*, *Candida* sp., *Lecytosphora* sp. NGV1, *Candida* sp. NGV9, *Aureobasidium pullulans* VR-8) have developed the ability to bioconvert them into stable, nontoxic, and bioavailable forms (Ramírez-Ramírez et al. 2004; Juvera-Espinosa et al. 2006; Cheung and Gu 2007; Villegas et al. 2008; Poljsak et al. 2010). It has been already elucidated that Cr(VI) is taken up via nonspecific anion carriers which, under normal growth conditions, are involved in sulfate and phosphate anions uptake (Poljsak et al. 2010). In the case of some chromium resistant genotypes, the ability to survive in contaminated wastewaters with this heavy metal would be related to decreased Cr(VI) uptake or reduction ability (Poljsak et al. 2010). The enzymatic biospeciation of Cr(VI) to Cr(III) with eukaryotic microorganisms was reported in *C. maltose* (Ramírez-Ramírez et al. 2004) and *C. utilis* (Muter et al. 2001) and fungi *Hypocrea tawa* (Morales-Barrera et al. 2008) and *Aspergillus* sp. (Srivastava and Thakur 2006). But no purification or characterization of the protein involved was studied in these cases.

These findings are the basis to propose heavy metal-resistant varieties of selected yeast species as promising candidates for bioremediation of chromate contaminated waters (Poljsak et al. 2011).

13.2 Case of Study: Removal of Hexavalent Chromium by Yeasts Isolated from Effluent Near a Textile-Dye Factory

13.2.1 Isolation and Characterization of Cr(VI)-Tolerant Yeasts

Chromium-tolerant eukariotyc microorganisms (yeasts and filamentous fungi) were isolated from the liquid effluents and the biofilm adhered to the wall of the textile-dye effluent channel drainage in the proximity of a textile factory (Famaillá, Tucumán, Argentina). Microcosm methodology with periodical Cr(VI)-pulses was used for the samples enrichment (Fernández et al. 2013). Results of viable cell count (CFU) seemed to indicate that the microbial population underwent an adaptation process to the metal presence. But at the end of cultivation, the number of fungal colonies showed a decrease, probably due to the accumulated toxic effects of metal at high concentrations and/or culture age (Fernández et al. 2013).

The finding of higher counts of Cr-resistant viable specimens from biofilm samples was not surprising. Usually, the high exopolymer content of biofilms results beneficial for both entrapping dispersed solids and biosorption of dissolved metals. This microenvironment with high pH or high CO_2 may favor metal precipitation. Besides, microbial interaction can promote or allow the survival of sensitive strains (Malik 2003). The observed adaptation of cells to progressively higher metal concentrations might be related to the constitutive synthesis of metallothionein/other metal binding proteins or changes in the genetic make up (Malik 2003).

Despite the absence of high-chromium concentrations in the textile-dye effluent (10 μg l^{-1}) (Fernández et al. 2013), it seemed not to affect the isolation of Cr-tolerant strains. Polti et al. (2007) have also reported isolation of chromium-resistant actinomycetes from polluted sites where this heavy metal was not abundant in the environment. Typical levels of Cr pollution have been reported in the order of 25–150 μg l^{-1} (Kaszycki et al. 2004). Markedly different values of Cr(VI) concentration, such as 0.03 and 60 μg l^{-1}, might be found in textile industry wastewaters (Juvera-Espinosa et al. 2006). Literature refers total Cr contents like 0.1–0.5 μg l^{-1} in seawater, 0.3–0.6 μg l^{-1} in non-polluted river and surface water, 5–50 μg l^{-1} in polluted river water, and up to 200 μg l^{-1} in severely polluted water systems. For the case of drinking water, Cr(VI) is regulated under the 50 μg l^{-1} maximum contaminant level (MCL) (Bobrowski et al. 2004). Microorganisms exposed to sublethal effects of certain metals or stress conditions can acquire resistance to lethal exposure of the same agent (adaptive response) or unrelated agents (cross-protection response). This behavior represents a microbial survival strategy when exposed to stressful environments. Even though the fungal isolates selected represent only a small fraction of the community, it is probable that these isolates are representatives of the tolerant microorganisms that could be cultivated.

Sequence analysis of ITS1-NL4 PCR-amplified fragments of the selected yeasts revealed that selected isolates M9 and M10 showed 99 % similarities with its close relatives, *Cyberlindnera jadinii* NRRL Y-1542^T (GenBank accesion number: EF550309, previously *Pichia jadinii*; Kurtzman et al. 2008) and *Wickerhamomyces anomalus* NRRL Y-366^T (GenBank accesion number: EF550341, previously *Pichia anomala*; Kurtzman et al. 2008), respectively. Accordingly, phylogenetic relationships were analyzed from amplified rDNA sequences of *Cy. jadinii* M9 (GenBank accession number FJ865435) and *W. anomalus* M10 (GenBank accession number FJ865436) strains and some close related strains (Fig. 13.1) (Fernández et al. 2013). Analysis of physiological and biochemical properties contribute to yeasts global characterization and results useful for diagnosis and identification of specific groups potentially interesting from a biotechnological perspective (Fernández et al. 2013).

13.2.2 Time Course of Cr(VI) Decrease and Cr(III) Production

The ability of the strains M9 and M10 to remove the initial 1 mM Cr(VI) concentration and produce Cr(III) in culture medium at 25 °C were analyzed. For both strains, Cr(VI)-reduction began during the first hours of cultivation. Cr(VI)-reduction rate became highest during exponential growth. At 24 h, most of soluble Cr(VI) was removed for *W. anomalus* M10. In the case of *Cy. jadinii* M9, the Cr(VI) concentration sharply decreased to 80 % at 24 h and was completely reduced (and not detected even as traces) at 48 h (Fig. 13.2a, b). Spontaneous Cr(VI) reduction did not occur in the uninoculated medium after 120 h of incubation, indicating that culture medium components (YNB′ medium) did not unspecifically reduce Cr(VI) levels (Fernández et al. 2009, 2013).

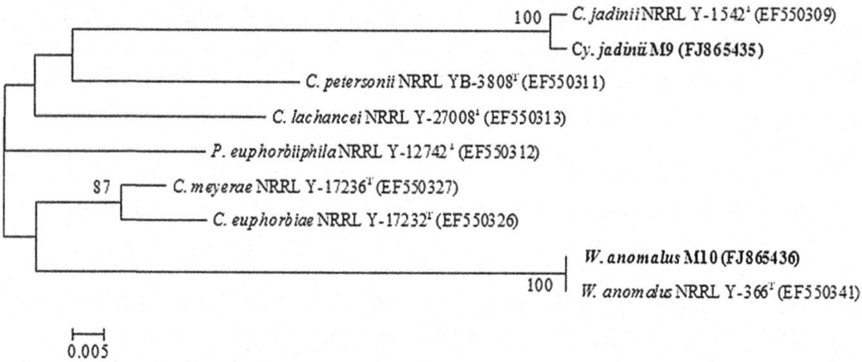

Fig. 13.1 Evolutionary relationships of M9 and M10 isolates based on D1/D2 domain of 28S, 5.8S rDNA; ITS1 and ITS2. The evolutionary history was inferred using the Neighbor-Joining method. The optimal tree with the sum of branch length = 0.32445281 is shown. The percentages of replicate trees in which the associated taxa clustered together in the bootstrap test (1,000 replicates) are shown next to the branches. The tree is drawn to scale, with branch lengths in the same units as those of the evolutionary distances used to infer the phylogenetic tree. The evolutionary distances were computed using the Maximum Composite Likelihood method and are in the units of the number of base substitutions per site. All positions containing gaps and missing data were eliminated from the dataset (Complete deletion option). There were a total of 575 positions in the final dataset. Phylogenetic analyses were conducted in MEGA4 (Fernández et al. 2013)

Total chromium concentration (determined by Flame Atomic Absorption Spectroscopy, FAAS) remained practically constant throughout the incubation time (most of the total chromium, initially added to culture media, was always present in solution), and a very little amount was accumulated in the cells (Fig. 13.2). Specific Cr accumulation (defined as mg of chromium accumulated per g of biomass dry wheigh, BDW) decreased with the biomass increasing. Obtained values between 0 and 120 h ranged from 1.196 to 0.786 mg Cr g^{-1} BDW and from 0.817 to 0.260 mg Cr g^{-1} BDW for *Cy. jadinii* M9 and *W. anomalus* M10, respectively. For *Cy. jadinii* M9 after 48 h of culture the value remained stable, while in *W. anomalus* M10 this occurred after 72 h, events most likely related to the beginning of the stationary growth phase. The high specific metal accumulation at the initial time of cultivation may be related to a rapid Cr adsorption to inoculum cells. Goyal et al. (2003), confirmed that Cr(VI) and Fe(III) biosorption by *Streptococcus equisimilis*, *S. cerevisiae*, and *A. niger* occurs in two stages: (1) immediate, passive and (2) active, much slower uptake. The first stage, as postulated would be related with the physical adsorption or ion exchange on the cell surface. Likewise, Ferraz et al. (2004) observed a similar phenomenon in the uptake of Cr(III) by *S. cerevisiae* in a period of 24 h. Therefore, small amount of chromium early accumulated in the cultures, could be better related to chromium bound to the cell surface instead of chromium accumulated into the cell structures.

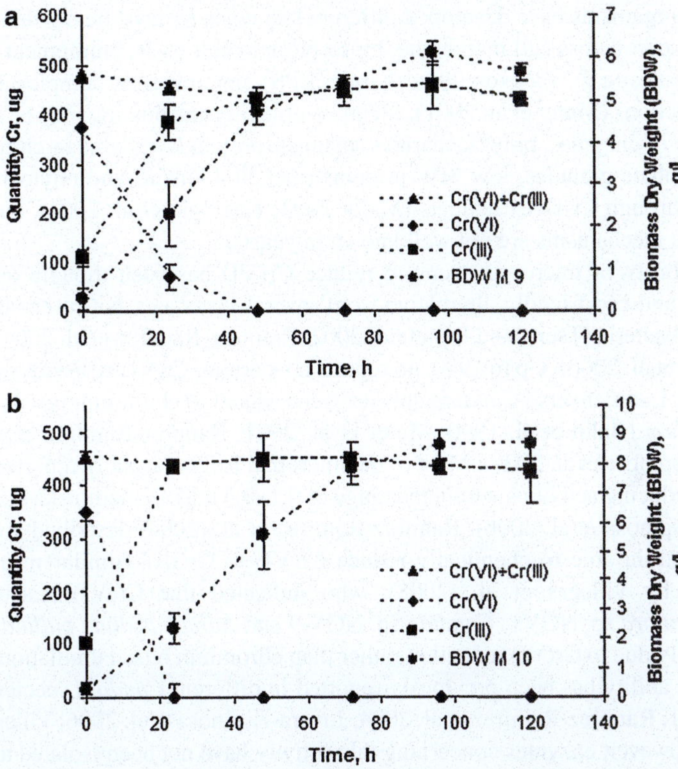

Fig. 13.2 Growth curves of *Cy. jadinii* M9 (**a**) and *W. anomalus* M10 (**b**) including determination of Cr(VI) and total chromium during 120 h of cultivation in YNB′ medium plus 1 mM Cr(VI) at 25 °C and 250 rpm. The amount of chromium is expressed to a volume of 10 ml^{-1}. Not visible standard deviation bars indicate that these do not exceed symbol size (Fernández et al. 2013)

Considering that Cr(III) is only a stable and soluble chromium state formed by Cr(VI) reduction, it could be inferred that the remaining chromium herein quantified in supernatant was Cr(III). The presence of Cr(III) in supernatant would be not surprising since Cr(III) is positively charged under physiological conditions and hence, cell membranes are impermeable to this metal. Therefore, Cr(III) requires binding to an organic compound to be able to pass the plasmatic membrane (Raspor et al. 2000; Paš et al. 2004). The high initial Cr(III) values observed with both strains (Fig. 13.2), would hardly represent an immediate reduction of Cr(VI) to Cr(III). Probably, these values result from the underestimation of Cr(VI) at time zero, because Cr(III) values were calculated from the difference between total chromium (by FAAS) and Cr(VI) (by *S*-diphenylcarbazide (DPC) method) (Urone 1955; Fernández et al. 2009). The Cr(III) concentration at time zero should have been nil or near zero, because the metal species added to culture medium was Cr(VI) and its immediate biotic reduction would be unfeasible.

Microorganisms have developed different strategies to resist heavy-metal stress, e.g., transport across cell membrane, biosorption to cell walls, entrapment into cellular capsules (e.g., via exopolysaccharide, EPS), precipitation, complexation, and redox reactions (Muter et al. 2001). Metal complexing and binding may be achieved by means of enzymes, polysaccharides (mannan-protein layer, glucan-chitin layer), polyphosphate granules, low MW proteins (metallothioneins and phytochelatins), or even through Cr-DNA adducts (Malik 2003; Kaszycki et al. 2004), but Cr(VI) resistance mechanisms are not well known in yeasts.

The ability of microorganisms to reduce Cr(VI) has been thought as a cost-effective and eco-friendly alternative from several years ago, but research usually involved bacteria (Tseng and Bielefeldt 2002; Ramírez-Ramírez et al. 2004; Juvera-Espinosa et al. 2006). Up to 1990, no reports concerning Cr(VI)-resistant yeasts had appeared. Later, diverse *Candida* species were related to chromium resistance and/or reduction (Baldi et al. 1990; Muter et al. 2001; Ramírez-Ramírez et al. 2004; Juvera-Espinosa et al. 2006). Most of the microbial isolates usually reported exhibit Cr(VI)-reducing activities within the range of 0.1–0.5 mM (Ksheminska et al. 2003; Juvera-Espinosa et al. 2006). Ramírez-Ramírez et al. (2004) described a *Candida maltosa* strain able to chemically reduce Cr(VI) to Cr(III). Similar results were obtained by Villegas et al. (2008), who indicated that Cr(VI) resistance by *Lecysthophora* sp. NGV1, *Candida* sp. NGV-9 and *Aureobasidium pullulans* VR-8 was mainly due to Cr(VI) reduction rather than chromium bioaccumulation. Cr(VI) reduction ability has been previously reported in different *Candida* species (Muter et al. 2001; Ramírez-Ramírez et al. 2004; Juvera-Espinosa et al. 2006; Villegas et al. 2008). However, enzymes concerning this activity have not been isolated and characterized so far.

On the other hand, some authors only obtained chromium accumulation with yeasts or filamentous fungi strains. They found that Cr(VI) bioaccumulation depends on the physiological state of the cells, cell density, incubation time and the strain used (Ksheminska et al. 2005; Srivastava and Thakur 2006). Chromium uptake strongly depends on a variety of factors, being the metal valence, concentration and nature of the chemical complex of Cr the most important.

The potential and kinetics for chromium remediation for a given yeast use to depend on various environmental and physiological factors such as Cr valence, concentration, chemical nature, medium pH and composition, inoculum size, mode of cultivation, growth phase, etc. (Kaszycki et al. 2004; Ksheminska et al. 2005; Juvera-Espinosa et al. 2006; Morales-Barrera and Cristiani-Urbina 2006). The critical influence of cell density on the yeast growth response to Cr supplementation has been also clearly demonstrated (Kaszycki et al. 2004; Fernández et al. 2010).

It was also herein noted that Cr(VI) concentrations at time zero as determined by DPC were lower than those obtained by FAAS (Fig. 13.2a, b). This difference could be related to the inoculum chemical composition, which might mask the Cr(VI) real value. The mentioned difference in initial Cr(VI) was almost the same for both strains, supporting the inoculum chemical composition influence on Cr(VI)-determination by DPC as the most probable interference. The observation that total chromium remained in solution almost constant while Cr(VI) was progressively removed, reinforced the idea of Cr(VI) reduction, as also found for other microbial processes (Morales-Barrera and Cristiani-Urbina 2006). Biospeciation to Cr(III) would be the most probable interpretation assumed, as it represents a more stable, less toxic and less mobile oxidation state. In addition, cells adsorbed minimal amounts of Cr. Based on these results and those from abiotic controls, the adsorptive removal of Cr(VI) as well as its removal by culture medium components could be neglected.

13.2.3 Characterization of Crude Chromate Reductase Activity

Considering Cr(VI) reduction ability of the strains *Cy. jadinii* M9 and *W. anomalus* M10, a characterization of crude chromate reductase (CChR) was evaluated. The protocol used was cell disruption using sonication with glass beads and fractionated precipitation with ammonium sulfate. The activity was measured using protocol described in Martorell et al. (2012).

The effect of temperature in the range 10–100 °C at pH 5 was evaluated. In the CChR of *Cy. jadinii* M9 a maximum at 60 °C was observed, in the case of *W. anomalus* M10 the higher chromate reductase activity was observed at 50 °C (Martorell et al. 2012). Among bacterial CChRs, the optimal temperature varies in the range 30–50 °C (Bae et al. 2005; Elangovan et al. 2006; Sarangi and Krishnan 2007), as no cell free extracts (CFEs) or CChRs with chromate reductase activity had been described in yeasts yet, there is nothing to compare with. In both CChRs assayed, another maximum value was observed at an unusually high temperature, 99 °C; being difficult to relate this value with an enzymatic activity, moreover it could be due to a non-protein metabolite (Martorell et al. 2012). Among the low molecular mass reductants, glutathione (GSH) is widespread in yeasts (Penninckx and Elsekens 1993) and seems to be the most important agent participating in the non-enzymatic reduction of Cr(VI).

Following incubation of the CChRs for different periods of time at various temperatures, their residual activity was measured at 30 °C. For *Cy. jadinii* M9 CChR, incubation at 55 °C produced a reduction in activity of a 55 %, after incubation at 73 °C the residual activity was reduced to 68 % compared with the CChR kept at −20 °C (control). In *W. anomalus* M10 CChR when incubated at 8 °C a decrease in activity of 25 % was observed and at 50 °C the activity was 50 % compared with the control (−20 °C). For *Escherichia coli* and *Bacillus* sp. CFEs, thermal stability was up to 30 °C (Bae et al. 2005; Elangovan et al. 2006). On the contrary, *Pseudomonas putida* CFE probed to be more resistant, keeping its stability up to 50 °C (Park et al. 2000).

To determine optimum pH, the chromate reductase activity of the CChRs was measured at different pH values using several buffers (CPB: 50 mM phosphate-citrate, pH 4.0–5.0; PB: 50 mM phosphate, pH 6.0–8.0, and 50 mM Tris–HCl, pH 8–9). In the CChR of *Cy. jadinii* M9, the higher activity was observed in CPB buffer pH 6.0; nevertheless a marked loss of activity was observed when the pH changed to higher or lower values. The CChR of *W. anomalus* exhibited its higher activity in PB pH 7.0. In this case there was a decrease in activity when increasing the pH value, however at pH values below the optimum, the decrease in activity was not as pronounced. The pH stability of CChRs was tested using different buffers (pH 4.0–9.0) at 8 °C for 24 h. Residual activity measured for *Cy. jadini* M9 CChR probed to be stable in the range of 5.0–9.0, at pH 4 and 10 a lost in activity of 55 and 20 % was observed. On the contrary, in *W. anomalus* M10 CChR, pH stability was moderately stable in the range of 4.0–6.0, nevertheless its optimum pH was 7.0. This may indicate more acidic pH stability. Other authors reported stability between 6.5 and 7.5 in *E. coli* CFE (Bae et al. 2005) and in the range 5.0–8.0 in *Bacillus* sp. (Elangovan et al. 2006).

The effects of electron donors, inhibitors and metal ions on chromate reduction by CChR of *Cy. jadinii* M9 and *W. anomalus* M10 were also evaluated. In both CChRs, the electron donors tested, NADH and NADPH alone had a specific activity. A slight abiotic reduction of Cr(VI) by both, NADH and NADPH (in a final concentration of 1 mM), was observed, but it was not substantial compared to Cr(VI) reduction with the CChRs plus either one of the electron donors. Subtracting NADH and NADPH unspecific effects, for *Cy. jadinii* CChR, specific activity was 17.5 and 13.5 nmol Cr(VI) min^{-1} mg $prot^{-1}$, respectively. *Cy. jadinii* CChR alone had a specific activity of 0.69 nmol Cr(VI) min^{-1} mg $prot^{-1}$, thus probing the necessity of an electron donor.

In *W. anomalus* CChR, discounting NADH and NADPH self-effects, chromate reductase activity was 19.2 nmol Cr(VI) min^{-1} mg $prot^{-1}$ and 18.98 nmol Cr(VI) min^{-1} mg $prot^{-1}$. CChR of *W. anomalus* alone had a specific activity of 2.74 nmol Cr(VI) min^{-1} mg $prot^{-1}$. Cr(VI) reduction by the CChRs of *Cy. jadinii* M9 and *W. anomalus* M10 was confirmed as a NAD(P)H dependent reaction. Other studies report that NADH or NADPH-dependent enzymatic reduction of Cr(VI) under aerobic conditions (Park et al. 2000; Mclean et al. 2000; Bae et al. 2005; Elangovan et al. 2006; Opperman et al. 2008). According to Ramirez-Díaz et al. (2008), the oxidation of NADH donates an electron to the chromate reductase enzyme and then, the electron is transferred to Cr(VI) reducing it to an intermediate form, Cr(V), which finally accepts two electrons from other organic substances to produce Cr(III).

Inhibition of the enzyme activity by selected metal ions (using 10 mM solutions of Na_2SO_4, $CaCl_2$, $CuCl_2$, $HgCl_2$, $MgCl_2$, $MnSO_4$, $ZnSO_4$, and $FeCl_3$) was also determined. In *P. jadinii* M9 CChR, only Cu^{2+} and Na^+ produced an augmentation in the activity of 63 and 30 %, respectively. All other ions tested had an inhibitory effect but in different levels. A decrease of 72 % was observed with Hg^{2+}, while addition of Mg^{2+}, Mn^{2+}, Ca^{2+}, Fe^{3+}, and Zn^{2+}, resulted in a decrease of activity between 38 and 58 %. In *W. anomalus* M10 CChR, only Cu^{2+} produced a raise in activity of a 31 %. Inhibition by Hg^{2+} and Mn^{2+} was higher than in *Cy. Jadinii* M9 CChR, with a decrease in activity of 85 and 90 %, respectively. Inhibition by Ca^{2+}, Mg^{2+}, and Zn^{2+} was approximately 60 %, while Fe^{2+} reduced the activity in a 32 %. These results agree with those reported for *Arthrobacter crystallopoietes* (Camargo et al. 2004) and *Bacillus* sp. (Camargo et al. 2003; Elangovan et al. 2006). Activation by Cu^{2+} can be explained considering that this transition metal may be acting as a prosthetic group for many reductases. Its main function would be related to the protection of electron transport, whether acting as a redox center or as an electron transporter between protein subunits. On the other hand, inhibition by Hg^{2+}, can be related with its affinity to –SH ligands, then suspecting the presence of this chemical group in the active site of the enzyme related to chromate reductase activity (Camargo et al. 2003).

In *Cy. jadinii* M9 CChR, for initial concentrations of 0.4, 0.7, and 1 mM, a total removal of Cr(VI) was observed. This removal was obtained after 2 h for 0.4 and 0.7 mM, and after 5 h for 1 mM. An initial concentration of 1.3 mM, can be considered as inhibitory based on the detention of the metal removal at 60 min and the maintenance of Cr(VI) concentration at 0.83 mM. In *W. anomalus* M10 CChR, initial concentrations of 0.4, 0.7, 1, and 1.3 mM a total removal of Cr(VI) was observed, at 2 h for 0.4 and 0.7 mM, at 3 h for 1 mM, and at 4 h for 1.3 mM. A Cr(VI) initial concentration of 1.7 mM was reduced to 0.8 mM in 3 h and remained constant for the rest of the incubation time, suggesting that this or a higher initial Cr(VI) concentration than this one may be inhibitory. The effect of initial Cr(VI) concentration was also evaluated for *B. sphaericus* AND 303 in the range 0–0.5 mM, the authors observed a fast increment on the Cr(VI) reduction up to a concentration of 0.3 mM (Pal et al. 2005). Also, Sarangi and Krishnan (2007) reported lower chromium initial concentration values, between 0.075 and 0.192 mM. All these concentrations were lower to that used with *Cy. jadinii* M9 and *W. anomalus* M10, showing the higher resistance of ChR activity in the CChR here described in reference with the initial Cr(VI) concentration, which was in the range of 0.4–2.0 mM.

References

Bae W, Lee H, Choe Y et al (2005) Purification and characterization of NADPH-dependent Cr(VI) reductase from *Escherichia coli* ATCC 33456. J Microbiol 43:21–27

Baldi F, Vaughan AM, Olson GJ (1990) Chromium(VI)-resistant yeast isolated from a sewage treatment plant receiving tannery wastes. Appl Environ Microbiol 56:913–918

Barceloux DG (1999) Chromium. Clin Toxicol 37:173–194

Bobrowski A, Mocak J, Dominik J et al (2004) Metrological characteristics and comparison of analytical methods for determination of chromium traces in water samples. Acta Chim Slov 51:77–93

Camargo FAO, Okeke BC, Bento FM et al (2003) *In vitro* reduction of hexavalent chromium by a cell-free extract of *Bacillus* sp. ES 29 stimulated by Cu^{2+}. Appl Microbiol Biot 65:569–573

Camargo FAO, Bento FM, Okeke BC et al (2004) Hexavalent chromium reduction by an actinomycete, *Arthrobacter crystallopoietes* ES 32. Biol Trace Elem Res 97:183–194

Cheung KH, Gu JD (2007) Mechanisms of hexavalent chromium detoxification by microorganisms and bioremediation application potential: a review. Inte Biodeter Biodegr 59:8–15

Elangovan R, Abhipsa S, Rohit B et al (2006) Reduction of Cr(VI) by a *Bacillus* sp. Biotechnol Lett 28:247–252

Fernández PM, Figueroa LIC, Fariña JI (2009) Critical influence of culture medium and Cr(III) quantification protocols on the interpretation of Cr(VI) bioremediation by environmental fungal isolates. Water Air Soil Pollut 206:283–293

Fernández PM, Fariña JI, Figueroa LIC (2010) The significance of inoculum standardization and cell density on the Cr(VI) bioremediation by environmental yeast isolates. Water Air Soil Pollut 212:275–279

Fernández PM, Cabral ME, Delgado OD et al (2013) Textile-dye polluted waters as a source for selecting chromate-reducing yeasts through Cr(VI)-enriched microcosms. Int Biodeter Biodegr 79:28–35

Ferraz AI, Tavares T, Texeira JA (2004) Cr(III) removal and recorvery from *Sacchoromyces cerevisiae*. Chem Eng J 105:11–20

Gonzalez CF, Ackerley DF, Park C et al (2003) A soluble flavoprotein contributes to chromate reduction and tolerance by *Pseudomonas putida*. Acta Biotechnol 2:233–239

Goyal N, Jain SC, Banerjee UC (2003) Comparative studies on the microbiol adsorption of heavy metals. Adv Environ Res 7:311–319

Jamnik P, Raspor P (2003) Stress response of yeast Candida intermedia to Cr(VI). J Biochem Mol Toxicol 17:316–323

Juvera-Espinosa J, Morales-Barrera L, Cristiani-Urbina E (2006) Isolation and characterization of a yeast strain capable of removing Cr(VI). Enzyme Microb Technol 40:114–121

Kamaludeen SPB, Megharaj M, Juhasz AL et al (2003) Chromium-microorganism interactions in soils: remediation implications. Rev Environ Contam T 178:93–164

Kaszycki P, Fedorovych D, Ksheminska H et al (2004) Chromium accumulation by living yeast at various environmental conditions. Microbiol Res 159:11–17

Ksheminska H, Jaglarz A, Fedorovych D et al (2003) Bioremediation of chromium by the yeast *Pichia guilliermondii*: toxicity and accumulation of Cr(III) and Cr(VI) and the influence of riboflavin on Cr tolerance. Microbiol Res 158:59–67

Ksheminska H, Fedorovych D, Babyak L et al (2005) Chromium(III) and (VI) tolerance and bioaccumulation in yeast: a survey of cellular chromium content in selected strains of representative genera. Process Biochem 40:1565–1572

Kurtzman CP, Robnett CJ, Basehoar-Powers E (2008) Phylogenetic relationships among species of *Pichia*, *Issatchenkia* and *Williopsis* determined from multigene sequence analysis, and the proposal of *Barnettozyma* gen. nov., *Lindnera* gen. nov. and *Wickerhamomyces* gen. nov. FEMS Yeast Res 8:939–954

Malik A (2003) Metal bioremediation through growing cells. Environ Int 30:261–278

Martorell MM, Fernández PM, Fariña JI et al (2012) Cr(VI) reduction by cell-free extracts of *Pichia jadinii* and *Pichia anomala* isolated from textile-dye factory effluents. Int Biodeter Biodegr 70:81–85

Mclean JS, Beveridge TJ, Phipps D (2000) Isolation and characterization of a chromium-reducing bacterium from a chromated copper arsenate contaminated site. Environ Microbiol 2:611–619

Molokwane PE, Meli CK, Evans and Chirwa MN (2008) Chromium (VI) reduction in activated sludge bacteria exposed to high chromium loading: Brits culture (South Africa). Water Res 42:4538–4548

Morales-Barrera L, Guillén-Jiménez F, Ortiz-Moreno A et al (2008) Isolation, identification and characterization of a Hypocrea tawa strain with high Cr(VI) reduction potential. Biochem Eng J 40:284–292

Morales-Barrera L, Cristiani-Urbina E (2006) Removal of hexavalent chromium by *Trichoderma viride* in an airlift bioreactor. Enzyme Microb Technol 40:107–113

Muter O, Patmalnieks A, Rapoport A (2001) Interrelations of the yeast *Candida utilis* and Cr(VI): metal reduction and its distribution in the cell and medium. Process Biochem 36:963–970

Opperman DJ, Piater LA, van Heerden E (2008) A novel chromate reductase from *Thermus scotoductus* SA-01 related to old yellow enzyme. J Bacteriol 190:3076–3082

Pal A, Dutta S, Mukherjee PK et al (2005) Occurrence of heavy metal resistance in microflora from serpentine soil of Andaman. J Basic Microbiol 45:207–218

Park CH, Keyhan M, Wielinga B et al (2000) Purification to homogeneity and characterization of a novel *Pseudomonas putida* chromate reductase. Appl Environ Microbiol 66:1788–1795

Paš M, Milačič R, Drašlar K et al (2004) Uptake of chromium(III) and chromium(VI) compounds in the yeast cell structure. Biometals 17:25–33

Pawlisz AV, Kent RA, Schnider UA et al (1997) Canadian water quality guidelines for chromium. Environ Toxic Water Qual 12:123

Penninckx MJ, Elsekens MT (1993) Modulation of chromium (VI) toxicity by organic and inorganic sulfur species in yeasts from industrial wastes. Biometals 5:170–185

Poljsak B, Pócsi I, Raspor P et al (2010) Interference of chromium with biological systems in yeasts and fungi: a review. J Basic Microbiol 50:21–36

Poljsak B, Pócsi I, Pesti M (2011) Interference of chromium with cellular functions. In: Banfalvi G (ed) Cellular effects of heavy metals. Springer, Amsterdam

Polti MA, Amoroso MJ, Abate CM (2007) Chromium(VI) resistance and removal by actinomycete strains isolated from sediments. Chemosphere 67:660–667

Ramirez-Díaz MI, Díaz-Pérez C, Vargas E et al (2008) Mechanisms of bacterial resistance to chromium compounds. Biometals 21:321–332

Ramírez-Ramírez R, Calvo-Méndez C, Avila-Rodríguez M et al (2004) Cr(VI) reduction in a chromate-resistant strain of *Candida maltosa* isolated from the leather industry. A van Leeuw 85:63–68

Raspor P, Batic M, Jamnik P et al (2000) The influence of chromium compounds on yeast physiology (a review). Acta Microbiol Imm H 47:143–173

Reynolds MF, Peterson Roth EC, Bespalov IA et al (2009) Rapid DNA double-strand breaks resulting from processing of Cr-DNA cross-links by both MutS dimers. Cancer Res 69: 1071–1079

Sarangi A, Krishnan C (2007) Comparison of *in vitro* Cr(VI) reduction by CFEs of chromate resistant bacteria isolated from chromate contaminated soil. Bioresour Technol 99:4130–4137

Srivastava S, Thakur IS (2006) Isolation and process parameter optimization of *Aspergillus* sp. for removal of chromium from tannery effluent. Bioresour Technol 97:1167–1173

Sultan S, Hasnain S (2005) Chromate reduction capability of a gram positive bacterium isolated from effluent of dying industry. Bull Environ Contam Tox 75:699–706

Thacker U, Parikh R, Schouche Y et al (2006) Hexavalent chromium reduction by *Providencia* sp. Process Biochem 41:1332–1337

Tseng JK, Bielefeldt AR (2002) Low-temperature chromium(VI) biotransformation in soil with varying electron acceptors. J Environ Qual 31:1831–1841

Urone PF (1955) Stability of colorimetric reagent for chromium, *S*-diphenylcarbazide, in various solvents. Anal Chem 27:1354–1355

Villegas LB, Fernández PM, Amoroso MJ et al (2008) Chromate removal by yeasts isolated from sediments of a tanning factory and a mine site in Argentina. Biometals 21:591–600

Zetic VG, Stehlik-Tomas V, Grba S et al (2001) Chromium uptake by *Saccharomyces cerevisiae* and isolation of glucose tolerance factor from yeast biomass. J Biosci 26:217–223

Chapter 14
Ecology of Dye Decolorizing Yeasts

Hipólito F. Pajot, María M. Martorell, and Lucía I.C. de Figueroa

Abstract Textile dyes are among the most recalcitrant pollutants. Industrial effluents containing textile dyes are usually disposed in large amounts into natural water bodies on a daily basis. Their pollution hazard is based on components which may be carcinogenic or toxic to living organisms. Also, because of their marked color, they affect light penetration, modifying photosynthetic activity in aquatic environments.

Physical and chemical methods may be used for dye removal from industrial effluents. These approaches are expensive, have operational problems and may lead to bigger problems.

Several studies have been reported on decolorization of numerous dyes using white rot fungi. These organisms could mineralize many types of synthetic dyes through their oxidative and nonspecific lignolytic system. However, the strict conditions for enzyme production and the jeopardy of bacterial contamination in non-sterile conditions in dye-containing wastewaters, made difficult the application of white rot fungi for textile dye effluents. Yeasts, on the other hand, have many advantages, not only because of their fast growth but also because of their ability to resist unfavorable environments.

Unfortunately, the ecology of dye-degrading yeasts is still poorly understood. In this chapter, we review several methods for textile dye effluent treatment. Then, we focus on dye decolorizing yeast, exploring the still fragmentary information on their ecology, taxonomy, and biotechnological applications in the field of textile dyes bioremediation.

H.F. Pajot (✉) • M.M. Martorell
PROIMI-CONICET (Planta Piloto de Procesos Industriales Microbiológicos), Av. Belgrano y Caseros, T4001MVB Tucumán, Argentina

Química Biológica, Calidad de productos IV, Gestión de Empresas Agroalimentarias, Universidad de San Pablo, Tucumán, Argentina
e-mail: hipolito_pajot@yahoo.com; hpajot@proimi.org.ar

L.I.C. de Figueroa
PROIMI-CONICET (Planta Piloto de Procesos Industriales Microbiológicos), Av. Belgrano y Caseros, T4001MVB Tucumán, Argentina

Cátedra de Microbiología Superior, Facultad de Bioquímica, Química y Farmacia, Universidad Nacional de Tucumán, Tucumán, Argentina

A. Alvarez, M.A. Polti (eds.), *Bioremediation in Latin America: Current Research and Perspectives*, DOI 10.1007/978-3-319-05738-5_14

14.1 Textile Dyes

Some industries, such as textile, paper and printing, use extensive amounts of textile dyestuffs. Before the discovery of "mauveine" by Perkin in 1856, dyes were extracted mainly from animals. Well-known examples are carminic acid (from *Dactylopius coccus*), Tyrian purple (from *Bolinus brandaris*), and Indigo (from *Indigofera tinctoria*). However, the use of natural dyes has several limitations, i.e., laborious and time consuming extractions with very low recovery yields (making them expensive), lack of purity, uniformity, and quality; poor fastness properties and easy loss of color under light conditions (Chatwal 2009).

Nowadays natural dyes have been almost completely replaced by synthetic ones. Even considering the huge improvement in dye quality, effluents of dyeing industries are markedly colored. Such effluents generate large amounts of dyes residues, which are directly released into water bodies, contaminating the environment.

14.1.1 Azo Dyes

More than 100,000 synthetic dyes exist commercially, with over 7×10^5 tons produced annually worldwide and half of them belong to the azo dyes compounds class. An estimated of 10–15 % of them is directly lost in the wastewater (Jarosz-Wilkolazka et al. 2002). Additionally, up to 50 % of dyes are lost through hydrolysis in the reactive dye process, and the environmental fate of their degradation products is still largely unknown (Singh 2006).

14.1.1.1 Chemical Structure

The azo dyes are by far the most important class, accounting for over 50 % of all commercial dyes, and having been studied more than any other class. Azo dyes contain at least one azo group (–N=N–) but can contain two (disazo), three (trisazo), or, more rarely, four (tetrakisazo) or even more (polyazo) azo groups. The azo group is attached to two groups, of which at least one, but more usually both, are aromatic (Hunger 2007). The phenyl and naphthyl radicals are extensively substituted with functional groups including: amino ($-NH_2$), chlorine (–Cl), hydroxyl (–OH), methyl ($-CH_3$), nitro ($-NO_2$), sulfonic acid, and sodium salts ($-SO_3Na$) (de Campos Ventura-Camargo and Marin-Morales 2013). The accumulation of such electron-withdrawing groups and some specific substitution patterns confers a xenobiotic character to synthetic azo dyes, making them extremely resistant to oxidation (Knackmuss 1996) (Fig. 14.1).

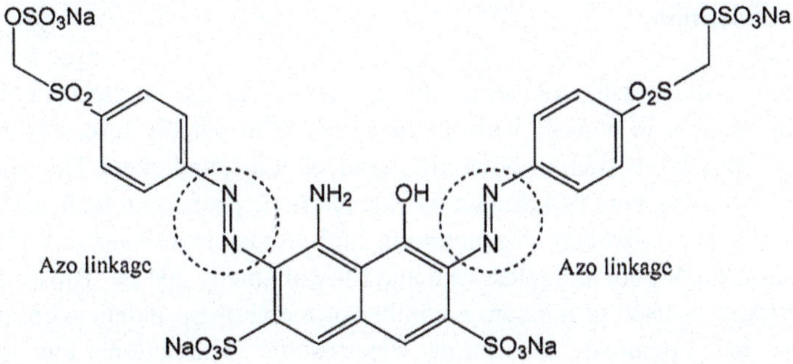

Fig. 14.1 C.I. Reactive Black 5 (CAS 17095-24-8), a highly sulfonated disazo dye, mainly used for dying cotton

14.1.1.2 Toxicity

Effluents of dyeing industries are usually markedly colored and their disposal into receiving waters causes damage to the environment. Since color reduces light penetration, dye pollution may significantly affect photosynthetic activity in aquatic life. In addition, due to their own toxicity as well as the presence of metals, chlorides, etc., they are harmful to the in situ aquatic life and also to the living organisms drinking from these waters (Pajot et al. 2011a, b; dos Santos et al. 2007).

14.2 Processes for Dye Removal

Through years, many processes have been employed in order to attempt color removal from industry effluents. Treatment methods are usually divided into two main categories: Biological treatments, which includes the use of living organisms in some stage and physicochemical processes, which do not include living organisms, even when they could involve the usage of dead biomass.

14.2.1 Physicochemical Processes

Physicochemical processes include ozonation, oxidation, coagulation/flocculation, ion exchange, irradiation, precipitation, filtration, and sorption. All these techniques have been shown to be effective, although with important limitations including: excess amount of chemical usage, accumulation of concentrated sludge, expensive plant requirements or operational costs; lack of effective color reduction; and sensitivity to a variable wastewater input (Aksu 2005).

14.2.1.1 Sorption

Adsorption is the transfer of solute dye molecules to the interface between two immiscible phases in contact with another. The term usually refers to the use of inorganic components such as bentonite, sand, zeolites, and clays. The selection of an appropriate adsorbent is based on its characteristic properties such as high affinity, retention, and adsorbent regeneration, high abrasion resistance, high thermal stability, and small pore diameters (Ali and El-Mohamedy 2012). "Biosorption," on the other hand, is used to indicate a number of metabolism-independent processes (physical and chemical adsorption, electrostatic interaction, ion exchange, complexation, chelation, and microprecipitation) taking place essentially in the cell surface. Both living and dead biomass can be used to remove hazardous organics, but the use of dead biomass is much more easy and advantageous (Aksu 2005). Activated granular carbon has been successfully used for color removal, although it is not very practical due to its high cost, compelling scientists toward the development of low cost adsorbents with considerable success.

Sorption is considered between the top wastewater treatment methods due to its universal nature, inexpensiveness, and ease of operation. However, the management of the exhausted adsorbent, the lack of knowledge on the issue of management of removed pollutants, pH requirements, contact time, and concentration of pollutants are major limitations of these techniques (Ali and El-Mohamedy 2012).

14.2.1.2 Filtration, Ultrafiltration and Reverse Osmosis

Filtration methods (ultrafiltration, nanofiltration and reverse osmosis) have also been explored for water reuse and chemical recovery. They simultaneously reduce color, BOD and COD of wastewater. Reverse osmosis is also the preferred method for effluent desalination, a common feature of textile effluents (Vijayaraghavan et al. 2013). However, high investment costs, potential membrane fouling and the production of secondary waste streams, which need further treatment, are major draw-backs (Rauf and Salman Ashraf 2012). To avoid membrane fouling it becomes essential to remove turbidity, suspended solids, colloids, and trace organics prior to treatment. Thus, these membrane methods are nowadays used mainly as finishing treatments (Lu and Liu 2010).

14.2.1.3 Coagulation, Flocculation and Sedimentation

According to Verma et al. (2012), chemical coagulation and flocculation involves the addition of chemicals to alter the physical state of dissolved and suspended solids and facilitate their removal by sedimentation.

Coagulation of dye-containing wastewater has been used for many years as main treatment or pretreatment due to its low capital cost. However, the generation of secondary pollution, in the form of sludge, and the ineffective decolorization of highly soluble dyes (mainly acid, direct, reactive, and vat dyes) are major limitation of this process. Color removal in such processes decreases with the increase in dye concentration and dye solubility (Zahrim et al. 2010). The effectiveness of the coagulation can be certainly improved by appropriate selection of coagulant and/or flocculant. However, due to the variability in textile dye effluents and the continuous development on synthesis technology, the selection of a universally suitable process has proven to be extremely difficult (Verma et al. 2012).

14.2.1.4 Advanced Oxidation Processes (AOP's)

Advanced oxidation processes constitute an emerging approach successfully employed in wastewater decolorization. These methods utilize oxidizing agents such as ozone (O_3), hydrogen peroxide (H_2O_2), and permanganate (MnO_4) in order to form stronger oxidizing species such as $^{\cdot}OH$ radicals. These strong oxidants oxidize big molecules into smaller and less harmful substances (de Campos Ventura-Camargo and Marin-Morales 2013).

Ozone is widely used because of its high reactivity with dyes and good removal efficiencies but is not efficient in decolorizing nonsoluble disperse and vat dyes. Moreover, the process is highly dependent on pH. The use of UV enhanced oxidation processes (O_3/UV and H_2O_2/UV), also proved to be effective methods, providing excellent effluent decolorization. Major drawbacks of such methods are an extremely poor COD reduction, and the low penetration of UV in deeply colored effluents since most of the UV light gets absorbed by the dyes, producing only very small amount of hydroxyl free radicals (Verma et al. 2012).

In Fenton and Fenton-like reactions, hydrogen peroxide is added in an acid solution (pH 2.0–3.0) containing Fe^{2+} ions:

$$Fe^{2+} + H_2O_2 \rightarrow Fe^{3+} + \cdot HO + HO^-$$

Compared with ozonation, this method results cheaper and gives higher COD reduction. However, because of the solubility of Fe ions, Fenton reactions are only possible in acidic environments. Moreover, they produce higher sludge amounts due to reagents flocculation (Verma et al. 2012). A further development is the use of so called Chelator-Mediated Fenton Reactions (CMFRs). Such reactions involves Fe^{3+}, H_2O_2, an iron chelator, and an iron reducing compound (most times chelator itself) to produce $^{\cdot}OH$. This process, allows Fenton reaction to take place in peri-neutral pHs, and helps to reduce the usage of Fe^{3+} and H_2O_2 and sludge production, but are hard to optimize.

14.2.2 Biological Methods

Synthetic textile dyes are considered xenobiotic compounds, highly recalcitrant against biodegradative processes. Nevertheless, during the last few years it has been demonstrated that several microorganisms are able to decolorize or even mineralize them (Stolz 2001).

14.2.2.1 Bioaccumulation and Biosorption

Bioaccumulation and biosorption possess good potential to replace conventional sorption methods, mainly because they are cost-effective and are perceived generally as eco-friendly alternatives. Bioaccumulation is defined as the phenomenon of active (energy consuming) uptake of toxicants by living cells; whereas biosorption can be defined as the passive (non-energy consuming) uptake of toxicants by dead or resting biological material. However, they clearly show also the same drawbacks that conventional sorption processes have: Sludge generation, lack of knowledge on the issue of management of removed pollutants, pH requirements, and extensive contact times (Vijayaraghavan et al. 2013).

14.2.2.2 Bioremediation/Biodegradation

Bioremediation/biodegradation is the process of degrading or transforming pollutants by biological methods, leading sometimes to complete mineralization of the pollutant. This sort of treatment causes the contaminant to be transformed into smaller molecules which are, usually, less harmful to the environment.

Biological methods have significant advantages over nonbiological remediation methods, mainly because of the adaptability of microorganisms to different conditions. Bioremediation processes can deal with lower concentration of contaminants, whereby the cleanup by physical or chemical methods would not be feasible. Additionally, biological methods are usually considered cost-effective and eco-friendly techniques, causing minimum environmental stress. However, biological processes may take longer and are intrinsically less predictable than conventional methods. Textile wastewater effluents have been successfully treated using microorganisms.

Many microorganisms including bacteria, fungi, and algae shown very promising results for dye degradation (Rauf and Salman Ashraf 2012).

Decolorization of Azo Dyes by Bacteria

Reductive cleavage of the –N=N– bond is usually the initial step of the bacterial degradation of azo dyes. Decolorization of azo dyes occurs under anaerobic (methanogenic), anoxic, and aerobic conditions by different trophic groups of bacteria.

Dye decolorization under methanogenic conditions requires an easily assimilable organic carbon/energy source. Glucose, starch, acetate, ethanol, whey, and tapioca have been employed with relative success. According to the review of Pandey et al. (2007), methanogens, acidogenic, and methanogenic bacteria contribute to dye decolorization. Most azo dyes could be decolorized under strict anaerobic conditions, with decolorization rates greatly depending on the carbon source and on dye structure. However, there is no correlation between decolorization rate and molecular weight, pointing to an unspecific cometabolic extracellular pathway of degradation (Stolz 2001; Van der Zee 2002). In this scenario, azo dyes are thought to act as electron acceptors, taking electrons directly from carriers of the electron transport chain. Alternatively, decolorization might be attributed to nonspecific extracellular reactions occurring between reduced compounds (i.e., quinones) generated by the anaerobic biomass (Van der Zee 2002).

Most frequently, decolorization of azo dyes occurs under conventional anaerobic, facultative anaerobic and aerobic conditions. The degradation mechanism usually involves the reductive cleavage of azo bonds (–N=N–) with the help of either specific or unspecific azoreductase enzymes (Saratale et al. 2011). However, treated and colorless effluents could still contain potentially hazardous-aromatic amines, generated directly from azo cleavage (Stolz 2001; Saratale et al. 2011; Rauf and Salman Ashraf 2012).

Degradation of Azo Dyes by Algae

Dye assimilation by algae seems to start with the expression of inducible azoreductases. The produced aromatic amines, then serves as C/N and energy sources (Acuner and Dilek 2004; Hanan Hafez 2008). The natural habitat of algae, makes them an interesting option when considering the removal of azo dyes and aromatic amines in stabilization ponds (Saratale et al. 2011).

Degradation of Azo Dyes by Fungi

Fungi are ubiquitous in the environment. They can exist and survive in almost every habitat, playing a vital role in all ecosystems, where they are known to degrade, or cause to deteriorate, a wide variety of materials and compounds (Singh 2006).

The obvious evolutionary success of fungi is evidenced by the number of species, the diversity of niches, and habitats occupied, the ability to survive under restrictive conditions, etc. In terms of biodiversity, since 1991 it has been estimated "in at least 1.5 million species but probably as many as 3 million" (Hawksworth 2012).

All fungi are heterotrophic and obtain the organic substance necessary for their growth through different ways including saprophytism, parasitism (pathosistic symbiosis), and mutualistic symbiosis. Virtually all natural organic compounds can be degraded by one or more fungal species thanks to the production of a variety of enzymes such as amylases, lipases, proteases, pectinases, cellulases, ligninases, etc. Due to the high unspecificity of the enzymes involved in the degradation of lignin, wood fungi can attack numerous aromatic and aliphatic xenobiotic compounds, including environmental pollutants such as polycyclic aromatic hydrocarbons (PAHs), polychlorinated biphenyls (PCBs), pesticides, and herbicides (Anastasi et al. 2013).

The first study reporting dyes biodegradation by the White Rot Fungus (WRFs) *Phanaerochaete. chrysosporium* was conducted almost 27 years ago. Since then, it has become the most widely studied fungy on bioremediation of textile dyes.

It has been reported that *P. chrysosporium* is able to decolorize all sort of dyes including Orange II, Azure B, Congo Red, Tropaeolin O, Reactofix Gold Yellow, Navinon Blue, Acid Violet 7, Acid Blue 25, Acid Black 4, Reactive Black 5, Reactive Blue, Indigo carmin, among others (Singh 2006). Other WRFs species have also received considerable attention, such as *Trametes* (*Coriolus*) *versicolor*, *Bjerkandera adusta*, *Aspergillus ochraceus*, several *Pleurotus* and *Phlebia* species, etc. (Saratale et al. 2011). White rot fungi (WRF) produce various isoforms of extracellular peroxidases including Lignin peroxidase (LiP), Variable peroxidase (VP), Dye Peroxidase (DyP), manganese peroxidase (MnP), etc. They also produce several phenoloxidases like laccases, catechol dioxygenases, tyrosinases, etc., which are involved in the degradation of lignin. This ligninolytic system of WRF is directly involved in the degradation of various xenobiotic compounds and dyes (Erkurt et al. 2010).

However WRFs present some inherent drawbacks such as the long growth cycle and the need for nitrogen limiting conditions. In addition, white-rot fungi are not naturally found in wastewater, and hence enzyme production may be unreliable. Moreover, the long hydraulic retention time required for complete decolorization, usually in the order of days or even weeks), also limits the performance of the fungal decolorization system (detailed reviews could be found in Singh 2006; Erkurt et al. 2010; Vijayaraghavan et al. 2013).The high oxygen requirements of most WRFs and the preservation of fungi in bioreactors is also a matter of concern when considering this bioremediation option (Stolz 2001; Saratale et al. 2011).

A lot less is known about biodegradation and/or decolorization of azo dyes by Brown Rot Fungi (BRF) an Soft Rot Fungi. Jarosz-Wilkolazka et al. (2002) reported the presence of oxidative factors, possibly laccases and peroxidases in three BRF, *Coprinus micaceus*, *Fomitopsis pinicola*, and *Gloeophyllum odoratum* in solid-agar media. However, degradation or decolorization in BRF may occur through a radically different mechanism, mainly driven by Chelator-Mediated Fenton Reactions (CMFRs) (Arantes et al 2012) (Fig. 14.2).

Degradation of Azo Dyes by Yeasts

The term "yeast" is usually employed, in a narrow sense, as a synonym of either *Saccharomyces cerevisiae* or *Schizosaccharomyces pombe* (Spencer et al. 2002). Any other yeast is usually referred as "non-conventional yeast." With the description of new ascomycetous and basidiomycetous yeasts, the term gained a broader meaning, and nowadays, the most accepted definition states that yeast are basidiomycetous or ascomycetous fungi that asexually reproduce by budding or fission, resulting in growth comprised mainly of single cells and sexual states are never enclosed in fruiting bodies (Kurtzman et al. 2011).

Very little work has been done to explore the decolorizing ability of yeast (Días et al. 2010). Initial studies dealt mainly with nonenzymatic decolorization processes such as biosorption or bioaccumulation but reduction of azo bounds through azoreductases and dye oxidation through ligninolytic enzymes have also been reported.

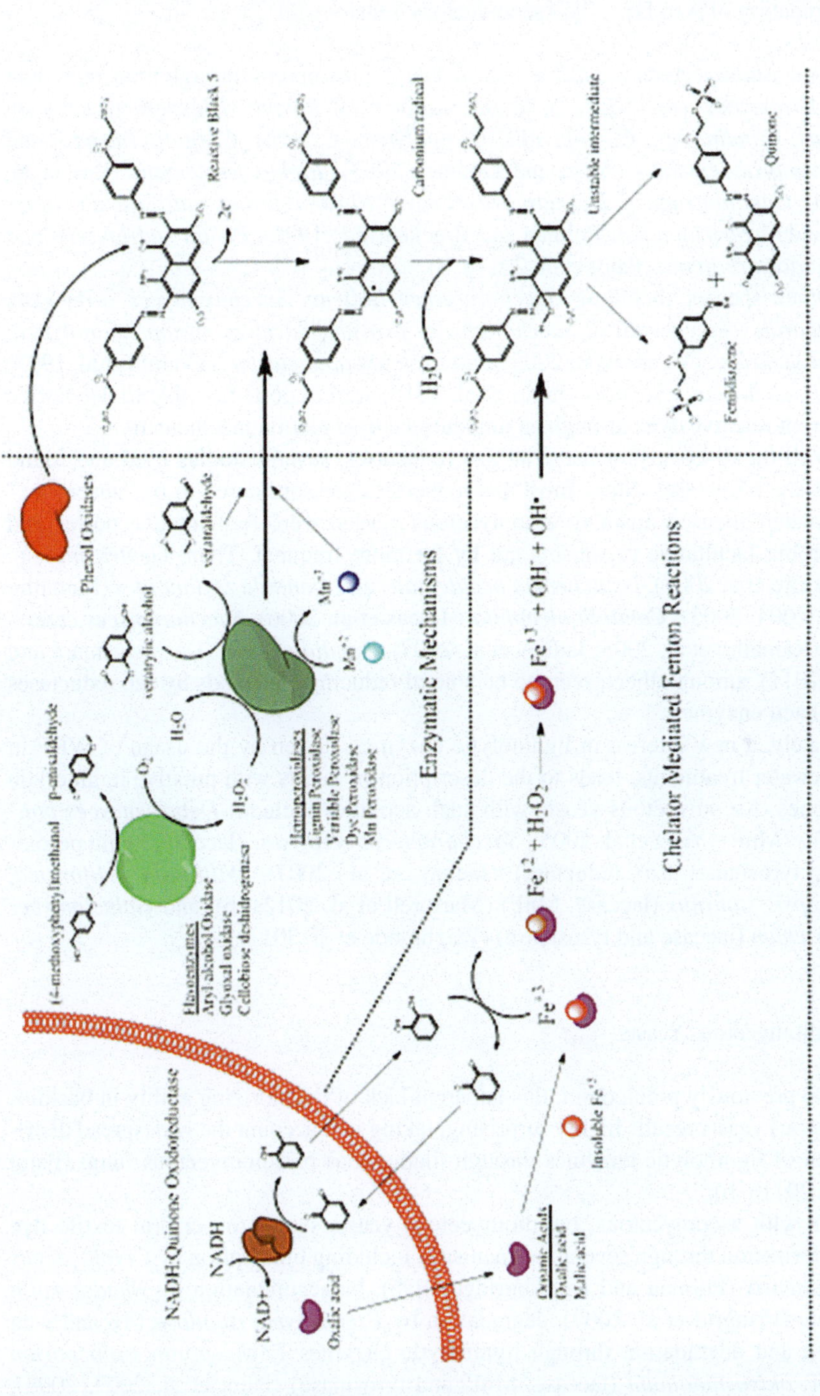

Fig. 14.2 Possible oxidative mechanisms of azo dye degradation by filamentous fungi and yeasts. Chemical modifications on Reactive Black 5 have been hypothesized from Chivukula and Renganathan (1995)

Degradation of Azo Dyes by Ascomycetous Yeast

Several ascomycetous yeasts has been tested as biosorbents through time, including *Saccharomyces cerevisiae*, *Schizosaccharomyces pombe*, *Kluyveromyces marxianus*, *Candida* sp., *C. tropicalis*, *C. lipolytica*, *C. utilis*, *C. quilliermendii*, and *C. membranaefaciens* (Aksu and Donmez 2003), *Pichia fermentans* (Das et al. 2010), *Candida rugosa*, *Dekkera bruxellensis*, *Kluyveromyces walti*, *Pichia carsonii*, and *Geotrichum fici* (Polman and Breckenridge 1996), *Pichia kudriavzevii* and *Candida sorbophila* (Pajot et al. 2011a, b).

Other species including *Saccharomyces italicus*, *Saccharomyces cerevisiae*, *Saccharomyces chevalieri*, *Saccharomyces uvarum*, *Saccharomycopsis lipolytica*, *Candida krusei* (Yu and Wen 2005), and *Candida aquaetextoris* (Vallini et al. 1997) and *Issatchenkia orientalis* (Jafari et al. 2013) were reported as able to decolorize different reactive dyes through an unidentified degradation mechanism.

With the discovery of bacterial azo reductases, several studies tried and found yeast azo reductases. Since most azo reductases are considered to be "anaerobic," the search focused again in ascomycetous yeasts, were fermentative power and anaerobic facultative organisms are by far more frequent. Then, *Geotrichum* sp. (Máximo et al 2003), *Issatchenkia occidentalis* and *Candida zeylanoides* (Ramalho et al. 2004, 2005), *Candida oleophila* (Lucas et al. 2006); *Saccharomyces cerevisiae* (Ramalho et al. 2005; Jadhav et al. 2007), *Candida krusei* (Deivasigamani and Das 2011), among others, seem to be able of reducing azo bonds by azo reductases or related enzymes.

Lately, a new interest in ligninolytic enzymes, driven by the usage of WRF in wastewater treatments, leads to the description of yeasts with putative ligninolytic enzymes. Ascomycetous yeasts with such activities includes *Debaryomyces polimorfus* (MnP) (Yang et al. 2005), *Saccharomyces cerevisiae* (laccase, lignin peroxidase, tyrosinase, azo reductase) (Jadhav et al. 2007), *Williopsis californica*, *Williopsis saturnus* (laccase, MnP) (Martorell et al. 2012a, b), and *Galactomyces geotrichum* (laccase and tyrosinase) (Waghmode et al. 2011).

Basidiomycetous Yeasts

As we previously pointed out, this apparent lack of decolorizing ability in basidiomycetous yeasts results highly surprising, taking into account the widespread distribution of ligninolytic enzymes through filamentous basidiomycetous fungi (Pajot et al. 2011a, b).

As with ascomycetous, basidiomycetous yeasts seems to perform textile dye decolorization through several mechanisms, including biosorption by *Cryptococcus heveanensis* (Polman and Breckenridge 1996), bioaccumulation by *Rhodotorulla minuta* (Ertuğrul et al. 2009), degradation by *Pseudozyma rugulosa* (Yu and wen 2005), and degradation through ligninolytic enzymes *Trichosporon multisporum* and T. *akiyoshidainum* (laccase, MnP, and tyrosinase) (Pajot et al. 2007, 2008),

Trichosporon beigelii (laccase, tyrosinase, LiP, NADH-DCIP reductase, and azo reductase (Saratale et al. 2009), *Trichosporon porosum* (laccase, MnP, and tyrosinase) (Martorell et al. 2012a, b).

14.3 State of the Art on Textile Effluent Treatment

There are many technologies currently available for treating wastewater from the textile industry. However, textile effluents are mainly treated by relatively traditional, low technology, cost-efficient methods. In China, most industries employs biological or chemico-biological treatments. According to Chen et al. 2007, the most typical biological treatment of textile effluent is the anaerobic–aerobic process, even when other techniques, such as bioaugmentation, immobilized microorganisms, and microorganism activity enhancement, are studied by Chinese researchers since the 1980s.

In Tamil Nadu, a major textile center in India, there exist 1,000 "textile units" along a 17-km-stretch on the banks of Amaravathi River. Out of which, 487 are bleaching and dyeing units, making water pollution a major issue. According to Rajamanickam and Nagan (2010) the standard effluent treatment system there consist on a collection well, an equalization tank, a flash mixer, a clariflocculator, an aeration tank, a clarifier, a pressure sand filter, a sludge thickener, and a centrifuge house. After treatment, effluents are discharged directly into the Amaravathi River.

Thus, even when some physicochemical, high technology methods, have proved to be efficient, bioremediation of textile dyeing effluents, either as main treatment or as complementary treatment, still remains as an easy-to-implement, cost-effective, and eco-friendly alternative.

14.4 Ecology of Dye Decolorizing Yeast

When studying the ecology of dye decolorizing yeasts, a clear distinction should be made between the habitat of a yeast and its niche. According to classical definitions (Hutchinson 1957 in Starmer and Lachance 2011) a niche is the group of all intrinsic characteristics (chemical, physical, and physiological) that describe an organism's ability to exist and persist. So defined, a niche is an abstract multidimensional space where optimal growth temperature, assimilation of several carbon or nitrogen sources, and stress tolerances are examples of dimensions. The actual habitat of a yeast is then defined as the physical place where yeasts live. Historically, habitat has been characterized as atmospheric, aquatic, or terrestrial, with further subdivisions into the biotic and abiotic realms.

Since it has been well established that anaerobic dye decolorization by yeasts is an unspecific process, involving Fe reductases (Ramalho et al. 2005) or poorly characterized azo reductases and NADH/DCIP reductases (Jadhav et al. 2007;

Saratale et al. 2009) the ecological significance of such degradation ability is destined to remain obscure. Consequently the following discussion is based on oxidative dye decolorizing yeasts.

14.4.1 The r-/K-Strategist Rationale

Ecology of dye-degrading yeasts remains poorly understood. We have previously proposed to adopt a r- to K-selection continuum concept, extended from traditional ecology, in order to understand the ecological role of dye decolorizing microorganisms (Pajot et al. 2011a, b). In general, r-strategists are adapted to maximize their intrinsic growth rate when resources are abundant, whilst K-strategists are adapted to compete and survive when populations are near the carrying capacity and resources become limited. Microbiologists are more likely to use the terms copiotroph and oligotroph to describe r- and K-strategist, respectively. Copiotrophs preferentially uptake easily assimilable substrates, have high nutritional requirements, and can exhibit high growth rates when resource conditions are abundant. In contrast, oligotrophs exhibit slower growth rates and are likely to outcompete copiotrophs in conditions of low nutrient availability due to their higher substrate affinities (Langer et al. 2004). Even when r- and K-selection theory is clearly an over-simplification, the general concept is well understood and provides a useful framework for comparing the ecological features of different taxa.

Oligotrophic yeasts can usually thrive in environments with relatively low nutrient concentrations, but generally at very low growth rates. When competing with copiotrophic organisms in rich culture media, oligotrophic population can result eliminated or hardly underestimated. So, even when most dye-decolorizing yeast belongs to the Ascomycota phylum, it could be assumed that this prevalence, far from reflecting a real trend, has been strongly influenced by the screening methodology adopted, mainly allowing only the selection of fast growing, bioaccumulating yeasts (Martorell et al. 2012a, b).

14.4.2 Wood Associate Yeasts on Dye Decolorization

The correlation between decolorization of certain dyes like Poly-R 478 or azure B and lignin degradation has been exploded for years as an easy screening method for Lignin degrading enzymes (Pointing 1999; Baldrian and Šnajdr 2006). This association is not accidental. Lignin synthesis is usually described as the combinatorial polymerization of three basic monomer types, often referred to as phenylpropane derivatives (Fig. 14.3), achieved through a radical coupling mechanisms. An important implication of these combinatorial is that, the chance of finding two identical lignin molecules is extremely low. Consequently, lignin degradation does not rely on classical hydrolytic enzymes but on less specific mechanisms. White Rot Fungi

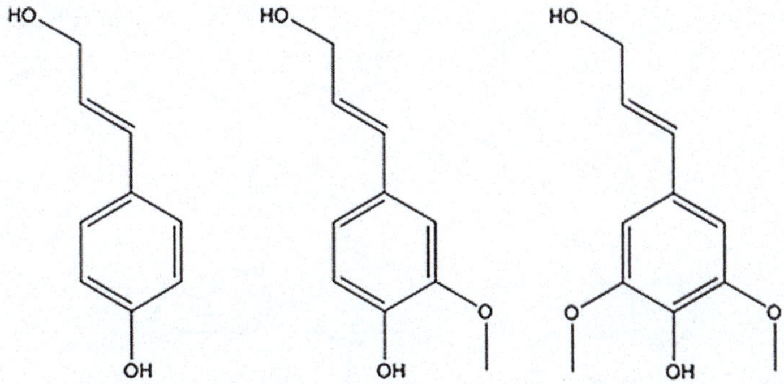

Fig. 14.3 Predominant monomers in lignin, hydroxycinnamyl alcohols

employs redox enzymes such as LiP, MnP, DyP, etc. (Solís et al. 2012; Koyani et al. 2013), while Brown and Soft Rot Fungi usually acts on lignin degradation through CMFR-like mechanisms. The nonspecific nature of these mechanisms makes them able to transform, and eventually mineralize, a variety of persistent environmental pollutants, including textile dyes (Martorell et al. 2012a, b).

An extension of the White/Brown/Soft Rot rationale, to yeast mediated dye decolorization could then sound reasonable. Thus, yeast expressing well known lignin degrading enzymes such as MnP, LiP or laccase, could initially be characterized as "White Rot Yeasts." Meanwhile, yeast performing dye degradation through Fenton like reactions could be described as "Brown/Soft Rot Yeasts." However, there exist a significant drawback to this simplistic approach. Even when many species have been isolated form wood related habitats like bark beetles gut (i.e., Rivera et al. 2009), phylloplane (Limtong and Koowadjanakul 2012) or rotting wood (Cadete et al. 2012) truly ligninolytic yeasts appears to be extremely rare (Dias et al. 2010; Kurtzman et al. 2011). Moreover, to our best knowledge, no nucleotide or amino-acid sequence of true lignin peroxidases is yet described in yeasts.

14.4.3 Soil Associated Yeasts on Dye Decolorization

Despite there exist some controversy about the synthesis mechanisms, Humic substances (HS) from different sources share common elements of structural organization (Fig. 14.4). According to Perminova and Hatfield (2005), the random nature of humus formation, involving a variety of reactions and precursors, may explain their resistance to biodegradation. Thus, humic substances- and litter-decomposers fungi usually produce MnP, laccase, tyrosinases, catechol oxidases, and other phenol oxidases. Lignin peroxidase activity, on the other hand, seems to be extremely uncommon between such soil habiting fungi (Valášková et al. 2007; Baldrian and Šnajdr 2006; Philippoussis 2009).

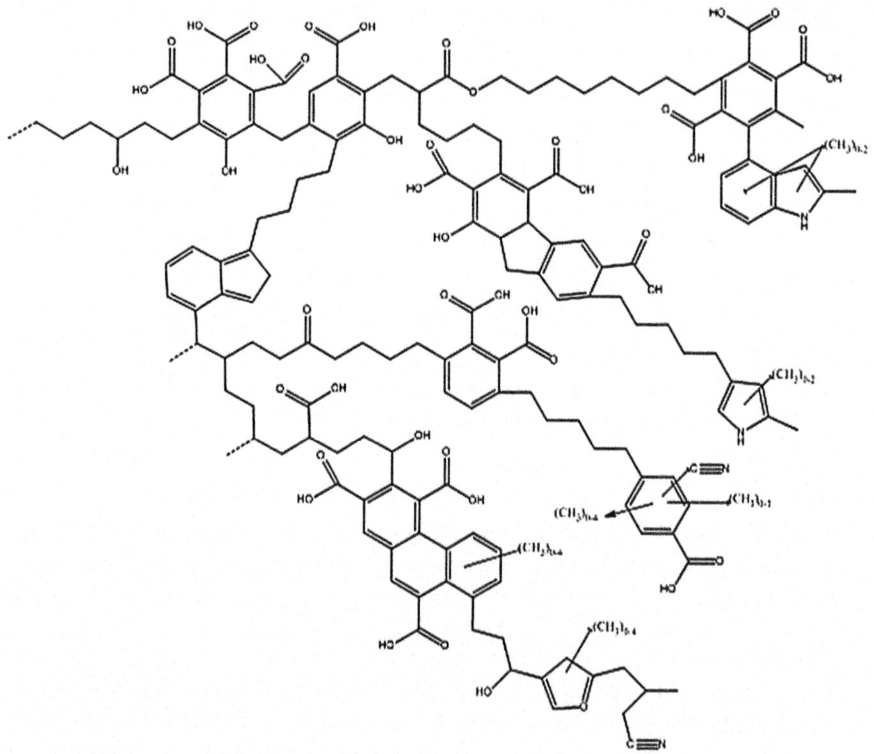

Fig. 14.4 Partial humic acid structure as proposed by Schulten and Schnitzer (1993)

It is widely accepted that soil associated yeasts occur in small numbers relative to bacteria and filamentous fungi and that they are limited to the upper surface of soil (5–15 cm) being rare in deeper layers (Starmer and Lachance 2011). However, most yeasts able to decolorize dyes through oxidative mechanisms have been isolated from soil, including *T. akiyoshidainum*, *T. multisporum*, *T. porosum* (Pajot et al. 2007; Martorell et al. 2012b) *Williopsis californica*, *Williopsis saturnus* (Martorell et al. 2012a, b) or are species usually founded in soils like *Debaryomyces polimorfus* (Yang et al. 2005) and *Galactomyces geotrichum* (Waghmode et al. 2011). Most of them seems to be related to litter or humic substances degradation in nature.

Accordingly, a positive correlation between humic substances degradation and dye decolorization ability could be demonstrated on 61 oligotrophic yeasts from Antarctic soils, with appreciable humic substances concentration (Rovati et al. 2013).

14.5 Concluding Remarks

Interesting and useful as they are, considerations about dye decolorizing yeasts ecology is still highly speculative. Most information is fragmentary, inconclusive or is based on oversimplified laboratory conditions. To illustrate this point *T. akiyoshidainum* cultures are able to decolorize several textile dyes in a variety of media and conditions. Under standard conditions, dye decolorization occurs concomitantly with the apparition of tyrosinase-like and manganese peroxidase activities in culture supernatants, pointing to a classical enzymatic driven dye decolorization process (Pajot et al. 2011a, b). However, by changing medium composition, it is possible to decouple dye decolorization from MnP and Tyr activities, without significantly increasing bioaccumulation (Martorell et al. 2012a, b). Unbuffered cultures from the closely related yeast *Trichosporon porosum*, also able to produce MnP and Tyr activities could remove up to 200 mg/l of Reactive Black 5 from culture media, through a biosorption mechanism with negligible MnP and Tyr titers (Viejobueno 2013). Obviously, under natural conditions, dye decolorization should be regarded as a complex process, with the participation of concomitant enzymatic and non-enzymatic subprocesses.

We are confident that the fast growing number of research articles in this field, based on the biotechnological potential of dye decolorizing microorganisms, will promote a better understanding of yeast ecology.

References

Acuner E, Dilek F (2004) Treatment of tectilon yellow 2G by *Chlorella vulgaris*. Process Biochem 39:623–631

Aksu Z (2005) Application of biosorption for the removal of organic pollutants: a review. Process Biochem 40:997–1026

Aksu Z, Donmez G (2003) A comparative study on the biosorption characteristics of some yeasts for Remazol Blue reactive dye. Chemosphere 50:1075–1083

Ali NF, El-Mohamedy RSR (2012) Microbial decolourization of textile waste water. J Saudi Chem Soc 16:117–123

Anastasi A, Tigini V, Varese GC (2013) The bioremediation potential of different ecophysiological groups of fungi. In: Goltapeh EM, Danesh YR, Varma A (eds) Fungi as bioremediators. Springer, Berlin, pp 29–50

Arantes V, Jellison J, Goodell B (2012) Peculiarities of brown-rot fungi and biochemical Fenton reaction with regard to their potential as a model for bioprocessing biomass. Appl Microbiol Biotechnol 94:323–338

Baldrian P, Šnajdr J (2006) Production of ligninolytic enzymes by litter-decomposing fungi and their ability to decolorize synthetic dyes. Enzyme Microb Technol 39:1023–1029

Cadete RM, Melo MA, Dussán KJ et al (2012) Diversity and physiological characterization of D-xylose-fermenting yeasts isolated from the Brazilian amazonian forest. PLoS ONE 7:e43135

Chatwal GR (2009) Introduction to dyes. In: Chatwal GR, Arora M (eds) Synthetic dyes. Himalaya Publishing House, Delhi, pp 11–19

Chen J, Wang Q, Hua Z et al (2007) Research and application of biotechnology in textile industries in China. Enzyme Microb Technol 40:1651–1655

Chivukula M, Renganathan V (1995) Phenolic azo dye oxidation by laccase from *Pyricularia oryzae*. Appl Environ Microbiol 61:4374–4377
Das D, Charumathi D, Das N (2010) Combined effects of sugarcane bagasse extract and synthetic dyes on the growth and bioaccumulation properties of *Pichia fermentans* MTCC 189. J Hazard Mater 183:497–505
de Campos Ventura-Camargo B, Marin-Morales MA (2013) Azo dyes: characterization and toxicity–a review. Text Light Ind Sci Technol 2:85–103
Deivasigamani C, Das N (2011) Biodegradation of Basic Violet 3 by *Candida krusei* isolated from textile wastewater. Biodegradation 22:1169–1180
Días A, Lucas M, et al (2010). Decolorization of Azo Dyes by Yeasts. In Erkurt HA (ed) Biodegradation of Azo Dyes. Springer-Verlag Berlin. pp 183–193
dos Santos AB, Cervantes FJ, van Lier JB (2007) Review paper on current technologies for decolourisation of textile wastewaters: perspectives for anaerobic biotechnology. Bioresour Technol 98:2369–2385
Erkurt EA, Erkurt HA, Unyayar A (2010) Decolorization of azo dyes by white rot fungi. In: Erkurt HA (ed) Biodegradation of azo dyes. Springer-Verlag, Berlin, pp 157–168
Ertuğrul S, San NO, Dönmez G (2009) Treatment of dye (Remazol Blue) and heavy metals using yeast cells with the purpose of managing polluted textile wastewaters. Ecol Eng 35:128–134
Hanan Hafez O (2008) Algal decolorization and degradation of monoazo and diazo dyes. Pakistan J Biol Sci 11:1310–1316
Hawksworth D (2012) Global species numbers of fungi: are tropical studies and molecular approaches contributing to a more robust estimate? Biodivers Conserv 21:2425–2433
Hunger K (2007) Important chemical chromophores of dye classes. In: Hunger K (ed) Industrial dyes. Chemistry, properties, applications. Wiley, Weinheim, pp 13–112
Jadhav JP, Parshetti GK, Kalme SD et al (2007) Decolourization of azo dye methyl red by *Saccharomyces cerevisiae* MTCC 463. Chemosphere 68:394–400
Jafari N, Soudi MR, Kasra-Kermanshahi R (2013) Biodecolorization of textile azo dyes by isolated yeast from activated sludge: Issatchenkia orientalis JKS6. Ann Microbiol 64:475–482, 10.1007/s13213-013-0677-y
Jarosz-Wilkolazka A, Kochmanska-Rdest J, Malarczyk E et al (2002) Fungi and their ability to decolourize azo and anthraquinonic dyes. Enzyme Microb Technol 30:566–572
Knackmuss HJ (1996) Basic knowledge and perspectives of bioelimination of xenobiotic compounds. J Biotechnol 51:287–295
Koyani RD, Sanghvi GV, Sharma RK et al (2013) Contribution of lignin degrading enzymes in decolourisation and degradation of reactive textile dyes. Int Biodeter Biodegr 77:1–9
Kurtzman CP, Fell JW, Boekhout T (2011) Definition, classification and nomenclature of the yeasts. In: Kurtzman CP, Fell JW, Boekhout T (eds) The yeast, a taxonomic study, vol 1. Elsevier, London, pp 3–8
Langer U, Bohme L, Bohme F (2004) Classification of soil microorganisms based on growth properties: a critical view of some commonly used terms. J Plant Nutr Soil Sci 167:267–269
Limtong S, Koowadjanakul N (2012) Yeasts from phylloplane and their capability to produce indole-3-acetic acid. World J Microbiol Biotechnol 28:3323–3335
Lu X, Liu R (2010) Treatment of azo dye-containing wastewater using integrated processes. In: Erkurt HA (ed) Biodegradation of azo dyes. Springer-Verlag, Berlin, pp 133–156
Lucas MS, Amaral C, Sampaio A et al (2006) Biodegradation of the diazo dye Reactive Black 5 by a wild isolate of *Candida oleophila*. Enzyme Microb Technol 39:51–55
Martorell MM, Pajot HF, de Figueroa LIC (2012a) Dye-decolourizing yeasts isolated from Las Yungas rainforest. Dye assimilation and removal used as selection criteria. Int Biodeter Biodegr 66:25–32
Martorell MM, Pajot HF, Rovati JI et al (2012b) Optimization of culture medium composition for manganese peroxidase and tyrosinase production during Reactive Black 5 decolourization by the yeast *Trichosporon akiyoshidainum*. Yeast 29:137–144
Máximo C, Pessoa Amorim MT, Costa-Ferreira M (2003) Biotransformation of industrial reactive azo dyes by *Geotrichum* sp. CCMI 1019. Enzyme Microb Technol 32:145–151

Pajot HF, de Figueroa LIC, Fariña JI (2007) Dye-decolorizing activity in isolated yeasts from the ecoregion of Las Yungas (Tucuman, Argentina). Enzyme Microb Technol 40:1503–1511

Pajot HF, de Figueroa LIC, Spencer JF et al (2008) Phenotypical and genetic characterization of *Trichosporon* sp. HP-2023. A yeast isolate from Las Yungas rainforest (Tucumán, Argentina) with dye-decolorizing ability. A Van Leeuw 94:233–244

Pajot HF, Fariña JI, de Figueroa LIC (2011a) Evidence on manganese peroxidase and tyrosinase expression during decolourization of textile industry dyes by *Trichosporon akiyoshidainum*. Int Biodeter Biodegr 65:1199–1207

Pajot HF, Delgado OD, de Figueroa LIC et al (2011b) Unraveling the decolourizing ability of yeast isolates from dye-polluted and virgin environments: an ecological and taxonomical overview. A Van Leeuw 99:443–456

Pandey A, Singh P, Iyengar L (2007) Bacterial decolorization and degradation of azo dyes. Int Biodeter Biodegr 59:73–84

Perminova I, Hatfield K (2005) Remediation chemistry of humic substances: theory and implications for technology. In: Perminova IV, Hatfield K, Hertkorn N (eds) Use of humic substances to remediate polluted environments: from theory to practice, vol 32. Springer, Berlin, pp 3–36

Philippoussis AN (2009) Production of mushrooms using agro-industrial residues as substrates. In: Nigam PSN, Pandey A (eds) Biotechnology for agro-industrial residues utilization. Springer, Netherlands, pp 163–196

Pointing SB (1999). Qualitative methods for the determination of lignocellulolytic enzyme production by tropical fungi. Fungal Diversity 2:17–33

Polman JK, Breckenridge CR (1996) Biomass-mediated binding and recovery of textile dyes from waste effluents. AATCC Rev 28:31–35

Rajamanickam R, Nagan S (2010) Groundwater quality modeling of Amaravathi River Basin of Karur District, Tamil Nadu, using Visual Modflow. Int J Environ Sci 1:91–108

Ramalho PA, Cardoso MH, Cavaco-Paulo A et al (2004) Characterization of azo reduction activity in a novel ascomycete yeast strain. Appl Environ Microbiol 70:279–2288

Ramalho PA, Paiva S, Cavaco-Paulo A et al (2005) Azo reductase activity of intact *Saccharomyces cerevisiae* cells is dependent on the Fre1p component of plasma membrane ferric reductase. Appl Environ Microbiol 71:3882–3888

Rauf MA, Salman Ashraf S (2012) Survey of recent trends in biochemically assisted degradation of dyes. Chem Eng J 209:520–530

Rivera FN, Gonzalez E, Gomez Z et al (2009) Gut associated yeast in bark beetles of the genus *Dendroctonus Erichson* (Coleoptera: Curculionidae: Scolytinae). Biol J Linn Soc 98:325–342

Rovati JI, Pajot HF, Ruberto L et al (2013) Polyphenolic substrates and dyes degradation by yeasts from 25 de Mayo/King George Island (Antarctica). Yeast 30:459–470

Saratale R, Saratale G, Chang J-S et al (2009) Decolorization and biodegradation of textile dye Navy blue HER by *Trichosporon beigelii* NCIM-3326. J Hazard Mater 166:1421–1428

Saratale RG, Saratale GD, Chang JS et al (2011) Bacterial decolorization and degradation of azo dyes: a review. J Taiwan Inst Chem Eng 42:138–157

Schulten HR, Schnitzer M (1993) A state of the art structural concept for humic substances. Naturwissenschaften 80:29–30

Singh H (2006) Fungal decolorization and degradation of dyes. In: Singh H (ed) Mycoremediation: fungal bioremediation. Wiley, Hoboken, NJ, pp 420–483

Solís M, Solís A, Pérez HI et al (2012) Microbial decolouration of azo dyes: a review. Process Biochem 47:1723–1748

Spencer J, de Spencer AR, Laluce C (2002) Non-conventional yeasts. Appl Environ Microbiol 58:147–156

Starmer WT, Lachance MA (2011) Yeast ecology. In: Kurtzman C, Fell JW, Boekhout T (eds) The yeasts: a taxonomic study, vol 1. Elsevier, London, pp 65–86

Stolz A (2001) Basic and applied aspects in the microbial degradation of azo dyes. Appl Microbiol Biotechnol 56:69–80

Valášková V, Šnajdr J, Bittner B et al (2007) Production of lignocellulose-degrading enzymes and degradation of leaf litter by saprotrophic basidiomycetes isolated from a *Quercus petraea* forest. Soil Biol Biochem 39:2651–2660

Vallini G, Frassinetti S, Scorzetti G (1997) *Candida aquaetextoris* sp. nov., a new species of yeast occurring in sludge from a textile industry wastewater treatment plant in Tuscany, Italy. Int J Syst Bacteriol 47:336–340

van der Zee FP (2002) Anaerobic azo dye reduction. Wageningen University, Doctoral Thesis

Verma AK, Dash RR, Bhunia P (2012) A review on chemical coagulation/flocculation technologies for removal of colour from textile wastewaters. J Environ Manage 93:154–168

Viejobueno J (2013) Optimización del medio de cultivo para la decoloración del colorante Negro Reactivo 5 con *Trichosporon porosum* MM-4029. Universidad Nacional de Tucumán (Argentina), Graduate thesis

Vijayaraghavan J, Basha SS, Jegan J (2013) A review on efficacious methods to decolorize reactive azo dye. J Urban Environ Eng 7:30–47

Waghmode TR, Kurade MB, Govindwar SP (2011) Time dependent degradation of mixture of structurally different azo and non azo dyes by using *Galactomyces geotrichum* MTCC 1360. Int Biodeter Biodegr 65:479–486

Yang Q, Yediler A, Yang M et al (2005) Decolorization of an azo dye, Reactive Black 5 and MnP production by yeast isolate: *Debaryomyces polymorphus*. Biochem Eng J 24:249–253

Yu Z, Wen X (2005) Screening and identification of yeasts for decolorizing synthetic dyes in industrial wastewater. Int Biodeter Biodegr 56:109–114

Zahrim AY, Tizaoui C et al (2010) Evaluation of several commercial synthetic polymers as flocculant aids for removal of highly concentrated C.I. Acid Black 210 dye. J Hazard Mater 182:624–630

Chapter 15
Copper Resistance and Oxidative Stress Response in *Rhodotorula mucilaginosa* RCL-11. Yeast Isolated from Contaminated Environments in Tucumán, Argentina

Verónica Irazusta and Lucía I.C. de Figueroa

Abstract The threat of heavy metal pollution to public health and wildlife has led to a great interest in the development of effective technologies for heavy metal immobilization in a non-bioavailable form or their conversion into less toxic forms. Organisms subjected to metal exposure in their natural environments have developed resistance mechanisms such us dedicated components and sophisticated homeostasis. *Rhodotorula mucilaginosa* RCL-11, a pigmented yeast isolated from a filter plant of a copper mine in the province of Tucumán, Argentina, supports high concentrations of the heavy metal Cu(II). In order to understand the mechanism involved in resistance to copper in this yeast, a proteomic study was conducted. Identification of differentially expressed proteins was performed. The results obtained show that when *R. mucilaginosa* RCL-11 was exposed to 0.5 mM copper, differential proteins, involved in cell resistance mechanisms, were expressed. Moreover, copper overload augmented carotenoid biosynthesis in this yeast, modifying at the same time the relative proportion of the pigments produced. Inhibition of the synthesis pathway with diphenylamine suggests an inverse relationship between carotenoid and copper biosorption by *R. mucilaginosa* RCL-11. The increased activity of superoxide dismutase and catalase measured under inhibition of carotenoid biosynthesis could explain these observations. The change in the relative proportion of the carotenoids torularhodin, torulene, and beta-carotene, as well as the detection of gamma-carotene in the presence of Cu(II) allows to hypothesize that the carotenoids produced by *R. mucilaginosa* RCL-11 play different roles in the oxidative stress response of this yeast.

V. Irazusta (✉)
PROIMI-CONICET, Av. Belgrano y Pje. Caseros, Tucumán T4001MVB, Argentina
e-mail: irazustaveronica@gmail.com

L.I.C. de Figueroa
PROIMI-CONICET, Av. Belgrano y Pje. Caseros, Tucumán T4001MVB, Argentina
Universidad Nacional de Tucumán, Tucumán, Argentina

A. Alvarez, M.A. Polti (eds.), *Bioremediation in Latin America: Current Research and Perspectives*, DOI 10.1007/978-3-319-05738-5_15

15.1 Introduction

Heavy metal pollution has become one of the most serious environmental problems today. Metal pollutants are generated through a wide range of industrial activities (including mining and smelting of metalliferous, surface finishing industry, energy and fuel production, fertilizer and pesticide industry, metallurgy, iron and steel, electroplating, electrolysis, electro-osmosis, leatherworking, photography, electric appliance manufacturing, metal surface treating, aerospace, and atomic energy installation) and continue to be released into the environment at potentially harmful levels (Avery 2001; Wang and Chen 2006).

The toxic characteristics of heavy meals are displayed as follows: (1) the toxicity can last for a long time in nature; (2) some heavy metals even could be transformed from relevant low toxic species into more toxic forms in a certain environment; (3) the bioaccumulation and bioaugmentation of heavy metal by food chain could damage normal physiological activity and endanger human life finally; (4) metals can only be transformed and changed in valence and species, but cannot be degraded by any methods; (5) the toxicity of heavy metals occurs even in low concentration (Wang 2002).

The threat of heavy metal pollution to public health and wildlife has lead to a great interest in the development of effective technologies for heavy metal immobilization in a non-bioavailable form or their conversion into less toxic forms. Conventional methods for removing metal ions from aqueous solution have been studied in detail, such as chemical precipitation, ion exchange, electrochemical treatment, membrane technologies, adsorption on activated carbon, etc. However, chemical precipitation and electrochemical treatment are ineffective and produce large amount of sludge to be treated with great difficulties. Ion exchange, membrane technologies and activated carbon adsorption processes are extremely expensive, especially when treating a large amount of water and wastewater containing heavy metal in low concentration, so they cannot be used at large scale.

Alternative process is biosorption or bioaccumulation, which utilizes various natural materials of biological origin, including bacteria, fungi, yeasts, and algae. These biosorbents possess metal-sequestering property and can be used to decrease the concentration of heavy metal ions in solution from parts per million (ppm) to parts per billion (ppb) level. It can effectively sequester dissolved metal ions out of dilute complex solutions with high efficiency and quickly, therefore it is an ideal candidate for the treatment of high volume and low concentration complex wastewaters (Machado et al. 2009).

Biosorption and bioaccumulation participate in the cycle of matter in the environment. Because living organisms bioaccumulate chemical substances, pollutants become toxicants, as well. The processes occur permanently and are performed by virtually all types of biomass. In natural environment human is not able to take control over these processes, although they can find an application in the industrial practice under controlled operation conditions (Chojnacka 2010). Biosorption and bioaccumulation also occur by accident in virtually all biological wastewater treatment processes and in all bioremediation technologies (Rehman et al. 2006).

Organisms subjected to metal exposure in their natural environments have developed resistance mechanisms such us dedicated components and sophisticated homeostasis allowing them to acquire and maintain adequate intracellular heavy metal concentrations, even under metal overload (Rensing et al. 2000). In this context, current researches on metal removal are focus on treatable sources to identify species of microorganisms that are capable of efficient uptake environmentally and economically important metals. Therefore, screening for autochthonous yeasts with high accumulation capacities and their stable resistance characteristics can help to understand cellular mechanisms as a response to high metal concentrations, such as copper.

15.2 Copper and Oxidative Stress

Copper (Cu) plays an essential role in cellular metabolism due to its versatility as a biological catalyst. It is required as a catalytic cofactor in many enzymes involved in diverse cellular processes, such as radical detoxification, oxidative phosphorylation, and iron metabolism (Puig and Thiele 2002). The two oxidation states of copper, Cu(I) and Cu(II), make it important to metalloenzymes in many redox-driven reactions (Solioz and Stoyanov 2003). While trace amounts of copper are essential for life, copper can easily react with oxygen or hydrogen peroxide generating reactive oxygen species (ROS).

Several mechanisms have been proposed to explain Cu-induced cellular toxicity. Most often, the basis for these theories is the propensity of free Cu ions to participate in the formation of ROS. Both cupric and cuprous Cu ions can participate in oxidation and reduction reactions. In the presence of superoxide ($O_2^{\cdot-}$) or reducing agents such as ascorbic acid or glutathione (GSH), Cu(II) can be reduced to Cu(I), which is capable of catalyzing the formation of hydroxyl radicals ($OH^{\cdot}$) from hydrogen peroxide (H_2O_2) via the Haber–Weiss reaction (Kadiiska et al. 1993; Bremner 1998). The hydroxyl radical is the most powerful oxidizing radical likely to arise in biological systems, and is capable of reacting with practically every biological molecule (Buettner 1993). It can initiate oxidative damage by abstracting the hydrogen from an amino-bearing carbon to form a carbon centered protein radical and from an unsaturated fatty acid to form a lipid radical (Powell 2000; Gaetke and Chow 2003).

Oxidative stress is described as a situation in which antioxidant defenses are insufficient to completely inactivate ROS. This imbalance between production and destruction of ROS can affect many cell components, including lipids, proteins, carbohydrates, and nucleic acids (Dalle-Donne et al. 2006). Both enzymatic and non-enzymatic systems are involved in the cell defense against these damaging oxidants (Raha and Robinson 2000). Key components in the enzymatic defense are the superoxide dismutases (SOD), which eliminate the superoxide radical, and the catalases (CAT) and peroxiredoxin, responsible for the H_2O_2 removal (D'Autreaux and Toledano 2007). Together with enzymatic antioxidant mechanisms, low molecular weight compounds also interact with oxidizing species, detoxifying them. The most significant non-enzymatic antioxidants involved in direct scavenging of ROS or recycling of oxidized compounds are ascorbate, glutathione, alpha-tocopherol, and carotenoids (Moore et al. 1989).

Microorganisms are known to have an ability to acquire tolerance to otherwise toxic levels of heavy metals. In this context, yeasts have previously been shown to acquire resistance toxic metals such as copper.

15.3 *Rhodotorula* Genus

Rhodotorula is common environmental yeast that is found in air, soil, lakes, ocean water, milk, and fruit juice. *Rhodotorula* species, part of the Basidiomycota phylum, colonize plants, humans, and other mammals. *Rhodotorula* produces pink to red colonies and blastoconidia that are unicellular lacking pseudohyphae and hyphae. Several authors describe the isolation of this fungus from different ecosystems, including sites with unfavorable conditions, such as the depths of the Baltic Sea, the high-altitude Lake Patagonia, the soil and vegetation of Antarctica and aquatic, hypersaline, and high-temperature environments such as the Dead Sea (Israel), Lake Enriquillo (Dominican Republic), the Great Salt Lake (USA), and beaches located in northern Brazil (Kutty and Rosamma 2008)

The genus *Rhodotorula* includes three active species: *Rhodotorula glutinis*, *Rhodotorula minuta*, and *Rhodotorula mucilaginosa*.

15.3.1 Rhodototula mucilaginosa *RCL-11*

Rhodotorula mucilaginosa RCL-11 was isolated from filter plant of a copper mine located in the NorthWest of Argentina (Villegas et al. 2005). This microorganism is capable of accumulating up to 44 % of the copper from a medium supplemented with 0.5 mM $CuSO_4$ as a Cu(II) source (Villegas et al. 2005). After 48 h of growth in the presence of Cu(II), electron microscopy showed dark grains in the cytoplasm of *R. mucilaginosa* RCL-11. The number of dark bodies in the cells increased with increasing incubation time. Scanning electron micrographs revealed that *R. mucilaginosa* RCL-11 cells grown in the presence of Cu(II) were larger than those grown in the absence of the heavy metal (Villegas et al. 2009). An increase in catalase (CAT) and superoxide dismutase (SOD) activity correlated with Cu(II) concentrations was observed in *R. mucilaginosa* RCL-11 cells (Villegas et al. 2009).

15.3.2 Proteomic Study of R. mucilaginosa *RCL-11*

As it was mentioned above, *R. mucilaginosa* RCL-11 showed increased rates of SOD and CAT probably as a consequence of oxidative stress due to elevated Cu(II) concentrations. However, so far no studies have been conducted to identify novel

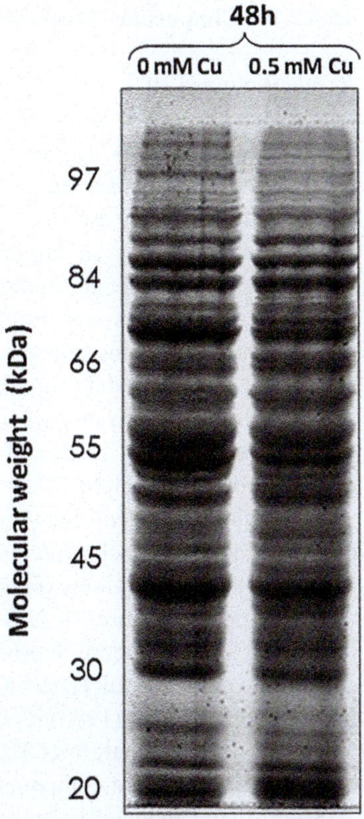

Fig. 15.1 Differential protein expression of *R. mucilaginosa* RCL-11 under copper overload. Crude extracts from cells grown during 48 h in medium supplemented or not with copper, were separated on a one-dimensional gel

proteins associated with this resistance. Proteomics is a formalized approach to obtain a rapid "snapshot" of the protein complement of a tissue, cell or cell components. It provides an excellent tool for studying variations in protein expression between different states and conditions. Undoubtedly, changes in protein expression are essential in any study aimed at examining cellular networks.

One-dimensional analysis of crude cell extracts revealed expression of different bands between cells with and without copper in *R. mucilaginosa* RCL-11 (Fig. 15.1). Most of the differentially expressed bands in yeast exposed to copper were between 55 and 90 kDa. The degree of difference in expression can be an indication of the type of proteins that cells need to support high copper concentrations. In the presence of Cu(II) the protein band profile also depended on the incubation time (Irazusta et al. 2012). The differentially expressed proteins may be responsible for the extraordinary cell capacity to resist copper toxicity.

Two-dimensional analysis (2D-analysis) of *R. mucilaginosa* RCL-11 revealed that when the yeast was exposed to copper for 48 h it produced an overexpression of 16 spots. *R. mucilaginosa* RCL-11 cells exposed to copper overload, overexpressed proteins which are involved in defense against oxidative stress (Irazusta et al. 2012).

15.3.2.1 Chaperones Overexpression

R. mucilaginosa RCL-11 cells exhibited an increased number of chaperons when copper was present. It known that a mechanism involved in stress resistance is the protection of proteins against misfolding and aggregation. Heat-shock proteins (Hsps) are highly conserved proteins within species and carry out essential functions such as protein translocation, folding and assembly under normal cellular conditions. Hsps play an important role in protection against multiple stressors (heat stress, toxic metals, ionizing and UV radiation, among others) and act as molecular chaperones assisting in ATP-dependent folding and stabilization of stress-damaged proteins (Parsell and Lindquist 1993). In most cell types, Hsps constitute 1–2 % of the total proteins even prior to stress, suggesting their importance in the biology and physiology of the unstressed cell. Hsps are distributed among diverse compartments inside cells and are classified into different groups depending on their molecular weight.

The majority of identified spots in 2D-analysis of *R. mucilaginosa* RCL-11 corresponded to heat shock proteins, belonging to Hsp88, Hsp70 and Hsp60 families (Irazusta et al. 2012). Overexpression of Hsps would be an adaptive response of the cells to copper exposure stress. There has seen a relationship between stress conditions (heat shock, glucose deprivation, exposure to free radicals and heavy metals, etc.) and expression of Hsps in different models and under adverse conditions. Parsell and Lindquist (1993) reported that, depending on the organism, heavy metals induced Hsps. Therefore, Hsps are commonly referred as stress proteins and constitute an important component of the stress response.

Most of the induced proteins identified as overexpressed in *R. mucilaginosa* RCL-11 in 0.5 mM Cu(II) belong to the 70 kDa heat shock protein family, which is critical in the cellular response to stress. In fact, these chaperones are among the most abundant cellular proteins protecting against stress-induced damage (Mayer and Bukau 2005). Hsp70 proteins are ubiquitous chaperones, which are essential ATP-binding proteins and involved in cellular functions under heat stress and non-heat stress conditions (Zhang et al. 2006). Hsp70 is the family of universal cytosolic chaperones involved in folding of damaged but repairable proteins and in the degradation of those that are damaged beyond repair (Mayer and Bukau 2005). The study found members of the Hsp70 subfamily, Ssa proteins, that are abundant cytosolic proteins and exhibit the highest identity of the group (76 %) to mammalian Hsp70. This class of chaperones was mainly present when cells were exposed to copper (Fig. 15.2). They play an essential role in cell viability and can be functionally substituted by one another even though they are differentially regulated (Jarosz and Lindquist 2010). An interesting member of the Hsp70 family is mitochondrial Hsp70 (mtHsp70), which resides in the mitochondrial matrix and is essential for cell viability (Yoneda et al. 2004). A key component, known as Ssc1, function as a molecular motor to drive translocation of proteins across the inner mitochondrial membrane (Mapa et al. 2010). Two different spots, 8 and 9, were identified as mtHsp70 SSc1 in *R. mucilaginosa* RCL-11 (Fig. 15.2).

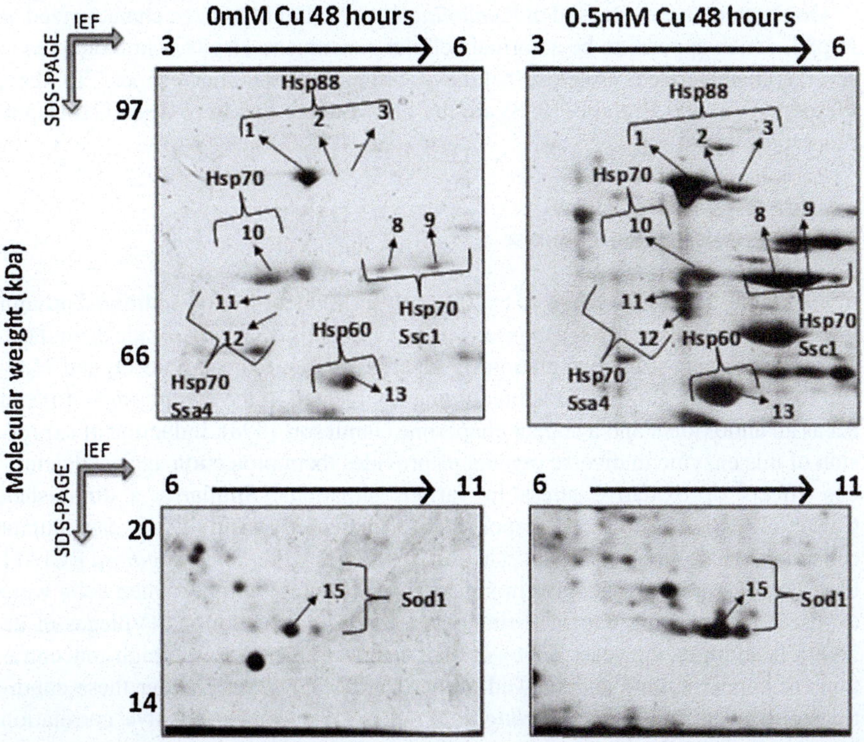

Fig. 15.2 Analysis of *R. mucilaginosa* RCL-11 proteome in the presence and absence of 0.5 mM Cu(II). Cells were grown in minimal medium in the presence or absence of copper and total cell lysate were separated by 2D gel electrophoresis

Another essential mitochondrial chaperone, Hsp60, was also overexpressed under copper overload (spot 13, Fig. 15.2). There exists a collective of double-ring assemblies that promote the folding of proteins to their native state, by importing the proteins into the mitochondrial matrix or into the intermembrane space (Raggam et al. 2011). Hsp60 has been directly associated with the response to oxidative stress. It is predominantly found in mitochondria and chloroplasts assisting in the protein folding and stress protection in these organelles. Cells containing different levels of Hsp60 were obtained by using mutant strains in which the Hsp60 gene expression was under control of a regulable promoter. In this model, cells displaying higher doses of Hsp60s were more resistant to both H_2O_2 and superoxide anions, whereas cells presenting lower levels of Hsp60 showed increased levels of intracellular ROS under oxidative stress (Cabiscol et al. 2002). Proteomic analysis of yeast cells showed that Hsps are induced in response to H_2O_2 stress, which is important for the protection of cells against this adverse condition. They help abnormal proteins that accumulate under stress conditions regain their proper folding or assist in their proteolytic degradation (Raggam et al. 2011).

One protein overexpressed in *R. mucilaginosa* RCL-11 has been characterized as Hsp88, which appears to be a normal cellular constituent. Hsp88 is homologous to several characterized proteins that have been related to heat shock. To date, Plesofsky-Vig and Brambl (1998) are the only authors that have described Hsp88 properties.

15.3.2.2 Superoxide Dismutase

R. mucilaginosa RCL-11 showed by 2D-analysis, induction of an additional protein involved in oxidative stress response, which was identified as superoxide dismutase. This enzyme catalyzes dismutation of superoxide free radicals into O_2 and H_2O_2 (Fridovich 1999). CuZnSOD is predominantly located in the cytoplasm and it can act as an antioxidant and a copper chaperone (Jamieson 1998). Induction of expression of this enzyme in diverse organisms provides them protection against deleterious effects of oxidative stress in various situations. Similarly, a diminished CuZnSOD activity is directly responsible for increased sensitivity of yeast mutant cells to oxidative stress (Irazusta et al. 2010). Studies with *R. mucilaginosa* RCL-11 have demonstrated an increase in superoxide dismutase activity when cells were exposed to increasing exposure times and copper concentrations (Villegas et al. 2009). In addition, the yeast *Yarrowia lipolytica* was able to grow at high concentrations of copper sulfate and showed higher CuZnSOD activity under these conditions (Ito et al. 2007). *R. mucilaginosa* RCL-11 showed a positive correlation between CuZnSOD activity and protein concentration when was exposed to copper overload.

15.3.3 *Carotenoids* Biosynthesis on *R. mucilaginosa* RCL-11

Similar to other colored species, the characteristic pigmentation of the colonies of *R. mucilaginosa* RCL-11 is due to the production of carotenoids. Carotenoids are one of the most important low molecular weight antioxidants. These compounds form a group of more than 600 molecules found in most life forms fulfilling diverse functions, from their original evolutionary role as photosynthetic or light-quenching pigments to antioxidants and precursors of vitamin A. The most significant part of the carotenoid molecule is the conjugated double bonds that determine their color and biological action (Sandmann 2001). A number of microorganisms, including bacteria, algae, molds, and yeasts, produce a broad range of carotenoids, including gamma- and beta-carotene, torulene, lycopene, and astaxanthin (Frengova and Beshkova 2009). Yeasts of the genera *Rhodotorula*, *Rhodosporidium*, *Sporobolomyces*, and *Phaffia* have been described to possess a remarkable ability to produce carotenoids when they are grown under unfavorable conditions, including UV radiation, temperature, solvents, and heavy metals (Buzzini 2001; Breierova et al. 2008).

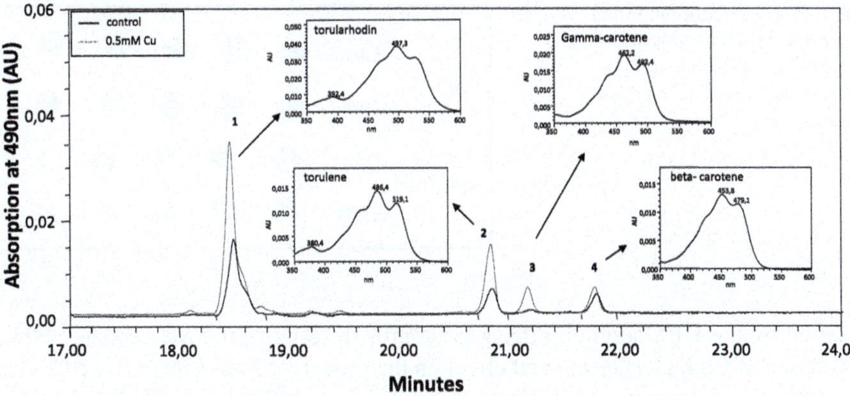

Fig. 15.3 Composition of carotenoids produced by *R. mucilaginosa* RCL-11 as determined by RP-HPLC coupled to an online diode-array detector under control conditions, in presence of Cu(II)

In *R. mucilagiona* RCL-11 the supplementation of culture media with Cu(II) as $CuSO_4$ produced a remarkable change in pigmentation intensity. Analysis revealed that *R. mucilaginosa* RCL-11 strain expressed three major carotenoids identified as torularhodin, torulene, and beta-carotene (Fig. 15.3). Under the assayed condition, gamma carotene was not detected. Cu(II) exposure produced an increase in the three compounds as well as the appearance of gamma-carotene, with a retention time between torularhodin and beta-carotene (Fig. 15.3). Torularhodin, torulene and beta-carotene are the main pigments produced in *Rodothorula* species (Weber et al. 2007). Libkind and Van Brook (2006) demonstrated a carotenoid proportion in decreasing order to torularhodin, beta-carotene, and torulene in *R. mucilaginosa* CRUB 006. Likewise, torularhodin, the most oxidized pigment, was the principal carotenoid produced in *R. mucilaginosa* RCL-11 under control conditions, followed by torulene and beta-carotene (Fig. 15.3). As other secondary metabolites, carotenoid biosynthesis is influenced by diverse factors such as light, temperature, aeration, solvents, and divalent cations (i.e., Ba, Fe, Mg, Ca, Zn, and Co). Interestingly, the heavy metal copper can increase the concentration of these pigments in *R. mucilaginosa* RCL-11 (Irazusta et al. 2013).

Carotenoid inhibition was analyzed in order to identify the role played by Cu(II) in pigment modification. Diphenylamine (DPA) is a compound largely utilized to interfere the carotenoid synthesis in both eukaryotic and prokaryotic microorganisms. Diphenylamine, blocks the sequence of de-saturation reactions by inhibiting phytoene synthase, leading to an accumulation of phytoene together with other, normally absent, saturated carotenoids (Britton et al. 1977; Squina and Mercadante 2005). As expected, 50 μM DPA produced almost the complete inhibition of carotenoid synthesis by *R. mucilaginosa* RCL-11 decreasing the pigment concentration (Irazusta et al. 2013).

DPA addition affected the growth of *R. mucilaginosa* after 48 h of incubation (Fig. 15.4). It is possible that DPA has a toxic effect per se, or the lack of carotenoids

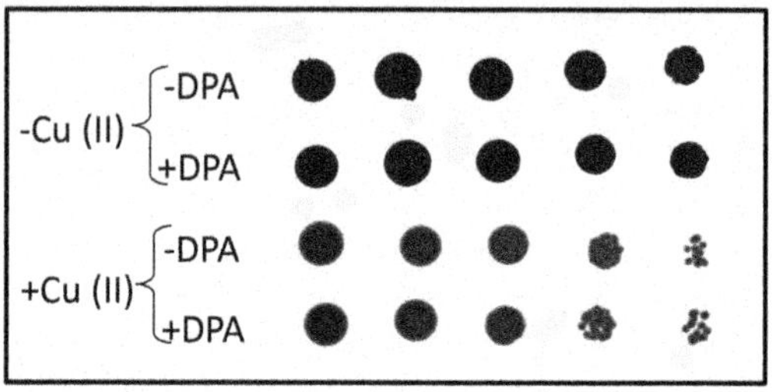

Fig. 15.4 Effect of inhibition of carotenoid biosynthesis and Cu(II) overload on *R. mucilaginosa* RCL-11 growth

caused by the DPA addition affects yeast growth (Squina and Mercadante 2005). Addition of Cu had a detrimental effect on the growth of *R. mucilaginosa* RCL-11. $CuSO_4$ at a concentration of 0.5 mM decreased the cell yield (Fig. 15.4).

15.3.3.1 Carotenoids Biosynthesis and Cu(II) Removal on *R. mucilaginosa* RCL-11

It has been postulated that the main resistance mechanism to copper overload of *R. mucilaginosa* RCL-11 is the sequestration of the heavy metal. However, the analysis of the influence of carotenoids on the bioremediation of copper showed an inverse relationship between pigment production and copper bioremediation (Irazusta et al. 2013). The increase from 0.053 to 0.083 mM A_{600}^{-1} in the relative copper removal after DPA addition, representing a rise of 55 % in the cell capability to bioremediate, suggests that the cell response to copper exposure is more complex. In addition, this finding is of biotechnological importance since it suggests that a non-pigmented mutant of *R. mucilaginosa* RCL-11 could be more effective to concentrate, remove, and recover metals from streams and could enhance the efficiency of wastewater treatment processes. Production of carotenoids could be part of a physiological response triggered to avoid, at least in part, the intracellular accumulation of heavy metals. When DPA is added, carotenoids are not produced and more copper is accumulated; when culture media are not supplemented with DPA, more pigments are produced and copper is also accumulated, although to a lesser extent (Irazusta et al. 2013).

15.3.3.2 Antioxidant Activity and Carotenoid Biosynthesis

A relationship between carotenoids and ROS has previously been established (Frengova and Beshkova 2009). For instance, these pigments have a demonstrated role in the prevention of oxidative injury due to ROS exposure in pigmented heterotrophic yeasts (Mendez-Alvarez et al. 2000). As was mentioned above, the exposure to heavy metals triggered the overexpression of SOD and CAT in *R. mucilaginosa* RCL-11, enzymes directly involved in responses to oxidative stresses. When the production of carotenoid pigments was inhibited with 50 μM DPA, the highest CAT activity could be measured (Irazusta et al. 2013). The maximum catalase activity in the absence of carotenoid synthesis shows that *R. mucilaginosa* RCL-11 responded to the lack of pigments induced by DPA, at least in part, activating other antioxidant mechanisms such as catalase and superoxide dismutase activity (Irazusta et al. 2013). Yan et al. (2011) showed that introduction of the beta-carotene gene into yeasts with a deficiency in cytosolic catalase activity increased cell resistance to H_2O_2 stress. The authors suggested that carotenoids would substitute catalase to protect the yeast from H_2O_2-induced oxidative damage (Yan et al. 2011). An increase in catalase activity in *R. mucilaginosa* RCL-11 could be explained by the same mechanisms, suggesting that increased catalase activity can substitute carotenoids in protecting the yeast from the oxidative damage caused by the heavy metal copper. Superoxide dismutase activity seems to exhibit a similar behavior showing highest activity under supplementation with copper. It has been suggested that the absence of CuZnSOD in pigmented yeasts is complemented by the presence of carotenoproteins that act as an extramitochondrial antioxidant (Moore et al. 1989).

Acknowledgments This work was supported by grants from Agencia Nacional de Promoción Científica Tecnológica, FONCYT (PICT 2010-0101), and Consejo Nacional de Investigaciones Científicas y Técnicas CONICET (PIP096 and PIP0391).

References

Avery SV (2001) Metal toxicity in yeasts and the role of oxidative stress. Adv Appl Microbiol 49:111–142

Breierova E, Gregor T, Marova I et al (2008) Enhanced antioxidant formula based on a selenium-supplemented carotenoid-producing yeast biomass. Chem Biodivers 5(3):440–446

Bremner I (1998) Manifestations of copper excess. Am J Clin Nutr 67:1069–1073

Britton G, Singh RK, Malhotra HC et al (1977) Biosynthesis of 1,2- dihydrocarotenoids in *Rhodopseudomonas viridis*: experiments with inhibitors. Phytochemistry 16:1561–1566

Buettner G (1993) The packing order of free radicals and antioxidants: lipid peroxidation, alpha-tocopherol and ascorbate. Arch Biochem Biophys 300:535–543

Buzzini P (2001) Batch and fed-batch carotenoid production by *Rhodotorula glutinis-Debaryomyces castellii* co-cultures in corn syrup. J Appl Microbiol 905:843–847

Cabiscol E, Belli G, Tamarit J et al (2002) Mitochondrial Hsp60, resistance to oxidative stress, and the labile iron pool are closely connected in *Saccharomyces cerevisiae*. J Biol Chem 277:44531–44538

Chojnacka K (2010) Biosorption and bioaccumulation-the prospects for practical applications. Environ Int 36:299–307

D'Autreaux B, Toledano MB (2007) ROS as signalling molecules: mechanisms that generate specificity in ROS homeostasis. Nat Rev Mol Cell Biol 8:813–824

Dalle-Donne I, Aldini G, Carini M et al (2006) Protein carbonylation, cellular dysfunction, and disease progression. J Cell Mol Med 10:389–406

Frengova GI, Beshkova DM (2009) Carotenoids from *Rhodotorula* and *Phaffia*: yeasts of biotechnological importance. J Ind Microbiol Biotechnol 36:163–180

Fridovich I (1999) Fundamental aspects of reactive oxygen species, or what's the matter with oxygen? Ann N Y Acad Sci 893:13–18

Gaetke LM, Chow CK (2003) Copper toxicity, oxidative stress, and antioxidant nutrients. Toxicology 189:147–163

Irazusta V, Obis E, Moreno-Cermeno A et al (2010) Yeast frataxin mutants display decreased superoxide dismutase activity crucial to promote protein oxidative damage. Free Radic Biol Med 48:411–420

Irazusta V, Estévez C, Amoroso MJ et al (2012) Proteomic study of the yeast *Rhodothorula mucilaginosa* RCL-11 under copper stress. BioMetals 25:517–527

Irazusta V, Nieto-Peñalver CG, Cabral MG et al (2013) Relationship among carotenoid production, copper bioremediation and oxidative stress in *Rhodotorula mucilaginosa* RCL-11. Process Biochem 48:803–809

Ito H, Inouhe M, Tohoyama H et al (2007) Characteristics of copper tolerance in Yarrowia lipolytica. BioMetals 20:773–780

Jamieson DJ (1998) Oxidative stress responses of the yeast *Saccharomyces cerevisiae*. Yeast 14:1511–1527

Jarosz DF, Lindquist S (2010) Hsp90 and environmental stress transform the adaptive value of natural genetic variation. Science 330:1820–1824

Kadiiska MB, Hanna PM, Jordan SJ et al (1993) Electron spin resonance evidence for free radical generation in copper-treated vitamin E- and selenium-deficient rats: in vivo spin-trapping investigation. Mol Pharmacol 44:222–227

Kutty SN, Rosamma P (2008) Marine yeasts—a review. Yeast 25:465–483

Libkind D, Van Brook M (2006) Biomass and carotenoid pigment production by patagonian native yeasts. World J Microbiol Biotechnol 22:687–692

Machado MD, Janssens S, Soares HM et al (2009) Removal of heavy metals using a brewer's yeast strain of *Saccharomyces cerevisiae*: advantages of using dead biomass. J Appl Microbiol 106:1792–1804

Mapa K, Sikor M, Kudryavtsev V et al (2010) The conformational dynamics of the mitochondrial Hsp70 chaperone. Mol Cell 38:89–100

Mayer MP, Bukau B (2005) Hsp70 chaperones: cellular functions and molecular mechanism. Cell Mol Life Sci 62:670–684

Mendez-Alvarez S, Rufenacht K, Eggen RI (2000) The oxidative stress-sensitive yap1 null strain of *Saccharomyces cerevisiae* becomes resistant due to increased carotenoid levels upon the introduction of the *Chlamydomonas reinhardtii* cDNA, coding for the 60S ribosomal protein L10a. Biochem Biophys Res Commun 267:953–959

Moore MM, Breedveld MW, Autor AP (1989) The role of carotenoids in preventing oxidative damage in the pigmented yeast, *Rhodotorula mucilaginosa*. Arch Biochem Biophys 270: 419–431

Parsell DA, Lindquist S (1993) The function of heat-shock proteins in stress tolerance: degradation and reactivation of damaged proteins. Annu Rev Genet 27:437–496

Plesofsky-Vig N, Brambl R (1998) Characterization of an 88-kDa heat shock protein of *Neurospora crassa* that interacts with Hsp30. J Biol Chem 273:11335–11341

Powell SR (2000) The antioxidant properties of zinc. J Nutr 130:1447–1454

Puig S, Thiele DJ (2002) Molecular mechanisms of copper uptake and distribution. Curr Opin Chem Biol 6:171–180

Raggam RB, Salzer HJ, Marth E et al (2011) Molecular detection and characterization of fungal heat shock protein 60. Mycoses 54:e394–e399

Raha S, Robinson BH (2000) Mitochondria, oxygen free radicals, disease and ageing. Trends Biochem Sci 25:502–508

Rehman A, Shakoori FR, Shakoori AR (2006) Uptake of heavy metals by a ciliate, *Tachysoma pellionella*, isolated from industrial effluents and its potential use in bioremediation of toxic wastewater. Bull Environ Contam Toxicol 77:469–476

Rensing C, Fan B, Sharma R, Mitra B, Rosen BP (2000) CopA: an *Escherichia coli* Cu(I)-translocating P-type ATPase. Proc Natl Acad Sci U S A 97:652–656

Sandmann G (2001) Carotenoid biosynthesis and biotechnological application. Arch Biochem Biophys 385:4–12

Solioz M, Stoyanov JV (2003) Copper homeostasis in *Enterococcus hirae*. FEMS Microbiol Rev 27:183–195

Squina FM, Mercadante AZ (2005) Influence of nicotine and diphenylamine on the carotenoid composition of *Rhodotorula* strains. J Food Biochem 29:638–652

Villegas LB, Amoroso MJ, de Figueroa LI (2005) Copper tolerant yeasts isolated from polluted area of Argentina. J Basic Microbiol 45:381–391

Villegas LB, Amoroso MJ, Figueroa LI (2009) Responses of *Candida fukuyamaensis* RCL-3 and *Rhodotorula mucilaginosa* RCL-11 to copper stress. J Basic Microbiol 49:395–403

Wang JL (2002) Biosorption of copper (II) by chemically modified biomassof *Saccharomyces cerevisiae*. Process Biochem 37:847–850

Wang J, Chen C (2006) Biosorption of heavy metals *by Saccharomyces cerevisiae*: a review. Biotechnol Adv 24:427–451

Weber RW, Anke H, Davoli R (2007) Simple method for the extraction and reversed-phase high-performance liquid chromatographic analysis of carotenoid pigments from red yeasts (*Basidiomycota*, Fungi). J Chromatogr A 1145:118–122

Yan GL, Liang HY, Wang Z et al (2011) Important role of catalase in the production of beta-carotene by recombinant *Saccharomyces cerevisiae* under H_2O_2 stress. Curr Microbiol 62:1056–1061

Yoneda T, Benedetti C, Urano F et al (2004) Compartment-specific perturbation of protein handling activates genes encoding mitochondrial chaperones. J Cell Sci 117:4055–4066

Zhang S, Hacham M, Panepinto J et al (2006) The Hsp70 member, Ssa1, acts as a DNA-binding transcriptional co-activator of laccase in *Cryptococcus neoformans*. Mol Microbiol 62:1090–1101

Chapter 16
Surface-Active Compounds of Microbial Origin and Their Potential Application in Technologies of Environmental Remediation

Verónica Leticia Colin, Álvaro Juárez Cortes, Analía Rodríguez, and María Julia Amoroso

Abstract The world is now confronted with serious problems of environmental contamination. Industrial and mining activities represent the main sources of heavy metal contamination, which provide unique challenges for their remediation, as they cannot be degraded into innocuous products. On the other hand, extensive production and use of hydrocarbons and pesticides in diverse fields has resulted in widespread environmental contamination by these chemicals. A variety of remediation technologies that include physicochemical methods are available to address contamination with organic and inorganic pollutants. However, these technologies have high cost and the risk of secondary environmental pollution. The microbial remediation includes the use of certain microorganisms or products derived to them to address the problems of environmental pollution. In this connection, the synthesis and release of exopolymers with surfactants activity is one of the strategies of the first microbial cell defense line against diverse toxics. These compounds can be successfully used to release hydrocarbons and other pollutants of low solubility from soils matrix. There are also reports regarding the release of heavy metals from soils

V.L. Colin (✉)
Planta Piloto de Procesos Industriales y Microbiológicos (PROIMI), CONICET,
Av. Belgrano y Pje. Caseros, 4000 Tucumán, Argentina

Universidad de San Pablo-Tucumán, Tucumán, Argentina
e-mail: veronicacollin@yahoo.com.ar

Á.J. Cortes • A. Rodríguez
Planta Piloto de Procesos Industriales y Microbiológicos (PROIMI), CONICET,
Av. Belgrano y Pje. Caseros, 4000 Tucumán, Argentina

M.J. Amoroso
Planta Piloto de Procesos Industriales Microbiológicos (PROIMI), CONICET,
Av. Belgrano y Pasaje Caseros, 4000 Tucumán, Argentina

Facultad de Bioquímica, Química y Farmacia, Universidad Nacional de Tucumán,
4000 Tucumán, Argentina

Universidad del Norte Santo Tomás de Aquino (UNSTA), 4000 Tucumán, Argentina

A. Alvarez, M.A. Polti (eds.), *Bioremediation in Latin America: Current Research and Perspectives*, DOI 10.1007/978-3-319-05738-5_16

and sediments using compounds of biological origin as surfactants microbial. This chapter considers the current advances on this topic, emphasizing on the properties and potential applications of microbial surfactants in environmental remediation technologies.

16.1 General Aspects, Classification, and Properties of the Biosurfactants

Biosurfactants comprise a structurally diverse group of surface-active substances of biological origin. They consist in amphiphiles molecules that present two parts, a polar moiety (hydrophilic) and nonpolar group (hydrophobic), which permit the partition preferentially at the interfaces such as liquid–liquid, gas–liquid, or solid–liquid interfaces. Most of these compounds are either anionic or neutral. Only a few are cationic such as those containing amine groups (Mulligan et al. 2001).

Synthetic surfactants are classified as anionic, cationic, or non-ionic according to predominant electric charge after contact with water. However, biological surfactants are typically categorized based on criteria as origin, chemical composition, molecular weight, and mode of action. Concerning to the origin, biosurfactants can derivate from plants or microorganisms, including in the last case, bacteria, filamentous fungi, and yeasts (Colin et al. 2010, 2013a; Luna et al. 2012). In comparison to those derivate from plants which to the far have been less explored, microbial surfactants are produced in substantial amounts during the exponential or stationary phase. In addition, those of microbial origin are characterized by be active at extreme conditions of temperatures, pH, and salinity, properties that could increase their scope of application in several biotechnological areas. Based upon these highlights, the next sections only it focuses on the production and potential application of surfactants of microbial origin.

Biosurfactants can be also categorize according to their molecular weight as low-molecular-mass biosurfactants (glycolipids, phospholipids and lipopeptides); and those of high-molecular-mass (proteins, polysaccharides, lipopolysaccharides, lipoproteins, or complex mixtures of these biopolymers). Based on mode of action, surface-active compounds are of two types: low-molecular-mass biosurfactants that are efficient in lowering interfacial tension at all levels, inclusive in air–water interface; and those of high-molecular-mass biosurfactants/bioemulsifiers that are more effective to reduce interfacial tension between immiscible liquids, stabilizing hydrocarbons–water emulsions.

The properties of the biosurfactants enable a wide potential of biotechnological and industrial applications, including their use in agriculture, the cosmetic and pharmaceutic industry, food production, etc. (Banat et al. 2010). Molecules as the biosurfactants can have also numerous environmental applications whose bases are considered in the next section.

16.2 Environmental Application of the Biosurfactants

The world is now confronted with serious problems of environmental contamination, which demand immediate solutions. A variety of remediation technologies that include mostly physicochemical methods are available to address to organic and inorganic pollutants. However, these technologies have several disadvantages including the high cost and the risk of secondary contamination. The situation is more critical in developing countries where there is no legislation to respect. As a result, it remains important to develop new technologies for the reduction of pollutants to acceptable levels, but at more manageable costs. Scientific evidence indicates that certain microbial activities and/or some products derived from microorganisms are among the constituents of the first cellular defense line against diverse toxics. Therefore, in the next paragraphs are considered the current advances on these topics. A particular emphasis is placed on the potential applications of microbial surfactants/emulsifiers for the removal of heavy metals and organic pollutants from soils and sediments.

16.2.1 Application of Microbial Surfactants/Emulsifiers in the Remediation of Heavy Metals

Due to the long term accumulation and lag in treatment, right now diverse countries are facing serious threats of mass heavy metal pollution. Industries diverse, including mining, smelting and metallurgical, chemical, textile printing and dyeing, leather tanning, etc., are involved in massive heavy metal discharge to the environments. There are reports in the literature regarding the removal of heavy metals from wastewater and soils using biological agents as biosurfactants/bioemulsifiers. The usefulness of biosurfactants for remediation of heavy metal-contaminated soils and sediments is mainly based on their ability to form complexes with metals, facilitating their subsequent desorption from soil matrix. Interestingly, has been pointed that metal removal effectiveness depends on the metallic species, but also of the biosurfactants's chemical nature and concentration. Juwarkar et al. (2007), for example, reported on a rhamnolipid surfactant produced by *Pseudomonas aeruginosa* strain BS2 which selectively favors mobilization of metals such cadmium and lead, being more effective in removing the first. Gutierrez et al. (2008) isolated and characterized an exopolymer with emulsifying activity from *Pseudoalteromonas* sp. strain TG12, which demonstrates a differential capacity to desorb metals from marine sediment according these are monovalent, divalent, or trivalent species. Wang and Mulligan (2009) evaluated the potential use of biosurfactants of microbial origin to mediate the arsenic removal from the mine tailings, noting that metal mobilization is enhanced as concentration of biosurfactant is increased. On the other hand, a screening of bioemulsifiers producer *Microbacterium*'s strains isolated from urban mangrove sediments was performed by Aniszewski et al. (2010).

They first evaluated the bioemulsifiers production using two carbon sources such as sucrose and glucose. Then, performance of the bioemulsifiers produced to remove cadmium and zinc from hazardous industrial residues were estimated. The authors noted that both the bioemulsifiers production as the removal metals from sediments depended of the producing strains and of the carbon substrate used for the production. Maximum cadmium removal (41 %) was reached during the washing experiments with *Microbacterium* Mc6b bioemulsifier produced from glucose. However, optimal zinc removal (68 %) was detected using as washed agent an emulsifier produced from glucose by *Microbacterium* MC1.

In our laboratory, recently it has been detected the spectacular able of two indigenous actinobacteria such as *Amycolatopsis tucumanensis* DSM 45259 and *Streptomyces* sp. MC1 (isolated from heavy metal-contaminated soils of the Tucumán province-Argentine) to produce compounds with emulsifying activity (Colin et al. 2013a, b). We estimated the emulsifier production according to Cooper and Goldenberg (1987) by measured of the emulsifying ability of the supernatants, using kerosene as substrate. The first studies of Cu(II) and Cr(VI) removal from soil using bioemulsifiers produced by *A. tucumanenis* DSM 45259 as washed agents were performed by Colin et al. (2013b). The authors noted that emulsifiers compounds produced in a minimal medium with different carbon and nitrogen source were unable to remove copper. However, these compounds effectively mediate the chromium-desorption from soil, with the removal percentage doubled compared to that seen when deionized water was used for washing (Colin et al. 2013b). These findings may be promising for the development of remediation technologies for hexavalent chromium compounds based upon direct use of these microbial emulsifiers.

Colin et al. (2013b) also noted that in presence of 0.5 g L^{-1} phosphate in the culture medium, maximun bioemulsifier production by indigenous actinobacteria was reached after 72 h of cultivation for both strains (Fig. 16.1a, b). In contrast, when the phosphate concentration in the medium was increased to 1.5 g L^{-1}, the maximun production was detected al 24 h for both microorganisms (Fig. 16.1c, d).

Similarly, Franzetti et al. (2009) reported the positive effect of phosphate ions in the biosynthesis of surface-active compounds. Diverse theories have been hypothesized in order to elucidate this effect. Lang and Philp (1998), for example, elucidate the synthetic pathway of surface-active compounds in *Rhodococcus*. These authors noted that one of the key reactions for the synthesis of a glycosidic biosurfactant was catalyzed by the trehalose-6-phosphate enzyme. Among the actinobacteria, it has been reported that a high phosphate concentration in liquid medium can to lead to an accumulation of glucose-6-phosphate, immediate precursor of trehalose-6-phosphate (Madry et al. 1979). Based upon this background, Pagilla et al. (2002) have suggested that the presence of phosphate ions could promote the synthesis of certain sugar residues, enhancing the production of surfactants of glycosidic nature. However, further studies are necessary to clear the specific role of phosphate ions on the biosynthesis and stability of these compounds.

Studies on the potential application of biosurfactant-producing microorganisms in bioremediation of metals were also proportioned by some authors. Gnanamani et al. (2010) for example, carried out experiments of hexavalent chromium

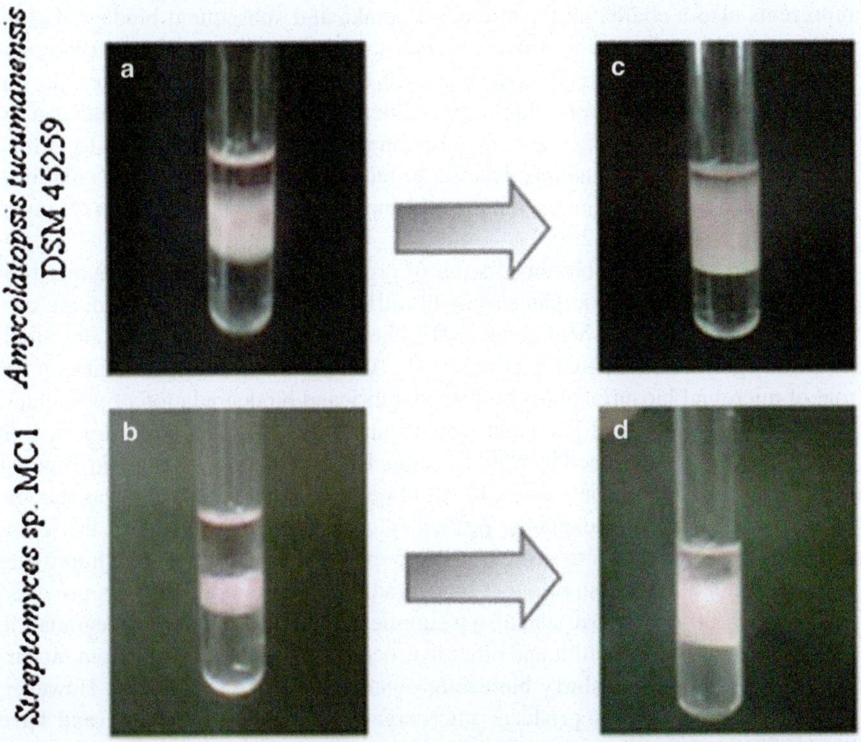

Fig. 16.1 Bioemulsifers production by indigenous actinobacteria in a minimal medium added with 0.5 (**a**, **b**) and 1.5 g L^{-1} of phosphate (**c**, **d**)

bioremediation mediated by biosurfactants produced by *Bacillus* sp. MTCC 5514. These authors hypothesized that chromium remediation is carried out through two processes: the first, consist in extracellular reduction of hexavalent chromium to trivalent specie via a chromium reductase. The second process could be consisting in the entrapment of trivalent chromium in the micelle of biosurfactants, which prevents microbial cells from exposure towards trivalent chromium. However, more studies on the application of biosurfactant-producing microorganisms are required in order to effectively estimate their use in technologies of heavy metals removal.

16.2.2 Application of Microbial Surfactant/Emulsifiers to Organic Pollutants Remediation

Organic pollutants, such as hydrocarbon, oils and pesticides present limited solubility in water and largely are absorbed on the soil matrix. As a consequence, their limited availability represents one of the greatest challenges for their recovery by physicochemical treatments. The low availability of these classes of pollutants represents also a challenge for microbial uptake and subsequent biodegradation, limiting the bioremediation processes. However, the environmental microorganisms have developed effective strategies to overcome these difficulties. In the presence of poorly water-soluble molecules, certain microorganisms can to produce and secrete surface active compounds. These compounds can to reduce surface and interfacial tensions, leading to increased mobility and bioavailability of water-insoluble compounds bound to the matrix of soils and sediments (Nguyen et al. 2008; Nievas et al. 2008).

Numerous reports on bioremediation of organic pollutants by using either bioemulsifier producer microorganisms or biosurfactants previously isolated, are currently available (Das and Mukherjee 2007; Nievas et al. 2008; Cameotra and Singh 2009; Kang et al. 2010; Reddy et al. 2010). Among the pioneering studies on the role of microbial biosurfactants in bioavailability and biodegradation of pesticides, we found the advances of Patel and Gopinathan (1986). These authors reported on a glycolipopeptide produced by *Bacillus* strain FE-2, capable of emulsifying immiscible organophosphorus pesticides. It was observed that this emulsifier was specific for immiscible organophosphorus pesticides. Also it was detected that this compounds was secreted during growth in the presence of such pesticides, improving their bioavailability and subsequent biodegradation. Laboratory findings provided by Kosaric (2001) demonstrated also a significant enhancing in the biodegradation of the herbicide metholachlor and other hydrocarbons when sophorose lipid biosurfactants was added to a slurry bioreactor containing soil in suspension. However, studies on biosurfactants producer microorganisms are very complex, and have been limited to laboratory scale. Thereby, its application perspective at a real scale is still uncertain. Direct application of biosurfactants in soil washing technologies has been used as one of the most pragmatic strategies for increasing the mass transfer of organic contaminants and their consequent removal (Calvo et al. 2009; Lai et al. 2009; Franzetti et al. 2009, 2012). It is important to remark that in this last case, it is not required that producer microorganisms are able to grow and survive in contaminated-environment (Pacwa-Plociniczak et al. 2011; Colin et al. 2013a). One of the most recent advances on the addition of surface-active compounds to enhance the plaguicide-contaminated soil remediation was provided by Jayashree et al. (2006). They reported on the potential of synthetic surfactants as Tween 80, Triton X-100, and of a biosurfactant produced by *Bacillus subtilis* (surfactin) for enhancing the release of endosulfan, a highly toxic compound to aquatic and human population. Interestingly, these authors found that *Bacillus subtilis* surfactin recorded the maximum endosulfan recovery (91.5 %), followed by Tween 80 and Triton X-100, ultimately. Despite these findings, the advances about the biosurfactant assisted-pesticide remediation are limited compared to the field of hydrocarbons. Thereby, more efforts are required to evaluate the prospects of the biosurfactants for the pesticides remediation.

In the field of the hydrocarbons, the application of biosurfactants as natural alternatives to those synthetically produced is an efficient strategy for enhanced their removal from soils and sediments. Franzetti et al. (2009), for example, evaluated the potential applications of bioemulsans producer *Gordonia* sp. BS29 for the bioremediation of soils contaminated by aliphatic and aromatic hydrocarbons. On the other hand, bioemulsans were used as soil washing agents for the removal of crude oil and polycyclic aromatic hydrocarbons (PAHs). Bioremediation experiments showed that the BS29 bioemulsans are able to enhance the biodegradation of recalcitrant branched hydrocarbons. Also, the authors detected an effective crude oil and PAHs removal from soil during the washing experiments.

Recently, Franzetti et al. (2012) have carried out the characterization and the evaluation studies on of potential environmental applications of the bioemulsifiers produced by *Variovorax paradoxus* 7bCT5. These bioemulsifiers were able to produce a stable oil/water emulsion and maintained the emulsification activity after boiling as well as at low temperatures. These authors showed also through different ecotoxicological tests that the bioemulsifiers produced did not have any toxic properties. In addition, the soil sorption affinity likely affected the bioemulsifier ability to remove hydrocarbons from contaminated soils. In fact, *V. paradoxus* 7bCT5 bioemulsifiers significantly increased the removal of crude-oil from sandy soil compared to water.

Currently, in our laboratory it is being explored the ability of the compounds produced by *A. tucumanensis* and *Streptomyces* sp. MC1 to emulsifying organic pollutants such as oils, pesticides and hydrocarbons in liquid medium. Likewise, we will assess the potential use of these compounds as soil washing agent in order to increase the desortion of these pollutants, and enhance the washing processes efficiency.

16.3 Production of Microbial Surfactants/Emulsifiers from Agro-industrial Residues

In comparison to the chemical surfactants, those of biological origin present unquestionable advantages since they are environmentally friendly, biodegradable, less toxic, and nonhazardous (Colin et al. 2010). However, their large scale application is currently restricted by high cost of production, due mainly to the use of synthetic culture media (Saharan et al. 2011). To overcome this limitation, greater emphasis is being laid on the use of various agro-industrial wastes, which can result appropriate substrates of cost low for the massive biosurfactants production. Table 16.1, provides some recent reports on the use of various agro-industrial wastes as substrates for the biosurfactants production at large scale.

While biofuels have emerged as promising substitutes for petroleum-based fuels, there is currently a great need for the proper management and utilization of the by-products such as crude glycerol derived from biodiesel synthesis; or the vinasse, derived from the bioethanol production process. De Souza Monteiro et al. (2012), for example, have reported on effective production of bioemulsifiers by *Trichosporon mycotoxinivorans* CLA2 using crude glycerol and other derived from biodiesel synthesis as sole carbon source. Recently, we have demonstrated the feasibility of using glycerol to support the growth and effective production of bioemulsifiers by *A. tucumanensis* DSM 45259 (Colin et al. 2013b).

Table 16.1 Agro-industrial residues used for the biosurfactants/bioemulsifiers production in scientific works published during the last 5 years

Agro-industrial substrates	Microorganism	Reference
Distilled waste	*Candida sphaerica*	Sobrinho et al. (2008)
Sugarcane molasses	*Candida bombicola*	Daverey and Pakshirajan (2010)
Molasses and olive oil	*Brevibacterium aureum* MSA13	Seghal Kiran et al. (2010)
Vegetable fat waste	*Candida glabrata*	de Gusmão et al. (2010)
Sugarcane molasses	*Starmerella bombicola* NBRC 10243	Takahashi et al. (2011)
Biodiesel residues	*Trichosporon mycotoxinivorans* CLA2	De Souza Monteiro et al. (2012)
Diesel oil	*Yarrowia lipolytica*	De Souza Monteiro et al. (2012)
Crude glycerol	*Yarrowia lipolytica*	Fontes et al. (2012)
Vinasse	*Bacillus pumilus*	Guerra de Oliveira and García-Cruz (2013)
Waste frying oil	*Bacillus pumilus*	Guerra de Oliveira and García-Cruz (2013)
Cassava wastewater	*Bacillus subtilis* sp.	Cavalcante Barros et al. (2013)

Currently, we being evaluated the ability of *Streptomyces* sp. MC1 to produce bioemulsifiers using vinasse as substrate. The growing use of the bioethanol in parallel with the oil price is the main source of vinasse generation. The vinasse reinstatement to the soil could be desirable and productive since organic matter and nutrients recycling to the soil is possible. Thereby, their application to the land does not generate pollution if it is well managed. In fact, it is considered as a sane concept that the soil, is a natural treatment system. However, the bioethanol industry has found in the drive of their effluents (12 L of vinasse per L of alcohol), the main limitant for their growth, since the vinasse is an effluent of high organic load, and represent a risk for the surrounding watersheds (Sanches Carrocci et al. 2012). The limiting "vinasse" is the shortest in countries of large areas or low population density as Brazil, which in those densely populated as the Valle del Cauca in Colombia or the Salí River Valley in Argentina (Province-Tucumán). But even in Brazil, the vinasse management became an obstacle to the development of business distillers because larger land is required to return the vinasse to the soil. Tucumán is a small agricultural province (22,500 km^2), with an important concentration of citrus industries and sugar mills with distilleries that discharging in liquid or solid form very high organic loads. In this context, is desirable to design effective strategies for the vinasse redirection, avoiding or mitigating the consequences of accidental spills or of an inadequate management. As an approximation to that effect, in our laboratory, we are evaluating the potential use of vinasse as substrate to support the growth and production of tenso-active compounds by regional actinobacteria. The concretion of this aim would allow proposing a strategy to achieve a more appropriate management of vinasse and minimize undesirable environmental impact. Finally, the redirecting of the final destination of this effluent could support the development of a profitable process for the production of microbial surfactants at industrial scale.

16.4 Concluding Remarks

Despite the limitations referred in this chapter concerning to the application of biosurfactants at large scale, the progressive entry of these biological molecules into the market and their application in remediation technologies is a concrete fact. Since cost of production is one of the key factors in the development of any biotechnological process, the use of agro-industrial waste as substrate for the biosurfactants synthesis appear to be promising in terms of supporting economical production of these biomolecules. In addition, the comprehensive utilization of waste derived from various industrial activities represents a new model of waste management, which reduces their negative environmental impact. It is hoped that the concerted efforts of the lines of research involved in this field would provide valuable progress on this topic in short-term. While these events occur, the first advances on the potential application of microbial emulsifiers in bioremediation technologies have already been achieved in our laboratory.

Acknowledgements We gratefully acknowledge the financial assistance of CIUNT D/401 and CONICET, Argentina.

References

Aniszewski E, Peixoto RS, Mota FF, Rosado A et al (2010) Bioemulsifier production by Microbacterium sp. strains isolated from mangrove and their application to remove cadmium and zinc from hazardous industrial residues. Braz J Microbiol 41:235–245

Banat IM, Franzetti A, Gandolfi I et al (2010) Microbial biosurfactants production, applications and future potential. Appl Microbiol Biotechnol 87:427–444

Calvo C, Manzanera M, Silva-Castro GA, Uad I, González-López J (2009) Application of bioemulsifiers in soil oil bioremediation processes. Future prospects. Sci Total Environ 407: 3634–3640

Cameotra SS, Singh P (2009) Synthesis of rhamnolipid biosurfactant and mode of hexadecane uptake by *Pseudomonas* species. Microbial Cell Fact 8:1–7

Cavalcante Barros FF, Resende Simiqueli AP, de Andrade CJ, Pastore GM (2013) Production of enzymes from agroindustrial wastes by biosurfactant-producing strains of *Bacillus subtilis*. Biotechnol Res Int. http://dx.doi.org/10.1155/2013/103960

Colin VL, Baigorí MD, Pera LM (2010) Bioemulsifier production by *Aspergillus niger* MYA 135: presumptive role of iron and phosphate on emulsifying ability. World J Microbiol Biotechnol 26:2291–2295

Colin VL, Castro MF, Amoroso MJ, Villegas LB (2013a) Bioemulifiers production by *Amycolatopsis tucumanensis* and their potential application for hexavalent chromium removal from soil. J Hazard Mater 261:577–583

Colin VL, Pereira CE, Villegas LB, Amoroso MJ, Abate CM (2013b) Production and partial characterization of bioemulsifer from a chromium-resistant actinobacterria. Chemosphera 90: 1372–1378

Cooper DG, Goldenberg BG (1987) Surface-active agents from two *Bacillus* species. Appl Environ Microbiol 53:224–229

Das K, Mukherjee AK (2007) Crude petroleum-oil biodegradation efficiency of *Bacillus subtilis* and *Pseudomonas aeruginosa* strains isolated from a petroleum-oil contaminated soil from North-East India. Bioresour Technol 98:1339–1345

Daverey A, Pakshirajan K (2010) Kinetics of growth and enhanced sophorolipids production by *Candida bombicola* using a low-cost fermentative medium. Appl Biochem Biotechnol 160: 2090–2101

de Gusmão C, Rufino R, Sarubbo L (2010) Laboratory production and characterization of a new biosurfactant from *Candida glabrata* UCP1002 cultivated in vegetable fat waste applied to the removal of hydrophobic contaminant. World J Microbiol Biotechnol 26:1683–1692

Fontes GC, Ramos NM, Amaral PFF, Nele M, Coelho MAZ (2012) Renewable resources for biosurfactant production by *Yarrowia lipolytica*. Braz J Chem Eng 29:483–493

Franzetti A, Caredda P, Ruggeri C, La Colla P, Tamburini E, Papacchini M, Bestetti G (2009) Potential applications of surface active compounds by *Gordonia* sp. strain BS29 in soil remediation technologies. Chemosphere 75:801–807

Franzetti A, Gandolfi I, Raimondi C, Bestetti G, Banat IM, Smyth TJ, Papacchini M, Cavallo M, Fracchia L (2012) Environmental fate, toxicity, characteristics and potential applications of novel bioemulsifiers produced by *Variovorax paradoxus* 7bCT5. Bioresour Technol 108: 245–251

Gnanamani A, Kavitha V, Radhakrishnan N, Rajakumar GS, Sekaran G, Mandal AB (2010) Microbial products (biosurfactant and extracellular chromate reductase) of marine microorganism is the potential agents reduce the oxidative stress induced by toxic heavy metals. Colloid Surf B 79:334–339

Guerra de Oliveira J, García-Cruz CH (2013) Properties of a biosurfactant produced by *Bacillus pumilus* using vinasse and waste frying oil as alternative carbon sources. Braz Arch Biol Technol 56:155–160

Gutierrez T, Shimmield T, Haidon C, Black K, Green DH (2008) Emulsifying and metal ion binding activity of a glycoprotein exopolimer produced by *Pseudoalteromonas* sp.strain TG12. Appl Environ Microbiol 15:4867–4876

Jayashree R, Vasudevan N, Chandrasekaran S (2006) Surfactants enhanced recovery of endosulfan from contaminated soils. Int J Environ Sci Technol 3:251–259

Juwarkar AA, Nair A, Dubey KV, Singh SK, Devotta S (2007) Biosurfactant technology for remediation of cadmium and lead contaminated soils. Chemosphere 68:1996–2002

Kang SW, Kim YB, Shin JD, Kim EK (2010) Enhanced biodegradation of hydrocarbons in soil by microbial biosurfactant, sophorolipid. Appl Biochem Biotechnol 160:780–790

Kosaric N (2001) Biosurfactants for soil bioremediation. Food Technol Biotechnol 39:295–304

Lai CC, Huang YC, Wei YH, Chang JS (2009) Biosurfactant-enhanced removal of total petroleum hydrocarbons from contaminated soil. J Hazard Mater 167:609–614

Lang S, Philp JC (1998) Surface-active lipids in *rhodococci*. A Van Leeuw Int J Genet Mol Microbiol 74:59–70

Luna JM, Rufino RD, Campos-Takaki GM, Leonie AS (2012) Properties of the biosurfactant produced by *Candida Sphaerica* cultivated in low-cost substrates. Chem Eng Trans 27:67–72

Madry N, Sprinkmeyer N, Pape H (1979) Regulation of tylosin synthesis in *Streptomyces*: effects of glucose analogs and inorganic phosphate. Eur J Appl Microbiol Biotechnol 7:365–370

Mulligan CN, Yong RN, Gibbs BF (2001) Surfactant-enhanced remediation of contaminated soil: a review. Eng Geol 60:371–380

Nguyen TT, Youssef NH, McInerney MJ, Sabatini DA (2008) Rhamnolipid biosurfactant mixtures for environmental remediation. Water Res 42:1735–1743

Nievas ML, Commendatore MG, Estevas JL, Bucalá V (2008) Biodegradation pattern of hydrocarbons from a fuel oil-type complex residue by an emulsifier-producing microbial consortium. J Hazard Mater 154:96–104

Pacwa-Plociniczak M, Plaza GA, Piotrowska-Seget Z, Cameotra SS (2011) Environmental applications of biosurfactants: recent advances. Int J Mol Sci 12:633–654

Pagilla KR, Sood A, Kim H (2002) *Gordonia* (*Nocardia*) *amarae* foaming due to biosurfactant production. Water Sci Technol 46:519–524

Patel MN, Gopinathan KP (1986) Lysozyme-sensitive bioemulsifier for immiscible organophosphorus pesticides. Appl Environ Microbiol 52:1224–1226

Reddy MS, Naresh B, Leela T, Prashanthi M, Madhusudhan NC, Dhanasri G, Devi P (2010) Biodegradation of phenanthrene with biosurfactant production by a new strain of *Brevibacillus* sp. Bioresour Technol 101:7980–7983

Saharan BS, Sahu RK, Sharma D (2011) A review on biosurfactants: fermentation. Current developments and perspectives. Genet Eng Biotechnol J 29:1–14

Sanches Carrocci J, Yuji Mori R, Cobra Guimarães OL, dos Santos Salazar RF, Fernandes de Oliveira M, de Castro Peixoto AL, Filho HJI (2012) Application of heterogenous catalysis with TiO_2 photo irradiated by sunlight and latter activated sludge system for the reduction of vinasse organic load. Engineering 4:46–76

Seghal Kiran G, Anto Thomas T, Selvin J, Sabarathnam B, Lipton AP (2010) Optimization and characterization of a new lipopeptide biosurfactant produced by marine *Brevibacterium aureum* MSA13 in solid state culture. Bioresour Technol 101:2389–2396

Sobrinho HBS, Rufino RD, Luna JM, Salgueiro AA, Campos-Takaki GM, Leite LFC, Sarubbo LA (2008) Utilization of two agroindustrial by-products for the production of a surfactant by *Candida sphaerica* UCP0995. Process Biochem 43:912–917

Souza FASD, Salgueiro AA, Albuquerque CDC (2012) Production of bioemulsifiers by *Yarrowia lipolytica* in sea water using diesel oil as the carbon source. Braz J Chem Eng 29:61–67

Takahashi M, Morita T, Wada K, Hirose N, Fukuoka T, Imura T, Kitamoto D (2011) Production of sophorolipid glycolipid biosurfactants from sugarcane molasses using *Starmerellabombicola* NBRC 10243. J Oleo Sci 60:267–273

Wang S, Mulligan CN (2009) Arsenic mobilization from mine tailings in the presence of a biosurfactant. Appl Geochem 24:928–935

Chapter 17
Use of Actinobacteria Consortia to Improve Methoxychlor Bioremediation in Different Contaminated Matrices

María S. Fuentes, Analía Alvarez, Juliana M. Saez, Claudia S. Benimeli, and María Julia Amoroso

Abstract Methoxychlor (MTX) is an organochlorine pesticide which has been banned in most countries; however, it is still being used in agricultural products and against mosquito. This pesticide has estrogenic activity and mimics endocrine hormone functions. Thus, it is important to analyze its behavior in different matrices.

Actinobacteria present the ability to degrade this pesticide, and its use in mixed cultures for bioremediation purposes can be advantageous.

Streptomyces spp. A3, A6, A12, A14, M7 and *S. coelicolor* A3 (2) were used as defined mixed cultures for MTX removal, after checking the absence of antagonistic effects among them. The consortium consisting of *Streptomyces* spp. A6, A12, A14,

M.S. Fuentes (✉) • J.M. Saez
Planta Piloto de Procesos Industriales Microbiológicos (PROIMI), CONICET, Av. Belgrano y Pasaje Caseros, 4000 Tucumán, Argentina
e-mail: soledadfs@gmail.com

A. Alvarez
Pilot Plant of Industrial and Microbiology, National Council of Scientific and Techniques, Natural Sciences College and Miguel Lillo Institute, Tucumán, Argentina
e-mail: alvanalia@gmail.com

C.S. Benimeli
Unidad de Administración Territorial, Centro Científico Tecnológico (CCT), CONICET, Crisóstomo Álvarez 722, 4000 Tucumán, Argentina

Universidad del Norte Santo Tomás de Aquino, 9 de Julio 165, 4000 Tucumán, Argentina

M.J. Amoroso
Planta Piloto de Procesos Industriales Microbiológicos (PROIMI), CONICET, Av. Belgrano y Pasaje Caseros, 4000 Tucumán, Argentina

Facultad de Bioquímica, Química y Farmacia, Universidad Nacional de Tucumán, 4000 Tucumán, Argentina

Universidad del Norte Santo Tomás de Aquino (UNSTA), 4000 Tucumán, Argentina

A. Alvarez, M.A. Polti (eds.), *Bioremediation in Latin America: Current Research and Perspectives*, DOI 10.1007/978-3-319-05738-5_17

and M7 was selected. This defined mixed culture was able to grow in slurry bioreactors with or without stimulation, in the presence of MTX and also to remove it from stimulated and non-stimulated bioreactors. An increase about 10 % in MTX removal was observed in stimulated slurry bioreactors.

MTX removal in soil was 56.4 %. Maximum microbial growth and the absence of stationary growth phase were both observed in soil when the consortium was grown in the presence of MTX, while the opposite was observed in soil without MTX.

When comparing soil and slurries, similar removal percentages values can be observed but at a time almost three times higher in soil. Indeed, when analyzing ex situ bioremediation by slurries bioreactors, reduced processing times can be achieved, compared to in situ bioremediation. However, it is important to analyze the costs and benefits involved in using either bioremediation technique.

17.1 Introduction

Among many pollutants of diverse origin and nature, organochlorine pesticides (OPs) have been of great concern, mainly because of their occurrence in high concentrations even in remote ecosystems, despite their bans on production and usage. These OPs, introduced for the first time to the environment after the Second World War, were extensively used in agriculture and industries. These compounds present chemical stability and slow biodegradation and, as a result of these properties, are ubiquitous pollutants (Borrell et al. 2010). In Argentina, the extensive use of this kind of compounds has left residues in the environment, as was detected in Salí River, Tucumán (Chaile et al. 1999), and in soil samples of an illegal OPs storage in the province of Santiago del Estero (Fuentes et al. 2010). On the other hand, Arias et al. (2011) found residues of OPs in superficial sediment samples, collected from Bahia Blanca Estuary, in the southwest of Buenos Aires (Argentina), in a period between April 2005 and March 2009. Among OPs, methoxychlor [2,2-bis(p-methoxyphenyl)-1,1,1-trichloroethane] (MTX) is developed to be used as a replacement of DDT [1,1,1-trichloro-2,2-bis(p-chlorophenyl) ethane], which was internationally prohibited since 1970s due to its high toxicity (Stuchal et al. 2006). MTX has been banned in most countries; however it is still being used on agricultural products against several insect pests (Basavarajappa et al. 2011) and also in tropical environments, where mosquito-borne malaria and typhus are serious health problems (Llados et al. 2002).

Although MTX has relatively low toxicity and short half-life (Metcalf 1976), there is considerable concern for its exposure because of its estrogenic activity (Staub et al. 2002; Fort et al. 2004). This toxic compound mimics endocrine hormone functions in the body acting as disrupting chemicals, interfering with the normal endocrine activity in humans and wildlife (Crisp et al. 1998). MTX is considered to be pro-estrogenic, but its metabolites mono- and bis-hydroxy exhibit greater estrogenic properties than the parental compound (Lee et al. 2006).

On the other hand, it is important to analyze the behavior of MTX in different systems such as water, slurry, and soil. In soil systems, the movement and fate of these chemical compounds involves complex mechanisms, including volatilization, adsorption, chemical and biological degradations. These mechanisms can differ depending on soil texture and organic matter content thereof. Of all these processes, adsorption, desorption, and degradation are the key processes regulating the concentrations of contaminants in soils (Dragun 1998). Also the fate of these compounds in a natural water system is highly dependent upon their sorption behavior. In addition to affecting the physical movement of pollutants, the sorption phenomenon can be involved directly in pollutant degradation via surface-associated chemical processes (Karickhoff et al. 1978).

Among gram-positive microorganisms, actinobacteria have a great potential for biodegradation of organic and inorganic toxic compounds. Previous studies demonstrated the ability of different genera of actinobacteria to degrade pesticides (Fuentes et al. 2010; Karn et al. 2011; Briceño et al. 2012). Fuentes et al. (2010) isolated 12 actinobacteria from OPs contaminated samples, with the ability to grow in the presence of MTX among other OPs, and also with the ability to remove or degrade them from culture medium. There is little information about the use of mixed defined cultures of actinobacteria for bioremediation of recalcitrant compounds as OPs (Fuentes et al. 2011, 2013a, b). The biodiversity of mixed cultures can enhance the environmental survival and increase the number of catabolic pathways available for the biodegradation of xenobiotic compounds (Siripattanakul et al. 2009; Yang et al. 2010). If a defined mixed culture is successfully constructed, it can be facilitate the examination of general characteristics of all its members in pure culture and the monitoring of the dynamics of the members in the community throughout a given cultivation period. Furthermore, by constructing a "knockout mixed culture," in which one member is eliminated from the original mixed culture, the role played by the eliminated member in situ and its impact on the other members of the community, could be evaluated (Kato et al. 2005).

17.2 Methoxychlor Biodegradation by Actinobacteria Defined Consortia in Liquid Systems

Twelve indigenous colonies belonging to the actinobacteria group were isolated from contaminated soil samples from a village named Argentina in the province of Santiago del Estero, where more than 30 t of obsolete pesticides (including MTX) were found in 1994. Eleven of these 12 isolates belonged to *Streptomyces* genus, and one belonged to *Micromonospora* genus (Fuentes et al. 2010). These strains and also *Streptomyces* sp. M7, isolated by Benimeli et al. (2003) with ability to degrade lindane (Benimeli et al. 2006) and *Streptomyces coelicolor* A3 (2), from the German Collection of Microorganisms and Cell Cultures (DSMZ), were cultured in liquid minimal medium (MM) containing MTX (1.66 mg L^{-1}). In these assay conditions, Fuentes et al. (2010) demonstrated that all these strains were able to grow and remove this pesticide. The release of chloride ions was analyzed as an indirect determination of MTX degradation. As a result, the presence of chloride ions was revealed in only two samples, although all the actinobacteria assayed were able to remove it from the culture medium (Fuentes et al. 2010). These results raise the hypothesis that some of the studied organisms would not be capable of degrading the pesticide, but to accumulate it inside. The bioaccumulation is a microbial transformation in which the pesticide is incorporated and accumulated by organisms both by active or passive uptake processes (Müller et al. 2007). Analyzing these results, no linear relationship was observed between microbial growth and MTX removal. Thus, four strains were then selected based on the ratio between these two parameters (MTX residual/dry weight) (Fuentes 2012). The selected strains (*Streptomyces* spp. A3, A6, A12, and A14), plus *Streptomyces* sp. M7 and *Streptomyces coelicolor* A3 (2) were used to evaluate their ability to grow as a defined mixed culture in the presence of MTX (Fuentes et al. 2013a).

From an applied perspective for bioremediation purposes, the use of a mixed culture rather than a pure culture in a polluted area is more advantageous. After checking the absence of antagonistic effects among the studied *Streptomyces* strains (Fuentes 2012), all of them were grown on pure and mixed cultures in MM added with MTX (1.66 mg L^{-1}), in order to evaluate their pesticide removal ability and the presence of specific dechlorinase activity (SDA). The six strains in study were assayed with all possible combinations, obtaining different consortia. All of them were able to grow and to remove MTX from the culture medium and/or degrade it, showing lower residual pesticide values in the culture supernatants than the initial concentration added to the medium. These diverse catabolic activities are due to the presence of catabolic genes and enzymes (Siripattanakul et al. 2009). Fuentes (2012) and Fuentes et al. (2013a) detected the presence of SDA in cell-free extracts in most of the consortia. On these bases, we assumed that MTX is predominantly degraded by dechlorination and dehydrogenation processes. Indeed, other studies also support the fact that actinobacteria were able to release chloride ions in liquid culture media contaminated with OPs (Benimeli et al. 2006; Fuentes et al. 2010). Moreover, the molecule of MTX has three chlorine atoms, and thus, the dechlorination is a very significant step in its degradation process (Fetzner and Lingens 1994). The MTX removal by the pure and mixed cultures was efficient, obtaining removal percentages between 72 and 100 % (Fuentes 2012). These results showed that there is not linear relationship between the residual MTX concentration (RM) and SDA. On this base it was analyzed the ratio between both parameters (RM/SDA) in order to select the best mixed culture. Table 17.1 shows the results of the ratio mentioned above for each consortium. Three mixed cultures presented the lesser values for this ratio, so they were found to be the most efficient consortia. Considering that no statistical significant differences were found in the RM/SDA among these three mixed culture ($P<0.05$), one of them consisting of *Streptomyces* spp. A6, A12, A14, and M7 was chosen. In addition this mixed culture showed the higher SDA among the three mixed cultures with the lesser values of the mentioned ratio (Fuentes 2012; Fuentes et al. 2013a).

Table 17.1 Ratio between residual methoxychlor concentration and specific dechlorinase activity in mixed cultures of *Streptomyces* spp.

Mixed cultures	RM/SDA × 10^2
A6-M7	3.62
A3-M7	0.58
A12-M7	1.12
A6-*Sc*	1.31
A3-*Sc*	0.51
A3-A12	1.10
A3-A6-A12	0.31
A3-A6-A14	0.35
A3-A12-M7	1.19
A3-A12-*Sc*	2.76
A3-A14-*Sc*	1.56
A6-A14-*Sc*	1.17
M7-A6-*Sc*	0.91
M7-A12-*Sc*	1.73
Sc-A14-M7	0.45
M7-A12-A14	1.56
A6-A12-*Sc*	2.05
A3-M7-*Sc*	0.42
A3-A6-*Sc*	2.26
A3-A14-M7	3.40
A6-A14-M7	0.14
A6-A12-M7	0.22
A3-A6-A12-A14	1.35
A3-A6-M7-*Sc*	0.13
A3-A6-*Sc*-A14	0.37
M7-A6-A14-*Sc*	0.28
A12-M7-A3-*Sc*	15.02
A12-A3-A14-*Sc*	0.44
A3-A6-A12-*Sc*	0.10
M7-A3-A14-*Sc*	0.17
A12-M7-A6-*Sc*	12.22
A3-A6-M7-A14	0.43
A3-A6-A12-M7	0.03
A12-A14-M7-A3	0
A6-A12-A14-M7[a]	0.04
A3-A12-A14-M7-*Sc*	0.21
A3-A6-A14-M7-*Sc*	0.11
A3-A6-A12-A14-M7	2.74
A3-A6-A12-A14-M7-Sc	1.69

Extracted from (Fuentes et al. 2013a)

RM residual methoxychlor concentration (mg L^{-1}), *SDA* specific dechlorinase activity (EU mg^{-1} protein), *EU* enzimatic unit, defined as the amount of chloride ions (μmol) released in 1 h, *Sc Streptomyces coelicolor* A3 (2)

[a]The most efficient mixed culture in methoxychlor biodegradation

17.3 Evaluation of the Ability of a *Streptomyces* spp. Consortium to Grow and Remove Methoxychlor on Slurry and Soil Systems

The bioremediation process depends on the metabolic potential of microorganisms to detoxify or transform the pollutant molecules, which is dependent of both accessibility and bioavailability (Antizar-Ladislao 2010). Following entry into the soil environment, pollutants rapidly bind to the mineral and organic matter (solid phases) via a combination of physical and chemical processes. The ability of soils to release (desorb) pollutants determines its susceptibility to microbial degradation, thereby influencing effectiveness of the treatment (Megharaj et al. 2011). Bioremediation processes have been classified in three broad categories, according to place and soil handling/conditioning: in situ, ad situ, and ex situ (Robles-González et al. 2008).

Besides, depending on the state of the contaminant to be removed, ex situ bioremediation is classified as solid phase system (including land treatment and soil piles) or as slurry phase systems (including solid/liquid suspensions in bioreactors) (Kumar et al. 2011). The slurry bioreactor has interesting and distinctive features, such as soil treatment in aqueous suspension and the mechanical or pneumatic mixing provided. These characteristics lead to several advantages in the process, such as the increased mass transfer rates, the increase in the contact in microorganisms/pollutant/nutrients and the increase of the rates of pollutant biodegradation compared to in situ bioremediation or to ad situ solid phase biotreatment. It can also achieve times of treatment significantly shorter, achieve the control and optimization of environmental parameters and increase the pollutant desorption and availability through the addition of surfactants and solvents (Robles-González et al. 2008). On these bases, we focused on the study of the ability of a microbial defined consortium to grow in the presence of MTX and to remove it from slurry and from soil systems, as a potential tool for its use in multiple bioremediation strategies.

The growth of the selected consortium (see above) and its MTX removal in slurry bioreactor (SB) operated in batch with different slurries formulations, allow seeing the different behaviors of the microbial consortium with or without stimulation. The soil/water (SB-water) composition was used for non-stimulated assays and soil/trypteine soy broth (SB-TSB) composition, for stimulated assays. In both cases, the systems were contaminated with MTX (1.66 mg L^{-1}) and incubated for 7 days at 30 °C (Fuentes et al. 2013a).

The *Streptomyces* spp. consortium was able to grow in both conditions (with or without stimulation) in the presence of MTX. In SB-TSB spiked with MTX the consortium showed maximum microbial growth at the third day of the experiment (Fig. 17.1a). The biomass value was statistically different to the biomass reached on SB-water ($P<0.05$) at the same culture time (Fig. 17.1b). Moreover, no statistically significant differences were found between biomass reached with and without nutrient supply and with and without MTX, at the end of the experiments ($P>0.05$) (Fig. 17.1a, b) (Fuentes et al. 2013a). It is possible that the rich culture medium (TSB) plus the organic matter own of soil and the pesticide may have provided extra carbon sources to

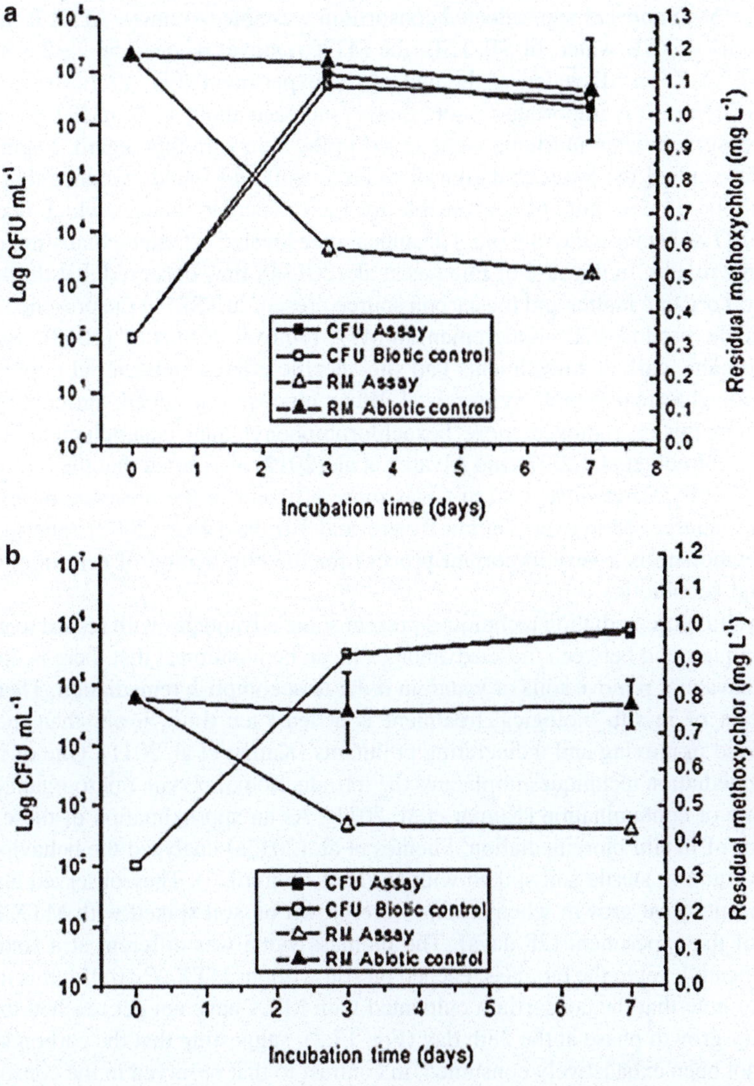

Fig. 17.1 Residual methoxychlor concentration (mg L^{-1}) and microbial enumeration of *Streptomyces* spp. consortium (CFU mL^{-1}) growing on slurry at different incubation times. RM: residual metoxychlor concentration; CFU: colony forming units; Assay: consortium cultivated with methoxychlor (MTX); Biotic control: consortium cultivated without MTX; Abiotic control: slurry spiked with MTX, without microorganisms. (**a**) Slurry formulated with TSB and (**b**) slurry formulated with water. (Extracted from Fuentes et al. 2013a)

support the higher microbial growth obtained at the third day, but having exhausted the extra carbon source the consortium growth stabilizes in both cases (Fuentes et al. 2013a). The favorable performance of the consortium to grow in the presence of MTX may be due to a selective pressure of the polluted environment from which these actinobacteria were isolated (Fuentes et al. 2010), as already noted in MM.

The *Streptomyces* spp. selected consortium was able to remove MTX from both SB-TSB and SB-water. In SB-TSB, the MTX removal reached 56.2 ± 2.3 %, and 45.6 ± 7.4 % on SB-water, at the end of the experiment (Fig. 17.1a, b) (Fuentes et al. 2013a). It is important to note that an increase about 10 % in MTX removal was observed when nutrients were added in the slurry. In this regard, Fogel et al. (1982) studied the biodegradation of MTX in soil and found evidence that a co-metabolic process may be responsible for the molecular changes which occurred with MTX, because the rate of its disappearance leveled off after exhaustion of soil organic matter. In the case of Fuentes et al. (2013a), they observed that the sum of the soil organic matter and the carbon source present in TSB, in the presence of the pesticide, produces an improvement in MTX removal. Luo et al. (2008) observed that organic carbon amendments had site-specific effects on bacterial populations and polychlorinated biphenyl removal. When glucose was added, agricultural soil microorganisms removed more hexachlorobiphenyl than unsupplemented samples. Benimeli et al. (2007) and Alvarez et al. (2012) also noted that the removal of other OP by *Streptomyces* strains was more efficient in the presence of an extra carbon source. Moreover, Garcia-Rivero and Peralta-Pérez (2008) reported that cometabolism is a very important process for the elimination of certain environmental xenobiotics.

In situ bioremediation techniques present some advantages with regard to ex situ techniques, and between these advantages it can be mentioned that there is no need to excavate or remove soils or water, in order to accomplish remediation. Therefore the cost of in situ biological treatment is at least ten times lower than the cost involved in moving and incinerating pollutants (Kumar et al. 2011). One of in situ bioremediation techniques implicates the introduction of certain microorganisms to the site of contamination (Kumar et al. 2011). As an approximation of these techniques of in situ bioremediation, Fuentes et al. (2013a) analyzed the behavior of a consortium in sterile soil spiked with MTX (1.66 mg L^{-1}). They observed a maximum microbial growth when strains were grown on soil spiked with MTX at the end of the experiment (28 days). The biomass value was different at a statistical significant level to the biomass reached on soil without MTX ($P < 0.05$). It is important to note that the consortium cultivated with MTX have not yet reached the stationary growth phase at the 28th day (Fig. 17.2), suggesting that the carbon source had not been exhaustively consumed, in contrast to that observed in the consortium cultivated in soil without MTX. It seems that pesticide provides an additional source of carbon to support the cells growth, besides to own soil organic matter (Fuentes et al. 2013a). Regarding the pesticide removal, Fuentes et al. (2013a) observed a substantial declination of the residual MTX along the assay with the consortium, reaching a 56.4 % of pesticide removal. Only a small fraction of MTX disappeared from the uninoculated control (Fig. 17.2) probably because of the aggregation and/or adsorption of the pesticide to soil particles. It is important to note that the soil has the ability to retain the pesticide principally by adsorption processes (Krishna and Philip 2008) which depends on both the chemical properties of the pesticide and the properties of the soil (Edwards 1975).

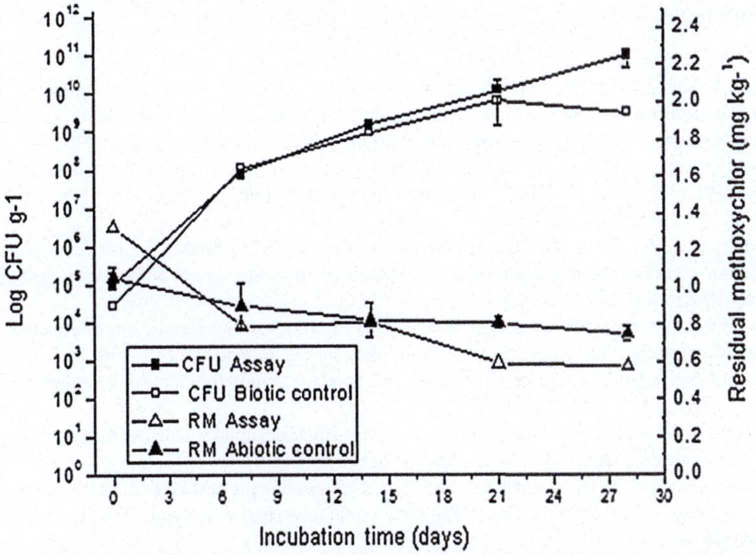

Fig. 17.2 Residual methoxychlor concentration (mg L^{-1}) and microbial enumeration of *Streptomyces* spp. consortium (CFU mL^{-1}) growing on sterile soil at different incubation times. (Extracted from Fuentes et al. 2013a)

When comparing both matrices, soil and slurry, it can be observed that SBs reached MTX removal percentages of 56 and 46 % in SB-TSB and SB-water, respectively at the seventh day of the assay, while in soil the pesticide removal reached a similar value at a time almost three times higher. In fact, the SBs have been described as appropriate for bioremediation of contaminated sites that require a short remediation time because of regulatory or other pressures. Indeed, Robles-González et al. (2008) also reported that SBs allow reducing the processing times, respect to the biodegradation in situ (soil). However, it is important to assess the costs and benefits involved in using either bioremediation technique.

17.4 Concluding Remarks

Four actinobacteria isolated from environments polluted with organochlorine pesticides, were selected to perform microbial consortia assays. These strains were able to grow and to remove MTX. A defined consortium consisting of *Streptomyces* spp. A6, A12, A14, and M7 was selected based on the lowest ratio between RM and SDA. This consortium was able to grow and remove MTX in slurries and soil systems. In the mentioned systems, MTX removal percentages ranged between 46 and 56 %. Hence, these results have given relevant evidence that actinobacteria defined consortia may be suitable for development of in situ and ex situ bioremediation methods; particularly it could be an excellent strategy for bioremediation of sites polluted with MTX.

References

Alvarez A, Yañez ML, Benimeli CS et al (2012) Maize plants (*Zea mays*) root exudates enhance lindane removal by native *Streptomyces* strains. Int Biodeter Biodegr 66:14–18

Antizar-Ladislao B (2010) Bioremediation: working with bacteria. Elements 6:389–394

Arias AH, Pereyra MT, Marcovecchio JE (2011) Multi-year monitoring of estuarine sediments as ultimate sink for DDT, HCH, and other organochlorinated pesticides in Argentina. Environ Monit Assess 172:17–32

Basavarajappa MS, Craig ZR, Hernández-Ochoa I et al (2011) Methoxychlor reduces estradiol levels by altering steroidogenesis and metabolism in mouse antral follicles *in vitro*. Toxicol Appl Pharmacol 253:161–169

Benimeli CS, Amoroso MJ, Chaile AP et al (2003) Isolation of four aquatic streptomycetes strains capable of growth on organochlorine pesticides. Biores Technol 89:133–138

Benimeli CS, Castro GR, Chaile AP et al (2006) Lindane removal induction by *Streptomyces* sp. M7. J Basic Microbiol 46:348–357

Benimeli CS, Castro GR, Chaile AP et al (2007) Lindane uptake and degradation by aquatic *Streptomyces* sp. strain M7. Int Biodeter Biodegr 59:148–155

Borrell A, Garcia-Solà A, Aguilar A et al (2010) Organochlorine residues in South American Sea Lions, *Otaria flavescens* (Shaw, 1800): bioaccumulation and time trends. Bull Environ Contam Toxicol 84:731–737

Briceño G, Fuentes MS, Palma G et al (2012) Chlorpyrifos biodegradation and 3,5,6-trichloro-2-pyridinol production by actinobacteria isolated from soil. Int Biodeter Biodegr 73:1–7

Chaile AP, Romero N, Amoroso MJ et al (1999) Organochlorine pesticides in Salí River. Tucumán-Argentina. Rev Bol Ecol 6:203–209

Crisp TM, Clegg ED, Cooper RL et al (1998) Environmental endocrine disruption: an effects assessment and analysis. Environ Health Perspect 106:11–56

Dragun J (1998) Adsorption and mobility of organic chemicals. In: Dragun J (ed) The soil chemistry of hazardous materials. Amherst Scientific Publishers, Amherst, MA

Edwards CA (1975) Factors that affect the persistence of pesticides in plants and soils. Pure Appl Chem 42:39–56

Fetzner S, Lingens F (1994) Bacterial dehalogenases: biochemistry, genetics and biotechnological applications. Microbiol Rev 58:641–685

Fogel S, Lancione RL, Sewall AE (1982) Enhanced biodegradation of methoxychlor in soil under sequential environmental conditions. Appl Environ Microbiol 44:113–120

Fort DJ, Guiney PD, Weeks JA et al (2004) Effect of methoxychlor on various life stages of *Xenopus laevis*. Toxicol Sci 81:454–466

Fuentes MS (2012) Degradación aeróbica de plaguicidas organoclorados (clordano, metoxicloro y lindano) por actinomycetes autóctonos en sistemas líquidos, suelos y fangos. Ph.D. thesis. Universidad Nacional de Tucumán, Argentina

Fuentes MS, Benimeli CS, Cuozzo SA et al (2010) Isolation of pesticide-degrading actinomycetes from a contaminated site: bacterial growth, removal and dechlorination of organochlorine pesticides. Int Biodeter Biodegr 64:434–441

Fuentes MS, Sáez JM, Benimeli CS et al (2011) Lindane biodegradation by defined consortia of indigenous *Streptomyces* strains. Water Air Soil Pollut 222:217–231

Fuentes MS, Alvarez A, Saez JM et al (2013a) Methoxychlor bioremediation by defined consortium of environmental *Streptomyces* strains. Int J Environ Sci Technol. doi:10.1007/s13762-013-0314-0

Fuentes MS, Briceño GE, Saez JM et al (2013b) Enhanced removal of a pesticides mixture by single cultures and consortia of free and immobilized *Streptomyces* strains. BioMed Res Int; doi:10.1155/2013/392573

Garcia-Rivero M, Peralta-Pérez MR (2008) Cometabolism in the biodegradation of hydrocarbons. Rev Mex Ing Biomed 7:1–12

Karickhoff SW, Brown DS, Scott TA (1978) Sorption of hydrophobic pollutants on natural sediments. Environmental Research Laboratory, US Environmental Protection Agency, Athens, GA

Karn SK, Chakrabarti SK, Reddy MS (2011) Degradation of pentachlorophenol by *Kocuria* sp. CL2 isolated from secondary sludge of pulp and papermill. Biodegradation 22:63–69

Kato S, Haruta S, Zong Jun C et al (2005) Stable coexistence of five bacterial strains as a cellulose-degrading community. Appl Environ Microbiol 71:7099–7106

Krishna KR, Philip L (2008) Adsorption and desorption characteristics of lindane, carbofuran and methyl parathion on various Indian soils. J Hazard Mater 160:559–567

Kumar A, Bisht BS, Joshi VD et al (2011) Review on bioremediation of polluted environment: a management tool. Int J Environ Sci 1:1079–1093

Lee SM, Lee JW, Park KR et al (2006) Biodegradation of methoxychlor and its metabolites by the white rot fungus *Stereum hirsutum* related to the inactivation of estrogenic activity. J Environ Sci Health B 41:385–397

Llados F, Sage G, Citra M et al (2002) Toxicological profile for DDT, DDE and DDD (Update), TP-35. Agency for toxic substances and disease registry. US Department of Health and Human Services, Atlanta, GA

Luo W, D'Angelo EM, Coyne MS (2008) Organic carbon effects on aerobic polychlorinated biphenyl removal and bacterial community composition in soils and sediments. Chemosphere 70:364–373

Megharaj M, Ramakrishnan B, Venkateswarlu K et al (2011) Bioremediation approaches for organic pollutants: a critical perspective. Environ Int 37:1362–1375

Metcalf RL (1976) The future of insecticides. In: McKelevey JJ, Metcalf RL (eds) Organochlorine insecticides, survey, and prospects. Wiley and Sons, New York, pp 223–285

Müller K, Magesan GN, Bolan NS (2007) A critical review of the influence of effluent irrigation on the fate of pesticides in soil. Agric Ecosyst Environ 120:93–116

Robles-González IV, Fava F, Poggi-Varaldo HM (2008) A review on slurry bioreactors for bioremediation of soils and sediments. Microb Cell Fact 7:1–16

Siripattanakul S, Wirojanagud W, McEvoy J et al (2009) Atrazine degradation by stable mixed cultures enriched from agricultural soil and their characterization. J Appl Microbiol 106: 986–992

Staub C, Hardy VB, Chapin RE et al (2002) The hidden effect of estrogenic/antiandrogenic methoxychlor on spermatogenesis. Toxicol Appl Pharmacol 180:129–135

Stuchal LD, Kleinow KM, Stegeman JJ et al (2006) Demethylation of the pesticide methoxychlor in liver and intestine from untreated, methoxychlor-treated, and 3-methylcholanthrene-treated channel catfish (*Ictalurus punctatus*): evidence for roles of cyp1 and cyp3a family isozymes. Drug Metab Dispos 34:932–938

Yang C, Li Y, Zhang K et al (2010) Atrazine degradation by a simple consortium of *Klebsiella* sp. A1 and *Comamonas* sp. A2 in nitrogen enriched medium. Biodegradation 21:97–105

Chapter 18
Biodegradation of α- and β-Hexachlorocyclohexane by Indigenous Actinobacteria

Pedro Sineli, Claudia S. Benimeli, María Julia Amoroso, and Sergio A. Cuozzo

Abstract The organochlorine pesticide lindane (γ-HCH) and its non-insecticidal isomers α-, β-, and δ- continue to pose serious environmental and health concerns, although their use has been restricted or completely banned for decades. The present chapter reports the first results on the ability of Actinobacteria strains, isolated from a HCH-polluted site, to grow in a minimal medium containing α- or β-HCH (8.3 mg L^{-1}) as sole source of carbon. Growth of cultures and HCHs degradation by *Streptomyces* sp. M7 was investigated after 1, 4, and 7 days of incubation by dry weight and GC with μECD detection, respectively. *Streptomyces* sp. M7 was able to metabolize the HCHs: removed up to 100 % of α-HCH and 55 % of β-HCH under optimal culture conditions: 30 °C, pH 7 and the isomers maxima concentration of

P. Sineli
Planta Piloto de Procesos Industriales Microbiológicos (PROIMI-CONICET), Avenida Belgrano y Pasaje Caseros, T40001MVB, Tucumán, Argentina

C.S. Benimeli
Unidad de Administración Territorial, Centro Científico Tecnológico, CONICET-Tucumán, Crisóstomo Álvarez 722, 4000, Tucumán, Argentina

Universidad del Norte Santo Tomás de Aquino (UNSTA), 9 de Julio 165, 4000 Tucumán, Argentina

M.J. Amoroso
Planta Piloto de Procesos Industriales Microbiológicos (PROIMI), CONICET, Av. Belgrano y Pasaje Caseros, 4000 Tucumán, Argentina

Facultad de Bioquímica, Química y Farmacia, Universidad Nacional de Tucumán, 4000 Tucumán, Argentina

Universidad del Norte Santo Tomás de Aquino (UNSTA), 4000 Tucumán, Argentina

S.A. Cuozzo (✉)
Planta Piloto de Procesos Industriales Microbiológicos (PROIMI-CONICET), Avenida Belgrano y Pasaje Caseros, T40001MVB, Tucumán, Argentina

Facultad de Ciencias Naturales e Instituto Miguel Lillo, Universidad Nacional de Tucumán, Miguel Lillo 205, 4000 Tucumán, Argentina
e-mail: scuozzo@proimi.org.ar

A. Alvarez, M.A. Polti (eds.), *Bioremediation in Latin America: Current Research and Perspectives*, DOI 10.1007/978-3-319-05738-5_18

8.3 mg L^{-1}. Also, *Streptomyces* sp. M7 showed greater overall growth in the presence of α-HCH than β-HCH, in concordance with the total or partial removal of the α-, β-, HCH isomers respectively.

18.1 Introduction

The organochlorine pesticide Lindane (γ-hexachlorocyclohexane, γ-HCH) is a widespread contaminant in several environments and food chain and the removal of pure γ-HCH was inefficient, because each ton of γ-HCH generated about 8–12 tons of waste HCH isomers (Vijgen 2006) and still persist as hazardous wastes in many locations. Therefore since August 2009 it has been included as new persistent organic pollutant in the Stockholm Convention list (POPs list) along with α- and β-HCH isomers (Vijgen et al. 2011). The theoretical HCH isomers are eight, but the technical formulation consists in α-, β-, γ-, and δ-HCH, where γ-HCH is the most abundant isomer and the only one that has specific insecticidal properties (Fig. 18.1).

Due to extensive use in the past for protection of crops and control of vector borne diseases; its residues have accumulated at the applied sites, where imparts toxicity (Willett et al. 1981). Although γ-HCH use has been restricted or banned in most countries after the 1990s, γ-HCH and its no insecticidal isomers (α-, β-, δ-, and ε-HCH) continue to pose serious environmental and health concerns (Willett et al. 1998; Pavlíková et al. 2012). Furthermore, in the environment there are abiotic and biotic interconversion between the different HCH isomers, which causes an accumulation and persistence of the more permanent forms: that is to say β and δ isomers (Engst et al. 1979; Phillips et al. 2005). Several HCH-degrading microorganisms have been characterized from different parts of world and the pathway for HCH aerobic degradation has been studied (Nagata et al. 2007; Lal et al. 2010; Fuentes et al. 2010; Cuozzo et al. 2009).

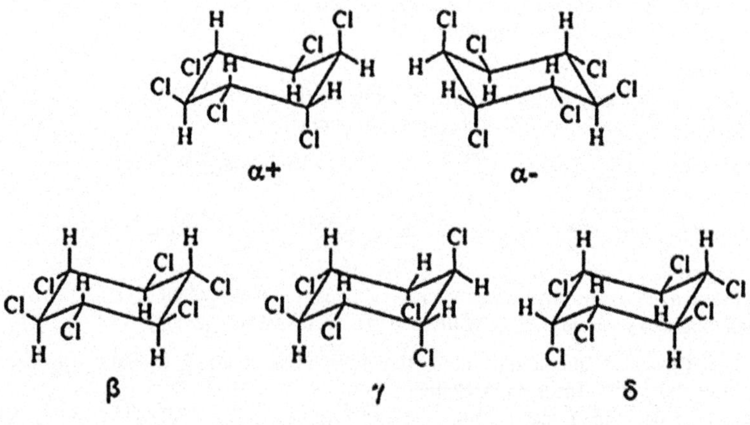

Fig. 18.1 Enantiomers and isomers of hexachlorocyclohexane (HCH)

Bioremediation by microbial activity is considered a potential strategy for the long-term in situ attenuation of HCH contamination and therefore, both single bacteria (Senoo and Wada 1989; Kaur et al. 2011) and bacterial consortia (Pesce and Wunderlin 2004; Elcey and Kunhi 2010) capable of degrading HCH isomers under both aerobic (Senoo and Wada 1989; Pesce and Wunderlin 2004; Elcey and Kunhi 2010; Nagata et al. 2007) and anaerobic (Buser and Muller 1995; van Doesburg et al. 2005; Jagnow et al. 1977) conditions have received considerable attention. Many HCH degrading aerobic bacteria have been isolated and characterized (Phillips et al. 2005; Lal et al. 2010; Camacho-Pérez et al. 2012; Benimeli et al. 2006) and most of them have been described to belong to the genus *Sphingomonas* (Senoo and Wada 1989; Johri et al. 1998; Manickam et al. 2008; Nagata et al. 1999, 2007; Lal et al. 2008; Raina et al. 2008) and the Actinobacteria Benimeli et al. (2006) . The aerobic metabolism of HCHs by *Sphingomonas* has been studied in several works and correlated to the enzymatic system of the bacteria species (Phillips et al. 2005; Camacho-Pérez et al. 2012; Raina et al. 2008; Nagata et al. 1999; Tabata et al. 2011). Therefore the HCH biodegradation consists in a progressive elimination of the chlorine and hydrogen atoms, with the introduction of double bonds; in some instances the chlorine atoms are substituted by hydroxyls (Raina et al. 2007, 2008; Wu et al. 2007). Because of their molecular geometry, α-, β-, γ-, and δ-HCH have different degradation pathways; with β- and δ-HCH being the most recalcitrant to bacterial attack. Pentachlorocyclohexenes are generally considered the initial metabolites, which are further degraded in different ways depending on the isomeric form and the bacterial strain. Several metabolites have been reported in aerobic bacterial cultures, among which the chlorobenzenes and chlorophenols are the more persistent compounds and in some cases are dead-end products. Some hydroxylated metabolites are also quite resistant to further bacterial degradation, as they have been detected in soil and water at a former pesticides production site (Raina et al. 2007).

The aim of this chapter was to study the isomers aerobic degradation of α- and β-HCH as a carbon source by regional actinobacteria. In this work were identified HCH isomers removal levels by *Streptomyces* sp. M7 grown in minimal medium (MM) supplemented with α- or β-HCH. The residual content in the supernatants were determinate after 7 incubation days. The methodologies employed in the different test were: Chloride ion released, growth of *Streptomyces* sp. M7 (dry weight) and HCH isomers residual by gas chromatography analysis with electron micro-capture.

18.2 Strains Selection Capable of HCH Isomers Degradation

The first analysis was focused on the selection of actinobacteria able to growth and degrades HCH isomers. Five actinobacteria strains, previously isolated from a pesticides contaminated soil from Argentina county, the province of Santiago del Estero, Argentina; and the reference strain *Streptomyces coelicolor* A3 (2) were

Table 18.1 Comparison of the growth rates in the presence of the α-/β-/δ-HCH in comparison to glucose as carbon source

Strain	Carbon source		
	α-HCH	β-HCH	δ-HCH
S. coelicolor A3 (2)	–	20	–
A8	50	50	–
A11	11	–	–
M7	38	11	–
A2	100	47	–
A5	20	–	–

selected (Fuentes et al. 2010). All strains were grown in Minimal Medium (MM), supplemented with each of the different HCH isomers (-α,-β, and δ-HCH) added aseptically to the autoclaved MM at a final concentration of 1.66 mg L^{-1}. All cultures were incubated on a rotatory shaker (100 rpm) at 30 °C, for 7 days. The growth was estimated by biomass, washing the pellets with 25 mM Tris–EDTA buffer (pH 8.0) and drying to constant weight at 105 °C.

After 7 day of incubation, it was observed that actinobacteria strains such as *Streptomyces* sp. A2, A8, and M7 were able to grow in the presence of α-HCH as the only carbon source. When the MM was supplemented with β-HCH, strains capable to grow in this condition were *Streptomyces* sp. A2 and A8 only. However, *Streptomyces* sp. A2, A8 and M7 did not show growth in the presence of δ-HCH. Therefore the best performances were A2 and A8 (Table 18.1). It is the first time evidence for the aerobic degradation of HCH-isomers especially β-HCH which is the most recalcitrant of the HCH isomers produced by indigenous actinobacteria.

According to the Table 18.1 the *Streptomyces* sp. A2 and A8 showed the best growth; however, *Streptomyces* sp. M7, was selected by continue the studies since it had the best performance with γ-HCH unlike the other strains studied (data not show).

18.3 Study of the Effect of the pH and HCH Isomers Initial Concentration on Microbial Growth

The growth of *Streptomyces* sp. M7 with β-HCH isomer as sole carbon source, was able to grow at rates of 0.41 mg mL^{-1} at 24 h incubation time and initial pH 9, overcome growth one time compared to that determined at initial pH 7, indicating that this was the optimal pH for *Streptomyces* sp. M7. On the other hand, when using the α-HCH isomer as a sole carbon source, the dry weight was 0.31 mg mL^{-1} to initial pH of 7 to fourth incubation day, the same values was obtained at pH 9 but after seventh incubation days, indicated that the best growth was at neutral pH (Fig. 18.2a).

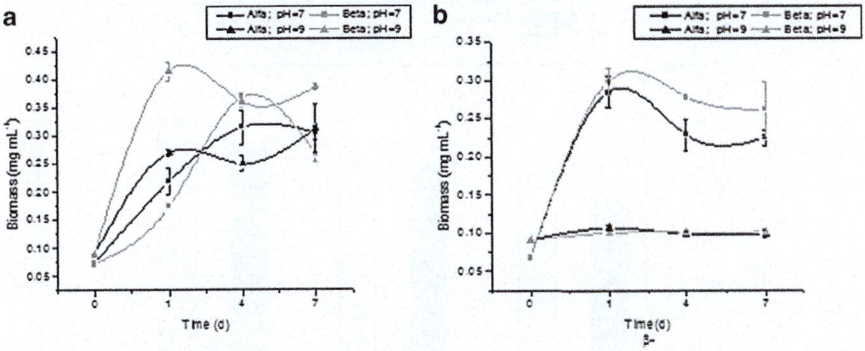

Fig. 18.2 Microbial growth determination by dry weight for two initial concentrations of **α**-HCH and β-HCH isomers: (**a**) 1.66 and (**b**) 8.33 mg L^{-1}. Error bars represent standard deviation

In the study of initial concentration of the different isomers was observed that at one concentration of 8.33 mg L^{-1} *Streptomyces* sp. M7 was able to grow rates of 0.28 mg mL^{-1} for α-HCH isomer and of 0.29 mg mL^{-1} for β-HCH isomer, both experiments were performed at pH 7. Exceeding five times in both cases the value observed at pH 9, which indicated that there was no inhibitory effect by concentration increasing of the different isomers analyzed (Fig. 18.2b).

18.4 Dechlorination Determination by Chloride Ion Release Detection at Different pH and Isomers Concentrations

One way to confirm that the HCH isomers are metabolized by actinobacteria was by chloride ion released (Phillips et al. 2001). Therefore the chloride ion release was determinate one initial concentration of 1.66 mg mL^{-1} for α-HCH isomer, where observed an increased ions release of at pH 7 after incubation 7 days (ΔAbs_{600} 0.0238 nm), exceeding three times compared to the chloride release at pH 9, while for β-HCH at pH 7 showed a greater chloride ions release with a absorbance value of ΔAbs_{600} 0.025 nm, which means five times higher than observed for pH9 (Fig. 18.3a).

When the test were performed at an initial concentration of 8.33 mg mL-1 of HCH isomer, the release of chloride ion was observed clearly favored to a pH7, whose value exceeds five times the value obtained at an alkaline pH of 9 for the test by α-HCH isomer, and three times for β-HCH isomer. In both experiments the values were determined on the seventh incubation day (Fig. 18.3b).

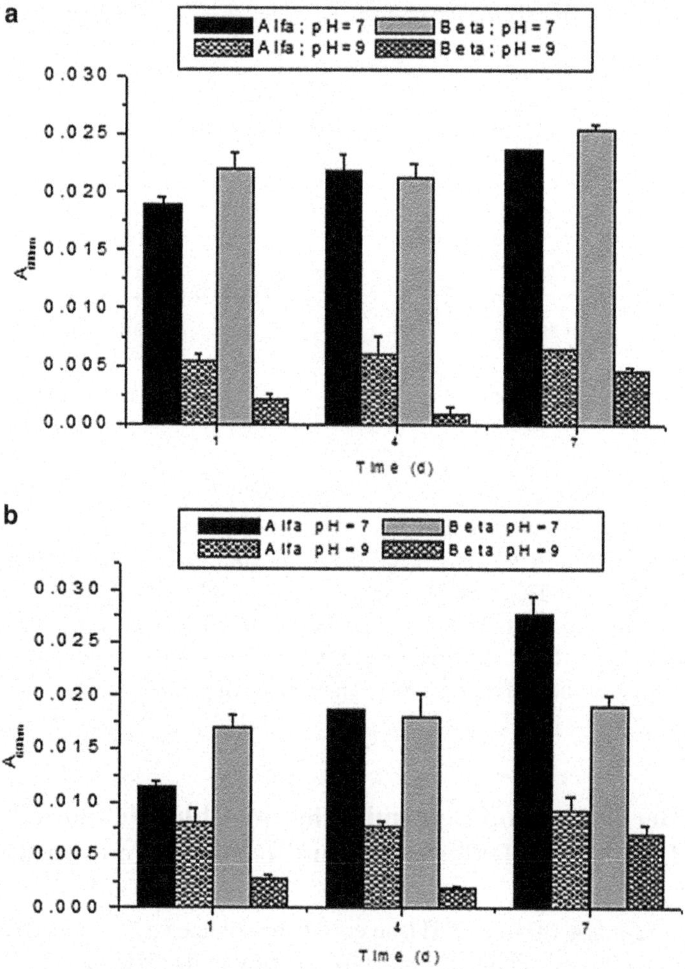

Fig. 18.3 Chloride ion released from chlordane after 1, 4, and 7 days of incubation by *Streptomyces* sp. M7 cultured in MM at two different initial concentrations of the isomers, **α**-HCH and **β**-HCH: (**a**) 1.66 and (**b**) 8.33 mg L^{-1}. Error bars represent standard deviation

18.5 Residual Determination of α and β-HCH Isomers

Another study was the determination of residual isomers present in the supernatant; it was observed that at an initial concentration of 1.66 mg L^{-1} of isomer α-HCH, removal reached 78.51 %, whereas with the isomer β-HCH it was of 65.30 %, while at initial concentration of 8.33 mg L^{-1}, it was observed that the α-HCH presented the maximum removal percentage of 84.62 %, exceeding eight times the value obtained for the isomer β-HCH at the same initial concentration. In all studied initial

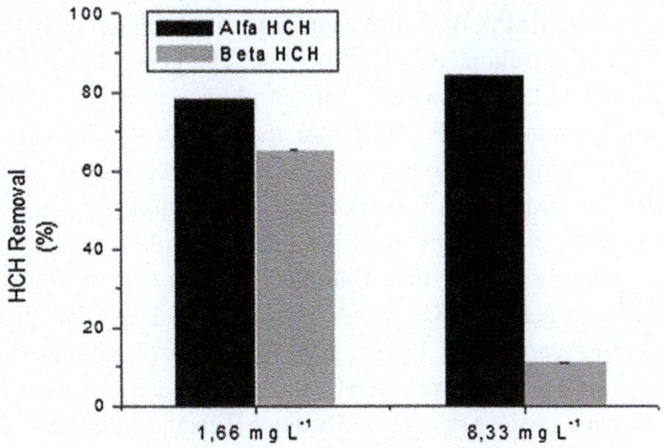

Fig. 18.4 Removal of the isomers in culture supernatants at pH7 on the seventh incubation day. Error bars represent standard deviation

concentrations no inhibitory effect was observed on growth by increasing concentration (Fig. 18.4), this behavior observed too was very similar to that presented by *Arthrobacter giacomelloi* after 72 h of incubation (De Paolis et al. 2013).

18.6 Concluding Remarks

In this chapter we put into evidence that *Streptomyces* sp. M7 has the ability to use α- and β-HCH isomers as only carbon source and subsequent removal. The results showed that optimal conditions for growth were pH 7 and initial concentration of both α-HCH and β-HCH isomers of 1.66 mg L^{-1}; as the sole carbon source. However at pH 9 and at a starting concentration of 1.66 mg L^{-1} with β-HCH isomer, *Streptomyces* sp. M7 showed the maximum growth.

Regarding the removal of the α-HCH and β-HCH isomers, it were show that the concentrations tested and initial pH of 7 is the optimum for both isomers, because the chlorides release values obtained were largely higher than those observed at pH 9 at both initial concentration of 1.66 and 8.33 mg L^{-1} of α-HCH and β-HCH, respectively, where for the value of 1.66 was removal plus 75 % and 8.33 mg L^{-1} was obtained the maximum removal percentage of 85 %. Demonstrating that removal is not inhibited when produces an increased concentration of the isomers.

In the same way the ion chlorides release process we are indicating that the HCH isomers are processing in the actinobacteria cells. Previously, Cuozzo et al. (2009) reported on dehalogenase activity in *Streptomyces* sp. M7 with HCH as specific substrate. It is important to note that the two first metabolites γ-pentachlorocyclohexene (γ-PCCH) and 1,3,4,6-tetrachloro-1,4-cyclohexadiene (1,4-TCDN) produced by the action of dechlorinase over lindane were detected in the cell free extract of *Streptomyces* sp. M7 at 48 and 96 h. This would allow postulating that the HCH isomers are processed by the LinA, according to prescriptive by Nagata et al. (2007). Moreover we know that in *Sphingomonas* sp. UT26 all the specific *lin* genes that are almost identical to those of UT26 are dispersed on four out of the five plasmids (linF on pISP0; linA, linC, and truncated linF on pISP1; linRED on pISP3; and linB, linC, and truncated linF on pISP4). The genome of *Sphingomonas* sp. MM-1 has 15 copies of IS6100, which is also highly associated with the specific lin genes in other strains, and all 15 copies are located on the five plasmids (Tabata et al. 2013). Whereas in *Streptomyces* sp. M7 were not plasmid detected but yes the IS6100 in the total M7 DNA considering is also highly associated with the specific lin genes (Cuozzo et al. 2009), so the lin genes would integrated at the chromosome.

Clearly demonstrate that the genomic organization and localization of the specific lin genes of *Sphingomonas* sp. MM-1 are significantly different from those of *Sphingomonas* sp. UT26, suggesting the two strains have independently acquired the ability to utilize -HCH (Tabata et al. 2013). Also most of the HCH-degrading isolates express dehydrochlorinase (LinA) and haloalkane dehalogenase (LinB, E.C. 3.8.1.5) activities, which are the first two enzymes of the γ-HCH degradation pathway (Lal et al. 2010). Experiments with pure enzyme preparations and with recombinant cells showed that a α-HCH, β-HCH, γ-HCH, δ-HCH, ε-HCH that is present in technical HCH are converted by LinA and LinB. Both enantiomers of α-HCH are transformed via β-pentachlorocyclohexene (β-PCCH) to 1,2,4-trichlorobenzene (1,2,4-TCB) by LinA and to trichlorocyclohexenediols (A4) by LinB (Suar et al. 2005; Raina et al. 2008). LinA cannot utilize β-HCH as substrate, but LinB converts it to partially hydroxylated metabolites (Geueke et al. 2013).

The results resumed here are the first evidence of aerobic biodegradation of HCH isomers by regional actinobacterias strains. Although *Streptomyces* sp. M7 seems to be a potential agent for bioremediation of environments contaminated with HCH isomers, many studies have shown that results from laboratory studies can greatly differ from results in field studies due to number of variables.

These partial results will lead to further explore different optimum removal of HCH isomers by *Streptomyces* sp. M7, to be used as bioremediator agent of HCH-contaminated environments.

Acknowledgments This research was financially supported by grants from CIUNT, ANPCYT, and CONICET.

References

Benimeli CS, Castro GR, Chaile AP et al (2006) Lindane removal induction by *Streptomyces* sp. M7. J Basic Microbiol 46:348–357

Buser HR, Muller MD (1995) Isomer and enantioselective degradation of hexachlorocyclohexane isomers in sewage sludge under anaerobic conditions. Environ Sci Technol 29:664–672

Camacho-Pérez B, Ríos-Leal E, Rinderknecht-Seijas N et al (2012) Enzymes involved in the biodegradation of hexachlorocyclohexane: a mini review. J Environ Manage 95:S306–S318

Cuozzo SA, Abate CM, Rollan GC et al (2009) Specific dechlorinase activity on lindane degradation by *Streptomyces* sp. M7. World J Microbiol Biotechnol 25:1539–1546

De Paolis MR, Lippi D, Guerriero E et al (2013) Biodegradation of α-, β-, and γ-hexachlorocyclohexane by *Arthrobacter fluorescens* and *Arthrobacter giacomelloi*. Appl Biochem Biotechnol 170:514–524

Elcey CD, Kunhi AAM (2010) Substantially enhanced degradation of hexachlorocyclohexane isomers by a microbial consortium on acclimation. J Agric Food Chem 58:1046–1054

Engst R, Fritsche W, Knoll R et al (1979) Interim result of studies of microbial isomerization of gamma-hexachlorocyclohexane. Bull Environ Contam Toxicol 22:699–707

Fuentes MS, Benimeli CS, Cuozzo SA et al (2010) Isolation of pesticide-degrading actinomycetes from a contaminated site: bacterial growth, removal and dechlorination of organochlorine pesticides. Int Biodeter Biodegr 64:434–441

Geueke B, Garg N, Ghosh S et al (2013) Metabolomics of hexachlorocyclohexane (HCH) transformation: ratio of LinA to LinB determines metabolic fate of HCH isomers. Environ Microbiol 15:1040–1049

Jagnow G, Haider K, Ellwardt PC (1977) Anaerobic dechlorination and degradation of hexachlorocyclohexane isomers by anaerobic and facultative anaerobic bacteria. Arch Microbiol 115:285–292

Johri AK, Dua M, Tuteja D et al (1998) Degradation of alpha, beta, gamma and delta-hexachlorocyclohexane by *Sphingomonas paucimobilis*. Biotechnol Lett 20:885–887

Kaur J, Verma M, Lal R (2011) *Rhizobium rosettiformans* sp. nov., isolated from a hexachlorocyclohexane dump site, and reclassification of *Blastobacter aggregatus* Hirsch and Müller 1986 as *Rhizobium aggregatum* comb. nov. Int J Syst Evol Microbiol 61:1218–1225

Lal R, Dadhwal M, Kumari K et al (2008) *Pseudomonas* sp. to *Sphingobium indicum*: a journey of microbial degradation and bioremediation of hexachlorocyclohexane. Indian J Microbiol 48:3–18

Lal R, Pandey G, Sharma P et al (2010) Biochemistry of microbial degradation of hexachlorocyclohexane and prospects for bioremediation. Microbiol Mol Biol Rev 74:58–80

Manickam N, Reddy MK, Saini HS et al (2008) Isolation of hexachlorocyclohexane degrading *Sphingomonas* sp. by dehalogenase assay and characterization of genes involved in γ-HCH degradation. J Appl Microbiol 104:952–960

Nagata Y, Miyauchi K, Takagi M (1999) Complete analysis of genes and enzymes for γ-hexachlorocyclohexane degradation in *Sphingomonas paucimobilis* UT26. J Ind Microbiol 23:380–390

Nagata Y, Endo R, Ito M et al (2007) Aerobic degradation of lindane (γ-hexachlorocyclohexane) in bacteria and its biochemical and molecular basis. Appl Microbiol Biotechnol 76:741–752

Pavlíková N, Bláhová L, Klán P et al (2012) Enantioselective effects of alpha-hexachlorocyclohexane (HCH) isomers on androgen receptor activity in vitro. Chemosphere 86:65–69

Pesce SF, Wunderlin DA (2004) Biodegradation of lindane by a native bacterial consortium isolated from contaminated river sediment. Int Biodeter Biodegr 54:255–260

Phillips TM, Seech AG, Lee H, Trevors JT (2001) Colorimetric assay for lindane dechlorination by bacteria. J Microbiol Methods 47:181–188

Phillips TM, Seech AG, Lee H et al (2005) Biodegradation of hexachlorocyclohexane (HCH) by microorganisms. Biodegradation 16:363–392

Raina V, Hauser A, Buser HR et al (2007) Hydroxylated metabolites of β- and δ-hexachlorocyclohexane: bacterial formation, stereochemical configuration, and occurrence in groundwater at a former production site. Environ Sci Technol 41:4292–4298

Raina V, Rentsch D, Geiger T et al (2008) New metabolites in the degradation of α- and γ-hexachlorocyclohexane (HCH): pentachlorocyclohexenes are hydroxylated to cyclohexenols and cyclohexenediols by the haloalkane dehalogenase LinB from *Sphingobium indicum* B90A. J Agric Food Chem 56:6594–6603

Senoo K, Wada H (1989) Isolation and identification of an aerobic γ-HCH decomposing bacterium from soil. Soil Sci Plant Nutr 35:79–87

Suar M, Hauser A, Poiger T, Buser HR, Muller MD et al (2005) Enantioselective transformation of alpha-hexachlorocyclohexane by the dehydrochlorinases LinA1 and LinA2 from the soil bacterium Sphingomonas paucimobilis B90A. Appl Environ Microbiol 71:8514–8518

Tabata M, Endo R, Ito M et al (2011) The lin genes for γ-hexachlorocyclohexane degradation in *Sphingomonas* sp. MM-1 proved to be dispersed across multiple plasmids. Biosci Biotechnol Biochem 75:466–472

Tabata M, Ohtsubo Y, Ohhata S et al (2013) Complete genome sequence of the γ-hexachlorocyclohexane-degrading bacterium *Sphingomonas* sp. Strain MM-1. GenomeA 1(3):e00247-13. doi:10.1128/genomeA.00247-13

van Doesburg W, van Eekert MH, Middeldorp PJ et al (2005) Reductive dechlorination of beta-hexachlorocyclohexane (beta-HCH) by a *Dehalobacter* species in coculture with a *Sedimentibacter* sp. FEMS Microbiol Ecol 54:87–95

Vijgen J (2006) The legacy of lindane HCH isomer production: a global overview of residue management, formulation and disposal. International HCH and pesticide association. http://www.ihpa.info/library_access.php

Vijgen J, Abhilash PC, Li YF et al (2011) Hexachlorocyclohexane (HCH) as new Stockholm Convention POPs—a global perspective on the management of Lindane and its waste isomers. Environ Sci Pollut Res 18:152–162

Willett KL, Ulrich EM, Hites RA (1998) Differential toxicity and environmental fates of hexachlorocyclohexane isomers. Environ Sci Technol 32:2197–2207

Wu J, Hong Q, Han P et al (2007) A gene linB2 responsible for the conversion of β-HCH and 2,3,4,5,6-pentachlorocyclohexanol in *Sphingomonas* sp. BHC-A. Appl Microbiol Biotechnol 73:1097–1105

Chapter 19
Cell Immobilization Technique for the Enhanced Removal of Lindane Using *Streptomyces* Strains Isolated from Northwestern Argentina

Juliana Maria Saez, Claudia Susana Benimeli, and María Julia Amoroso

Abstract Lindane (γ-HCH) is an organochlorine insecticide which has a negative effect as a pollutant agent of soil, water, and sediments. Nowadays it has been banned in almost all countries of the world, but its residues still remain in the environment. In this context, bioremediation, involving the use of microorganisms to degrade environmental contaminants, has received much attention as an effective biotechnological approach to clean up this kind of pollutants. Moreover, cell immobilization has been shown to present diverse advantages over conventional systems using free cells, such as the possibility of employing higher cell density, easier separation of cells from the system, repeated use of cells, and better protection of cells from harsh environments.

Thereby, this chapter compiles information about (1) the advantages and limitations of the use of immobilized cells; (2) the comparison between free or

J.M. Saez (✉)
Planta Piloto de Procesos Industriales Microbiológicos (PROIMI-CONICET),
Avenida Belgrano y Pasaje Caseros, 4000 Tucumán, Argentina
e-mail: julianasaez@hotmail.com

C.S. Benimeli
Planta Piloto de Procesos Industriales Microbiológicos (PROIMI-CONICET),
Avenida Belgrano y Pasaje Caseros, 4000 Tucumán, Argentina

Universidad del Norte Santo Tomás de Aquino (UNSTA),
9 de Julio 165, 4000 Tucumán, Argentina

Unidad de Administración Territorial, Centro Científico Tecnológico (CCT),
CONICET-Tucumán, Crisóstomo Álvarez 722, 4000 Tucumán, Argentina

M.J. Amoroso
Planta Piloto de Procesos Industriales Microbiológicos (PROIMI), CONICET,
Av. Belgrano y Pasaje Caseros, 4000 Tucumán, Argentina

Facultad de Bioquímica, Química y Farmacia, Universidad Nacional de Tucumán,
4000 Tucumán, Argentina

Universidad del Norte Santo Tomás de Aquino (UNSTA),
4000 Tucumán, Argentina

A. Alvarez, M.A. Polti (eds.), *Bioremediation in Latin America: Current Research and Perspectives*, DOI 10.1007/978-3-319-05738-5_19

immobilized cells for lindane removal by single cultures of actinobacteria, isolated from polluted environments in the northwest of Argentina; and (3) lindane removal by free and immobilized consortia of *Streptomyces* spp.

19.1 Introduction

Organochlorine pesticides are semivolatile compounds that exhibit resistance to photochemical, biological, and chemical degradation for a long period of time (Carvalho et al. 2009). These compounds, with their high persistence in the environment and their liposolubility, are accumulated primarily in adipose tissue and are biomagnified along food chain (Cid et al. 2007). Among organochlorine pesticides, lindane (γ-1,2,3,4,5,6-hexachlorocyclohexane, γ-HCH) is a highly chlorinated, recalcitrant compound, which has been used worldwide as a broad-spectrum insecticide for a variety of crops (Phillips et al. 2005). It has also been applied for human health as scabicide and pediculicide, in the form of lotions and shampoos (Bidlan et al. 2004).

Although the γ-isomer is the only one which exhibits insecticide properties, all HCH isomers have been shown to cause serious health effects in short and long term. They are toxic, carcinogenic, endocrine disrupters, and known to exert damaging effects on the reproductive and nervous systems in mammals (Salam and Das 2012). Nowadays, most countries, including Argentina, have banned or highly restricted the use of this xenobiotic. However, some developing countries are still using it because of economic reasons and consequently new sites are being contaminated (Pesce and Wunderlin 2004). In Argentina, residues of lindane and other HCH isomers have been found in diverse matrices and environments (Table 19.1).

Because contamination with organochlorine pesticides is still a serious problem in the agricultural environment, an efficient method for remediation is needed. In this context, bioremediation has received considerable attention as an effective biotechnological approach to clean up polluted environments. Bioremediation is the

Table 19.1 Lindane and HCHs[a] residues detected in Argentina

Compound	Concentration	Sample	Reference
Lindane	2 mg L^{-1}	Water	Chaile et al. (1999)
Lindane	Not specified	Wastewater treatment plant	Pesce and Wunderlin (2000)
HCHs	25.44 ng g^{-1}	Soils	Miglioranza et al. (2003)
Lindane	2.5–16.8 ng g^{-1}	Leek plants	González et al. (2003)
Lindane	1.3 ng g^{-1}	Untreated soils	González et al. (2003)
Lindane	371.71–1544.23 ng g^{-1}	Bird fat	Cid et al. (2007)
Lindane	1.9 μg g^{-1}	Soils	Fuentes et al. (2010)
HCHs	4.5–149.5 ng g^{-1}	Soils	Gonzalez et al. (2010)

[a]HCHs: Hexachlorocyclohexane residues without differentiating which isomer/s

exploitation of biological activities for the mitigation (and wherever possible, complete elimination) of the noxious effects caused by environmental pollutants in given sites (De Lorenzo 2008).

Bioremediation may be carried out by using different organisms, from bacteria to plants, or their derivatives, and its chief advantage is its reduced cost compared to conventional techniques. In addition, bioremediation is often a permanent solution, providing complete transformation of the pollutant to its molecular constituents like carbon dioxide and water, while other remediation methods only transfer wastes from one phase to another (Wood 2008).

Many microorganisms have been reported to present a great potential as pesticide bioremediators. Particularly lindane-degrading bacteria have been isolated and thoroughly studied (Nagata et al. 1999, 2007; Phillips et al. 2005; van Doesburg et al. 2005; Lal et al. 2006; Manickam et al. 2007, 2008).

Actinobacteria are Gram-positive microorganisms with a great potential for bioremediating toxic compounds. They constitute the main group of bacteria present in soils and sediments. In addition to their potential metabolic diversity, strains of *Streptomyces* genus may be well suited for soil inoculation, as a consequence of their mycelial growth habit, relatively rapid rates of growth, colonization of semi-selective substrates, and their ability to be genetically manipulated. Moreover, they are already adapted to this habitat (Shelton et al. 1996).

Benimeli et al. (2003) and Fuentes et al. (2010) isolated and selected wild-type actinobacteria strains, which are able to tolerate and remove lindane from culture media and soil. Cuozzo et al. (2009) reported the release of γ-2,3,4,5,6-pentachlorocyclohexene and 1,3,4,6-tetrachloro-1,4-cyclohexadiene, the first and second products of lindane degradation by *Streptomyces* sp. M7 according to the catabolic pathway proposed by Nagata et al. (2007) for a Gram-negative bacterium.

A very promising technique that has been attracting attention worldwide is cell immobilization. It may be used for several purposes, including bioremediation. For this reason, this chapter consists on a short compilation regarding the ability of actinobacteria, isolated from the northwestern Argentina region, to remove lindane from a liquid medium, either as pure or as mixed cultures, using four different immobilization techniques.

19.2 Cell Immobilization: Advantages and Limitations

Cell immobilization is a promising technique which consists of restricting the cellular mobility within a defined space, thereby retaining catalytic activity and enhancing both biological and physical stability of microorganisms (Siripattanakul et al. 2008; Yáñez-Ocampo et al. 2009). The immobilization of microbial cells is actually getting more importance than immobilized enzymes in the production of useful compounds because it eliminates the need for release of intracellular enzymes and the succeeding purification steps (Nigam et al. 2009).

Cell immobilization can be accomplished by using a wide variety of matrices or supports, whether natural, such as agar, alginate, sugarcane bagasse, and even volcanic stones, or synthetic, such as polyurethane, polystyrene, and nylon (Ahamad and Kunhi 2011; León-Santiestebán et al. 2011; Abdel-Razek et al. 2013). Natural matrices have the disadvantage of being prone to abrasion and are biodegradable. In fact, the choice of the immobilization support is a key factor in environmental application of immobilized biocatalyst (Poopal and Laxman 2008).

Immobilized cells present several advantages over free suspended cells, such as higher retention of microorganisms in the reactor, enhanced cellular viability (weeks or months), and easier solid-liquid separation (Poopal and Laxman 2009; Ahamad and Kunhi 2011). Besides all this, immobilized cells are easily reusable (Nigam et al. 2009). In addition, immobilized cultures tend to have a higher level of activity and higher tolerance to environmental perturbations, such as pH, temperature, or the exposure to toxic compounds, than suspended biomass cultures (León-Santiestebán et al. 2011).

Consequently, immobilization of microbial cells is increasingly applied in biotechnological processes for chemical conversion and bioremediation, obtaining higher biodegradation, higher dilution rate without washout in continuous processes, and tolerance against high concentrations of toxic compounds (Wang et al. 2004).

Despite all above, in some cases, immobilized cell systems have limitations such as lower substrate diffusion and/or lower oxygen diffusion, compared to free cells. In the case of alginate-immobilized cells, the pores are necessary for entrapping the cells and transporting substrate and oxygen. According to Li et al. (2005), the beads with 2-μm pores should not have oxygen and substrate transfer problems. On the other hand, cell-to-matrix ratio is an important point to consider too. Siripattanakul et al. (2008) reported that lower cell-to-matrix ratios resulted in faster and higher biodegradation of atrazine, due to less mass transfer limitation of oxygen and substrates.

The use of immobilized cells has been investigated for different purposes such as metabolite production—antibiotics and enzymes (Srivastava and Kundu 1998; Anisha and Prema 2008)—and removal of toxic compounds—colorants, heavy metals, among others (Boon et al. 2002; Poopal and Laxman 2009; Ahamad and Kunhi 2011). Also, the biodegradation of several pesticides has been assessed by using immobilized cells. Diuron (Bazot and Lebeau 2009), atrazine (Siripattanakul et al. 2009), methyl-parathion and tetrachlorvinphos (Yáñez-Ocampo et al. 2011), pentachlorophenol (León-Santiestebán et al. 2011), and lindane (Saez et al. 2012) are a few pesticides which removal and/or biodegradation were studied through the use of different cell immobilization techniques.

19.3 Microbial Growth and Lindane Removal in Pure Cultures of Free and Immobilized *Streptomyces* spp.

Actinobacteria are Gram-positive microorganisms with a great potential for biodegradation of organic and inorganic toxic compounds. Several studies have demonstrated removal and in some cases even oxidation and partial dechlorination

Table 19.2 Specific dechlorinase activity in cell-free extracts of *Streptomyces* sp.

Strains	Specific dechlorinase activity (EU[a] mg^{-1} protein)
Streptomyces sp. A2	5.88±0.03
Streptomyces sp. A5	18.29±0.03
Streptomyces sp. A11	8.15±0.03
Streptomyces sp. M7	16.06±0.05

Adapted from Fuentes et al. (2011)

[a]Enzymatic unit (EU) was defined as the amount of chloride ions (μmol) released in 1 h

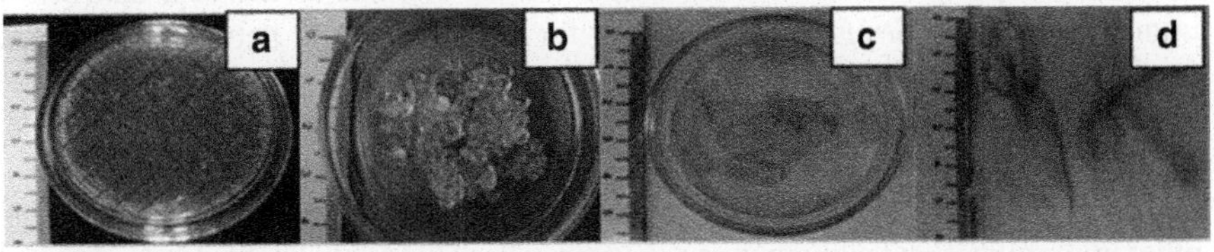

Fig. 19.1 Immobilization matrices. (**a**) Agar cubes, (**b**) PVA-alginate beads, (**c**) silicone tubes, and (**d**) cloth sachets

and dealkylation of pesticides such as diuron, methoxychlor, and alachlor by actinobacteria (Durães Sette et al. 2004; Castillo et al. 2006; Fuentes et al. 2013).

Benimeli et al. (2003) and Fuentes et al. (2010) isolated actinobacteria belonging to the *Streptomyces* and *Micromonospora* genera able to tolerate and/or degrade organochlorine pesticides. *Streptomyces* sp. M7 was isolated from wastewater sediment of a copper filter plant located in an agricultural area of Tucuman, in the northwest of Argentina. The sampling area was contaminated with heavy metals and organochlorine pesticides, including lindane (Benimeli et al. 2003). *Streptomyces* spp. A2, A5, and A11 were isolated by Fuentes et al. (2010), from a village located in the southeast of Santiago del Estero, a province in northwestern Argentina, where more than 30 t of obsolete pesticides, including lindane, were found in 1994 (see Table 19.1). The four strains were studied on their abilities to grow in the presence of lindane as carbon source and to remove it from culture medium, as pure and mixed cultures of free cells (Fuentes et al. 2011). All of them presented specific dechlorinase activity (Table 19.2).

Immobilized microbial systems have shown several advantages over free cells, such as a higher degradation rate, enhanced tolerance to high concentrations of toxic compounds, and the possibility of reusing the cells. For all the reasons mentioned, Saez et al. (2012) assayed single cultures of the four actinobacteria immobilized in four different matrices: agar cubes, polyvinyl alcohol (PVA)-alginate beads, silicone tubes, and cloth sachets for the removal of 1.66 mg L^{-1} of lindane (Fig. 19.1). As a result, the four *Streptomyces* spp. strains were able to grow in MM supplemented with lindane as carbon and energy source, both free and immobilized. Nevertheless, all the strains showed lowest growth values when immobilized in PVA-alginate beads compared to the other three supports. PVA is a cheap synthetic organic polymer, which is widely used for cell immobilization, mainly combined with alginate, because of its high mechanical strength, durability, and innocuousness to both microorganisms and environment (Siripattanakul et al. 2008). However, it is possible that cells have experienced physical stress during the PVA-alginate immobilization process or may have suffered low oxygen and/or substrate diffusion due to the matrix, although the size of the pore was not specified.

Siripattanakul et al. (2008) also reported lower proportion of viable cells when it were immobilized in phosphorylated-PVA, compared to free cells, indicating that immobilization could have produced a negative effect on cell viability. In contrast, Poopal and Laxman (2008) found PVA-alginate as the best matrix for immobilization of *Streptomyces griseus* for Cr^{6+} reduction, between other five matrices tested. Besides, the beads did not suffer disintegration during the experiment.

On the other hand, *Streptomyces* spp. A2, A5, and A11 showed significantly higher growth values when they were entrapped in silicone tubes compared to the other immobilization techniques and the free cells. This could be due to an adequate diffusion of both oxygen and substrates provided by the medium through the agar in the silicone tubes, since both ends were open. Moreover, the tubes contained nutritive agar inside, absent in the other immobilization modes, which may have helped sustain cell growth.

Saez et al. (2012) also showed that both lindane removal (Fig 19.2) and microbial growth were significantly higher in *Streptomyces* spp. A11 and M7 immobilized in any of the four matrices, compared to free cells. This suggests that these strains would be metabolically more active when they were immobilized than when they were free. A possible explanation is that the support could be acting as a protection for the cells against the toxicity of lindane in the medium. However, there is no clear understanding of how the immobilization process protects microorganisms (Boon et al. 2002). Mertens et al. (2006) and Siripattanakul et al. (2008) have also observed increased metabolic activity with encapsulated cells but the underlying mechanism has not yet elucidated.

Regarding the different matrices, lindane removal was most efficient for all strains when immobilized in silicone tubes followed by cloth sachets (Fig. 19.2). However, since the lindane removal of the abiotic control (i.e., the medium with lindane in contact with the silicone tubes) was not significantly different from the strains immobilized in this support, most removal was likely caused by adsorption of γ-HCH to the tubes. Mertens et al. (2006) obtained similar results when they estimated the amount of γ-HCH that could be adsorbed to agar-filled silicone tubes used for encapsulation, revealing that the tubes were responsible for the removal of a high percentage of the initially spiked lindane.

In contrast, when *Streptomyces* spp. strains were immobilized in cloth bags, lindane removal improved in approximately 25 % compared to the abiotic control, although pesticide adsorption contributed too. The difference with all the immobilized strains was significant, evidencing microbial action (Fig 19.2).

Several researchers have demonstrated that immobilization enhanced tolerance to toxic compounds and the degrading ability of the cells. Ahamad and

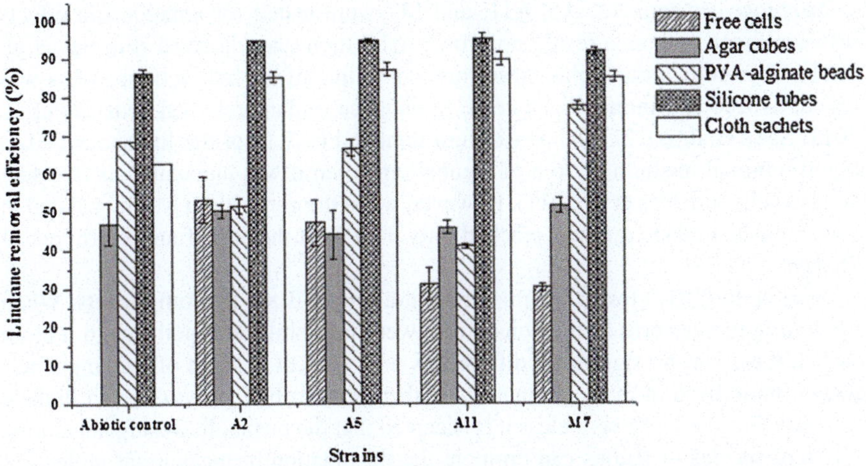

Fig. 19.2 Lindane removal efficiency (%) of free and immobilized cultures of *Streptomyces* spp. A2, A5, A11, and M7. The *bar* showing *one asterisk* indicates that it presented significant differences with the *bars* showing *two asterisks*. *Bars* showing *three asterisks* were significantly different from all others ($P \leq 0.05$) (adapted from Saez et al. 2012)

Kunhi (2011) reported that *Pseudomonas* sp. CP4 remarkably improved the removal of phenol when it was immobilized in calcium alginate and agar beads, and removal by agar-entrapped cells was far superior to Ca-alginate-immobilized microorganisms. Other researchers found that calcium alginate was a better matrix than agar-agar regarding degradation activity (Dey and Roy 2009). In contrast, Saez et al. (2012) reported that cells immobilized in agar and in PVA-alginate did not show in all cases an improvement in the lindane removal compared to free cells.

19.4 Microbial Growth and Lindane Removal by Mixed Cultures of *Streptomyces* spp.

Mixed cultures are considered potential agents in biodegradation of recalcitrant compounds, because in some cases they have proven to be more efficient than pure cultures. This may be due to an increased number of catabolic pathways available to degrade the contaminants (Siripattanakul et al. 2009). Nevertheless, it is important to consider that two different situations may occur when the behavior of mixed cultures for specific pollutant biodegradation is considered. These involve either the employment of a culture in which a complementary sequence of catabolic activities from several associated strains is harnessed in order to generate a complete degradative pathway for the pollutant under consideration or the employment of a culture in which the fastidiousness of the primary pollutant-degrading strain is diminished by the actions of several associated ancillary strains (Hamer 1997).

Streptomyces spp. A2, A5, A11, and M7 were tested for antagonistic effects between them, and, as a result, no growth inhibition was observed (Fuentes et al. 2011). Therefore, a microbial consortium consisting of the four actinobacteria was tested, as free and immobilized cells, to enhance lindane removal (Fuentes et al. 2011; Saez et al. 2012). The consortium removed 61.7 % of the lindane added to liquid minimal medium at 96 h of incubation, when it was inoculated as free suspended cells (Fuentes et al. 2011). However, when the consortium was inoculated in soil samples, lindane removal reached only 31.4 % at the end of an experiment of 28 days.

Saez et al. (2012) tested the mixed culture immobilized in cloth sachets. When the four actinobacteria of the consortium were immobilized together into a cloth sachet, there was no significant differences with respect to some of the single cultures immobilized in the same matrix, neither in microbial growth, nor in lindane removal (Fig. 19.3). Nevertheless it is accepted that the metabolic synergism among different species or strains can promote the degradation of recalcitrant molecules and in fact, this process can be further enhanced by cellular immobilization (Yáñez-Ocampo et al. 2009).

The microbial growth of the consortium immobilized in the sachet was similar to that presented by *Streptomyces* spp. A5, A11, and M7 when they were individually immobilized in the cloth sachets. The biomass of the cells released from the support was also examined as an inverse measure of the stability of the immobilization and matrix resistance. The results revealed that only 1.2 % of the initial biomass

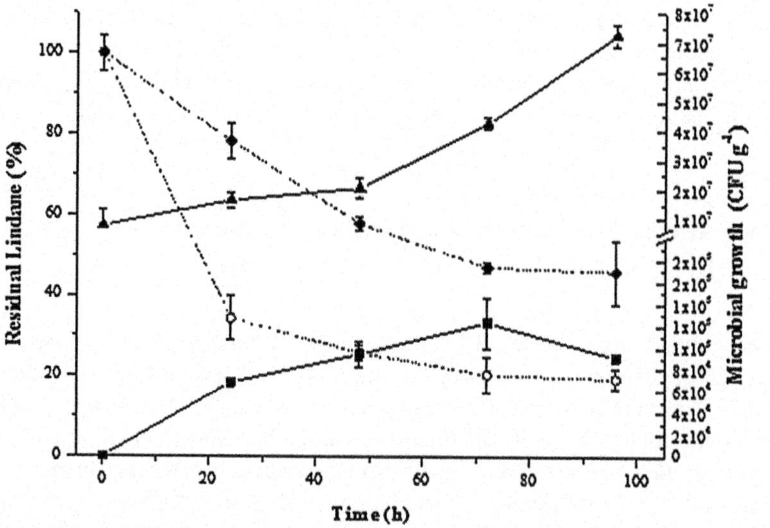

Fig. 19.3 Residual lindane (%) (*dashed line*) and microbial growth (CFU g^{-1}) (*solid line*) of a defined mixed culture of actinobacteria immobilized in cloth sachets. *Solid line with closed triangles* immobilized cells; *solid line with closed squares* released cells; *dotted line with open diamonds* immobilized cells; *dotted line with closed diamonds* abiotic control (adapted from Saez et al. 2012)

inoculated into the bag was released from it at the end of the experiment, demonstrating that the immobilization technique was efficient and the support did not disintegrate during the process.

On the other hand, the maximum pesticide removal obtained by the immobilized consortium was 81 % after 96 h of incubation, which means no significant difference to the single cultures immobilized in the cloth bags. However, it is important to note that immobilized cells could be reused for two additional cycles of 96 h each, obtaining a removal efficiency of 59 % in each case.

Other studies have reported that cells entrapped in calcium alginate could be used five times and agar-agar-entrapped cells three times (Dey and Roy 2009). Nigam et al. (2009) informed that they reused agar-immobilized *Streptomyces* sp. 20 times without much loss of activity.

19.5 Concluding Remarks

This review suggests that the use of immobilized indigenous actinobacteria is a promising alternative to bioremediate-polluted sites. The advantages of immobilized *Streptomyces* strains compared to free cells make them suitable candidates for the development of techniques for the removal of lindane and other organochlorine pesticides from polluted environments.

References

Abdel-Razek MARS, Folch-Mallol JL, Perezgasga-Ciscomani L et al (2013) Optimization of methyl parathion biodegradation and detoxification by cells in suspension or immobilized on tezontle expressing the *opd* gene. J Environ Sci Health B 48:449–461

Ahamad PYA, Kunhi AAM (2011) Enhanced degradation of phenol by *Pseudomonas* sp. CP4 entrapped in agar and calcium alginate beads in batch and continuous processes. Biodegradation 22:253–265

Anisha GS, Prema P (2008) Cell immobilization technique for the enhanced production of α-galactosidase by *Streptomyces griseoloalbus*. Bioresour Technol 99:3325–3330

Bazot S, Lebeau T (2009) Effect of immobilization of a bacterial consortium on diuron dissipation and community dynamics. Bioresour Technol 100:4257–4261

Benimeli CS, Amoroso MJ, Chaile AP et al (2003) Isolation of four aquatic streptomycetes strains capable of growth on organochlorine pesticides. Bioresour Technol 89:348–357

Bidlan R, Afsar M, Manonmani HK (2004) Bioremediation of HCH-contaminated soil: elimination of inhibitory effects of the insecticide on radish and green gram seed germination. Chemosphere 56:803–811

Boon N, De Gelder L, Lievens H et al (2002) Bioaugmenting bioreactors for the continuous removal of 3-chloroaniline by a slow release approach. Environ Sci Technol 21:4698–4704

Carvalho PN, Rodrigues PNR, Basto MCP et al (2009) Organochlorine pesticides levels in Portuguese coastal areas. Chemosphere 75:595–600

Castillo MA, Felis N, Aragón P et al (2006) Biodegradation of the herbicide diuron by streptomycetes isolated from soil. Int Biodeter Biodegr 58:196–202

Chaile AP, Romero N, Amoroso MJ et al (1999) Organochlorine pesticides in Sali River. Tucuman-Argentina (in Spanish). Rev Bol Ecol 6:203–209

Cid FD, Antón RS, Caviedes-Vidal E (2007) Organochlorine pesticide contamination in three bird species of the Embalse La Florida water reservoir in the semiarid midwest of Argentina. Sci Total Environ 385:86–96

Cuozzo SA, Rollán GG, Abate CM et al (2009) Specific dechlorinase activity in lindane degradation by *Streptomyces* sp. M7. World J Microbiol Biotechnol 25:1539–1546

De Lorenzo V (2008) Systems biology approaches to bioremediation. Curr Opin Biotech 19:579–589

Dey K, Roy P (2009) Degradation of trichloroethylene by *Bacillus* sp.: isolation strategy, strain characteristics, and cell immobilization. Curr Microbiol 59:256–260

Durães Sette L, da Mendonça Costa Alves LA, Marsaioli AJ et al (2004) Biodegradation of alachlor by soil streptomycetes. Appl Microbiol Biotechnol 64:712–717

Fuentes MS, Benimeli CS, Cuozzo SA et al (2010) Isolation of pesticide-degrading actinomycetes from a contaminated site: bacterial growth, removal and dechlorination of organochlorine pesticides. Int Biodeterior Biodegradation 64:434–441

Fuentes MS, Saez JM, Benimeli CS et al (2011) Lindane biodegradation by defined consortia of indigenous *Streptomyces* strains. Water Air Soil Pollut 222:217–231

Fuentes MS, Alvarez A, Saez JM et al (2013) Methoxychlor bioremediation by defined consortium of environmental *Streptomyces* strains. Int J Environ Sci Technol. doi:10.1007/s13762-013-0314-0

González M, Miglioranza KSB, Aizpún de Moreno JE et al (2003) Organochlorine pesticide residues in leek (*Allium porrum*) crops grown on untreated soils from an agricultural environment. J Agric Food Chem 51:5024–5029

Gonzalez M, Miglioranza KSB, Aizpún JE et al (2010) Assessing pesticide leaching and desorption in soils with different agricultural activities from Argentina (Pampa and Patagonia). Chemosphere 81:351–358

Hamer G (1997) Microbial consortia for multiple pollutant biodegradation. Pure Appl Chem 69:2343–2356

Lal R, Dogra C, Malhotra S et al (2006) Diversity, distribution and divergence of *lin* genes in hexachlorocyclohexane-degrading sphingomonads. Trends Biotechnol 24:121–130

León-Santiestebán H, Meraz M, Wrobel K et al (2011) Pentachlorophenol sorption in nylon fiber and removal by immobilized *Rhizopus oryzae* ENHE. J Hazard Mater 190:707–712

Li H, Li P, Hua T et al (2005) Bioremediation of contaminated surface water by immobilized *Micrococcus roseus*. EnvironTechnol 26:931–939

Manickam N, Misra R, Mayilraj S (2007) A novel pathway for the biodegradation of γ-hexachlorocyclohexane by a *Xanthomonas* sp. strain ICH12. J Appl Microbiol 102: 1468–1478

Manickam N, Reddy MK, Saini HS et al (2008) Isolation of hexachlorocyclohexane-degrading *Sphingomonas* sp. by dehalogenase assay and characterization of genes involved in γ-HCH degradation. J Appl Microbiol 104:952–960

Mertens B, Boon N, Verstraete W (2006) Slow-release inoculation allows sustained biodegradation of γ-hexachlorocyclohexane. Appl Environ Microbiol 72:622–627

Miglioranza KSB, Aizpún de Moreno JE, Moreno VJ (2003) Trends in soil science: organochlorine pesticides in Argentinean soils. J Soil Sediment 4:264–265

Nagata Y, Futamura A, Miyauchi K et al (1999) Two different types of dehalogenases, LinA and LinB, involved in γ-hexachlorocyclohexane degradation in *Sphingomonas paucimobilis* UT26 are localized in the periplasmic space without molecular processing. J Bacteriol 17: 5409–5413

Nagata Y, Endo R, Itro M et al (2007) Aerobic degradation of lindane (γ-hexachlorocyclohexane) in bacteria and its biochemical and molecular basis. Appl Microbiol Biotechnol 76:741–752

Nigam VK, Khandelwal AK, Gothwal RK et al (2009) Nitrilase-catalysed conversion of acrylonitrile by free and immobilized cells of *Streptomyces* sp. J Biosci 34:21–26

Pesce SF, Wunderlin DA (2000) Use of water quality indices to verify the impact of Córdoba city (Argentina) on Suquía River. Water Res 34:2915–2926

Pesce SF, Wunderlin DA (2004) Biodegradation of lindane by a native bacterial consortium isolated from contaminated river sediment. Int Biodeter Biodegr 54:255–260

Phillips TM, Seech AG, Lee H et al (2005) Biodegradation of hexachlorocyclohexane (HCH) by microorganisms. Biodegradation 16:363–392

Poopal AC, Laxman RS (2008) Hexavalent chromate reduction by immobilized *Streptomyces griseus*. Biotechnol Lett 30:1005–1010

Poopal AC, Laxman RS (2009) Chromate reduction by PVA-alginate immobilized *Streptomyces griseus* in a bioreactor. Biotechnol Lett 31:71–76

Saez JM, Benimeli CS, Amoroso MJ (2012) Lindane removal by pure and mixed cultures of immobilized actinobacteria. Chemosphere 89:982–987

Salam JA, Das N (2012) Remediation of lindane from environment-an overview. Int J Adv Biol Res 2:9–15

Shelton DR, Khader S, Karns JS et al (1996) Metabolism of twelve herbicides by *Streptomyces*. Biodegradation 7:129–136

Siripattanakul S, Wirojanagud W, McEvoy J et al (2008) Effect of cell-to-matrix ratio in polyvinyl alcohol immobilized pure and mixed cultures on atrazine degradation. Water Air Soil Pollut 8:257–266

Siripattanakul S, Wirojanagud W, McEvoy JM et al (2009) Atrazine removal in agricultural infiltrate by bioaugmented polyvinyl alcohol immobilized and free *Agrobacterium radiobacter* J14a: a sand column study. Chemosphere 74:308–313

Srivastava P, Kundu S (1998) A comparative evaluation of cephalosporin C production using various immobilization modes. J Gen Appl Microbiol 44:113–117

van Doesburg W, van Eekert MHA, Middeldorp PJM et al (2005) Reductive dechlorination of β-hexachlorocyclohexane (β-HCH) by a *Dehalobacter* species in coculture with a *Sedimentibacter* sp. FEMS Microbiol Ecol 54:87–95

Wang Y, Fana Y, Gua JD (2004) Dimethyl phthalate ester degradation by two planktonic and immobilized bacterial consortia. Int Biodeter Biodegr 53:93–101

Wood TK (2008) Molecular approaches in bioremediation. Curr Opin Biotech 19:572–578

Yáñez-Ocampo G, Sanchez-Salinas E, Jimenez-Tobon GA et al (2009) Removal of two organophosphate pesticides by a bacterial consortium immobilized in alginate or tezontle. J Hazard Mater 168:1554–1561

Yáñez-Ocampo G, Sanchez-Salinas E, Ortiz-Hernández ML (2011) Removal of methyl parathion and tetrachlorvinphos by a bacterial consortium immobilized on tezontle-packed up-flow reactor. Biodegradation 22:1203–1213

Index

A. Alvarez, M.A. Polti (eds.), *Bioremediation in Latin America: Current Research and Perspectives*, DOI 10.1007/978-3-319-05738-5

The manufacturer's authorised representative in the EU is Springer
Nature Customer Service Centre GmbH, Europaplatz 3, 69115 Heidelberg,
Germany. If you have any concerns regarding our products, please
contact ProductSafety@springernature.com

Printed and bound by CPI Group (UK) Ltd, Croydon, CR0 4YY
22/09/2026
02221065-0004